# Springer Tracts in Modern Physics
# Volume 146

**Springer-Verlag Berlin Heidelberg GmbH**

# Springer Tracts in Modern Physics

Springer Tracts in Modern Physics provides comprehensive and critical reviews of topics of current interest in physics. The following fields are emphasized: elementary particle physics, solid-state physics, complex systems, and fundamental astrophysics.

Suitable reviews of other fields can also be accepted. The editors encourage prospective authors to correspond with them in advance of submitting an article. For reviews of topics belonging to the above mentioned fields, they should address the responsible editor, otherwise the managing editor. See also http://www.springer.de/phys/books/stmp.html

## Managing Editor

### Gerhard Höhler

Institut für Theoretische Teilchenphysik
Universität Karlsruhe
Postfach 69 80
D-76128 Karlsruhe, Germany
Phone: +49 (7 21) 6 08 33 75
Fax: +49 (7 21) 37 07 26
Email: gerhard.hoehler@physik.uni-karlsruhe.de
http://www-ttp.physik.uni-karlsruhe.de/

## Elementary Particle Physics, Editors

### Johann H. Kühn

Institut für Theoretische Teilchenphysik
Universität Karlsruhe
Postfach 69 80
D-76128 Karlsruhe, Germany
Phone: +49 (7 21) 6 08 33 72
Fax: +49 (7 21) 37 07 26
Email: johann.kuehn@physik.uni-karlsruhe.de
http://www-ttp.physik.uni-karlsruhe.de/~jk

### Thomas Müller

Institut für Experimentelle Kernphysik
Fakultät für Physik
Universität Karlsruhe
Postfach 69 80
D-76128 Karlsruhe, Germany
Phone: +49 (7 21) 6 08 35 24
Fax: +49 (7 21) 6 07 26 21
Email: thomas.muller@physik.uni-karlsruhe.de
http://www-ekp.physik.uni-karlsruhe.de

### Roberto Peccei

Department of Physics
University of California, Los Angeles
405 Hilgard Avenue
Los Angeles, CA 90024-1547, USA
Phone: +1 310 825 1042
Fax: +1 310 825 9368
Email: peccei@physics.ucla.edu
http://www.physics.ucla.edu/faculty/ladder/
peccei.html

## Solid-State Physics, Editor

### Peter Wölfle

Institut für Theorie der Kondensierten Materie
Universität Karlsruhe
Postfach 69 80
D-76128 Karlsruhe, Germany
Phone: +49 (7 21) 6 08 35 90
Fax: +49 (7 21) 69 81 50
Email: woelfle@tkm.physik.uni-karlsruhe.de
http://www-tkm.physik.uni-karlsruhe.de

## Complex Systems, Editor

### Frank Steiner

Abteilung Theoretische Physik
Universität Ulm
Albert-Einstein-Allee 11
D-89069 Ulm, Germany
Phone: +49 (7 31) 5 02 29 10
Fax: +49 (7 31) 5 02 29 24
Email: steiner@physik.uni-ulm.de
http://www.physik.uni-ulm.de/theo/theophys.html

## Fundamental Astrophysics, Editor

### Joachim Trümper

Max-Planck-Institut für Extraterrestrische Physik
Postfach 16 03
D-85740 Garching, Germany
Phone: +49 (89) 32 99 35 59
Fax: +49 (89) 32 99 35 69
Email: jtrumper@mpe-garching.mpg.de
http://www.mpe-garching.mpg.de/index.html

Hubert Gnaser

# Low-Energy
# Ion Irradiation
# of Solid Surfaces

With 93 Figures

Springer

Dr. Hubert Gnaser
Universität Kaiserslautern
Fachbereich Physik
D-67663 Kaiserslautern
Email: gnaser@rhrk.uni-kl.de

Physics and Astronomy Classification Scheme (PACS):
79.20.-m, 79.20.Rf, 61.80.Ih, 61.82.Bg, 61.82.Fk, 68.35.Dv, 68.35.Bs


ISSN 0081-3869


Library of Congress Cataloging-in-Publication Data.

Gnaser, Hubert, 1953–. Low-energy ion irradiation of solid surfaces/Hubert Gnaser. p. cm.– (Springer tracts in modern physics, ISSN 0081-3869; vol. 146). Includes bibliographical references and index.
1. Solids–Effect of radiation on. 2. Ion bombardment. I. Title. II. Series: Springer tracts in modern physics;
146. QC1.S797 vol. 146 [QC176.8.R3] 539 s–dc21 [530.4'16] 98-44834

ISBN 978-3-662-14739-9      ISBN 978-3-540-49773-8 (eBook)
DOI 10.1007/978-3-540-49773-8
© Springer-Verlag Berlin Heidelberg 1999
Originally published by Springer-Verlag Berlin Heidelberg New York in 1999.
Softcover reprint of the hardcover 1st edition 1999

Typesetting: Data conversion by Satztechnik Katharina Steingraeber, Heidelberg
Cover design: *design & production* GmbH, Heidelberg

SPIN: 10670695      56/3144 - 5 4 3 2 1 0 – Printed on acid-free paper

# Preface

The interaction of energetic ions with solids gives rise to a variety of physical and chemical phenomena. This book examines specifically processes at the surface or in the near-surface region of a solid following the impact of low-energy ions. *Low energy*, in the present context, refers to ion energies from some 10 eV to about 10 keV, with a strong emphasis on sub-keV irradiations. The main themes treated are the slowing down of the impinging ion in the solid, the generation of bulk and surface damage due to the dissipation of the ion's energy, the ejection of atoms and molecules from the surface and the ionization processes involved in this emission, the composition changes multicomponent targets may experience due to ion irradiation and ion-bombardment effects on crystalline semiconductors, the latter ranging from the formation of isolated defects (adatoms and surface vacancies) by single-ion impacts to multilayer removal and surface roughening. Although metals and semiconductors are considered almost exclusively, the major fraction of the phenomena described appears to be relevant also for other materials. The examples chosen to illustrate the processes of low-energy ion bombardment of surfaces are taken both from experiments and from computer simulations. Recent findings from these approaches have contributed enormously, over the last decade, to an improved understanding of the processes induced by ion irradiation.

The modifications effected by low-energy ion irradiation at the surface or in the near-surface region of the solid have considerable impact on many applications: low-energy ion beams are ubiquitously employed, for example, in sputter deposition and thin-film growth, in ion implantation, in surface cleaning procedures and in various types of surface analysis techniques. Hence, this monograph emphasizes strongly the *surface*-related aspects of ion–solid interaction. Owing to the broad range of the subject, this book cannot (and does not) attempt to provide a comprehensive coverage of all individual phenomena that one may encounter in this field. Rather, some of the more prominent processes are highlighted by illustrative (and, mostly, recent) examples. The book, therefore, may serve as an overview of the field of ion–surface interactions; the extensive references included can provide guidance to further information on the subject.

It is a great pleasure to acknowledge the assistance this work has received from many colleagues. I am particularly grateful to H. Oechsner for his scientific advice and continuous support. The outcome of very fruitful cooperations with several coworkers is partly reflected in this work; I wish to thank especially W. Bieck, W. Bock, W. Hofer, I. Hutcheon and D. Weathers for these joint endeavors. I am very much indebted to P. Bedrossian, G. Betz, D. Cahill, W. Eckstein, N. Lam, T. Michely, M. Müller, P. Sigmund, H. Urbassek, P. Varga, M. Wahl, J. Weaver, A. Wucher, M. Yu and H. Zandvliet for permission to reproduce some of their published results and for promptly supplying me with the appropriate material. I wish to express my gratitude to Herbert Urbassek, who read major portions of the manuscript. I am grateful to Springer-Verlag for the opportunity to publish this book and the pleasant collaboration during its completion. Last, but not least, I thank my family for their patience in sparing me many evening and weekend hours and for their continuous encouragement during the writing of this book.

Kaiserslautern, September 1998                              *Hubert Gnaser*

# Contents

# 1. Introduction

A wide range of physical and chemical phenomena is associated with the interaction of low-energy ions with surfaces [1.1]. Such an ion irradiation often results in pronounced modifications of the surface and the near-surface region of the solid: it may alter the composition, induce defects at the surface, remove atoms and molecules from the outermost layer(s) or create substantial morphological changes, to name but a few of the possible processes. Many reviews have described the evolution of research in this field over the last three decades [1.1–13]. The interactions of low-energy ion beams with surfaces are also of significant technological importance, in such diverse applications as ion-beam assisted growth in thin-film deposition [1.14–17], surface characterization by means of surface analytical techniques [1.18–21] and sputter-induced cleaning of surfaces [1.22]. These well-established and/or evolving uses of ion beams have provided a stimulus for active research investigations into the fundamental processes, as they require a rather rigorous understanding of the relevant phenomena. This book constitutes an attempt to elucidate some of the more prominent processes in low-energy ion irradiation of surfaces and to highlight the recent progress made in this field.

The regime considered as *"low energy"* in this book is defined as follows: A natural *lower limit* corresponds to a kinetic energy of the order of the bond energies at the surface or in the bulk, i.e. an energy of about 5–15 eV [1.1]. Such an energy is typically required to displace an atom from its lattice site, either by ejection from the surface or by relocation within the solid. Although this very low-energy range appears to have important implications for beam-enhanced thin-film deposition, the number of well-characterized experiments is still rather limited. Computer simulations in this range may also face serious difficulties as the binary-collision approximations utilized to emulate atom–atom interactions are of limited use and a description in terms of many-body interactions is required. For the *upper energy limit* an ion kinetic energy of a few keV appears to be a reasonable choice in that ion–surface interactions in this range are still restricted to the near-surface region of the solid (some tens of nanometers, with the possible exception of very light ions), but are energetic enough to result in phenomena that typically occur also at much higher energies (albeit over a more extended depth range), such as collisional mixing, defect production and sputtering.

At these low irradiation energies, the surface of the solid [1.23, 24] is expected to play a significant role in the interactions that occur. Furthermore, at the lower end of this regime ions are near the threshold for creating atomic displacements in crystals. Owing to the lower coordination of surface atoms, an energy region might exist where surface displacements can occur without concurrent production of bulk defects. While bulk defects and the associated displacement energies are rather well studied [1.25], there is very little information available on the creation of surface defects and surface displacement energies [1.26]. The controlled production of surface defects which mediate epitaxial growth processes and the minimization (or even elimination) of bulk defects might often be the ultimate goal. Part of this book (see Chaps. 2 and 4) is devoted to elucidating this transition from pure surface effects to phenomena that extend some atomic layers from the surface into the bulk. Establishing this transition (in terms of beam energy, ion species, etc.) might provide the possibility to fine-tune ion irradiation effects to specific needs, e.g. in ion-beam-assisted deposition processes [1.15, 16].

Apart from the limitation in ion energy range, this book is subject also to a restriction in terms of the materials that are considered. With very few exceptions, only ion bombardment effects in metals and semiconductors are discussed. In view of the enormously large number of investigations in the field of ion bombardent, such a confinement is dictated not only by space limitations but also by the author's experience in the field. This selection does not, of course, indicate that ion irradiation of other materials is considered less important. On the contrary, it appears that encouraging progress [1.27–30] has been achieved in understanding ion bombardment effects on other (inorganic and organic) materials not covered in this work.

An energetic particle impinging on and penetrating the surface of a solid or liquid may trigger, while being slowed down through its interactions with the target species, a variety of processes; they may be grouped into the following broad classes.

(i)   Projectiles are incorporated into the solid, a process usually called ion implantation [1.31].

(ii)  Target atoms can experience a temporary or permanent relocation from their original lattice site, causing an accumulation of defects which may eventually transform an existing crystalline structure into an amorphous state [1.3–5, 8, 9].

(iii) Atoms at or close to the surface may receive sufficient energy to surmount the surface potential barrier and to leave the solid; this process, usually termed sputtering, results in a flux of ejected atomic and molecular species, which can be neutral as well as in various charge states [1.6, 7, 11, 12, 32]; their distributions in terms of emission angle and energy relay details of the energy-sharing mechanisms in the near-surface region.

(iv) The electronic excitation of target atoms may produce defects in the lattice and create excited (ionized) species which can be emitted in this state if they survive the passage to the solid's surface [1.2, 33, 34].

While these various processes are intrinsically coupled by their common origin (to the extent that the isolation of any single one may be difficult if not impossible), their occurrence and relevance in particle–solid interactions depend on a variety of parameters related to the incident projectile and to the bombarded sample.

The results presented in this work cover some aspects of the afore-mentioned processes: The creation of defects both at the surface and in the bulk, the sputtering of elemental and multicomponent systems and the associated compositional changes in the near-surface region of the latter are examined. For reasons outlined below, some emphasis is laid here on the preferential sputtering from isotopic mixtures and, both for elemental and for nonelemental samples, on very low impact energies. The importance of isotope effects is also considered for the ionization of sputtered species and sheds light on the relevant ionization mechanisms. The production of defects by (low-energy) ion bombardment in crystalline semiconductors and the gradual formation of an amorphous layer near the surface, as investigated by means of various experimental techniques and by computer simulations, is discussed. The following introductory remarks summarize the experiments, the computations and the theoretical concepts considered in the respective chapters of this work; specifically, these chapters address the following topics.

Chapter 2 describes first, in rather general terms, the interaction processes of low-energy ions with elemental solids. After briefly defining some fundamental parameters (such as the stopping cross section) governing the slowing down of the incident ion, ion ranges and the amount of energy transferred to the surface are discussed. The concept of a collision (or displacement) cascade initiated by the incoming projectile is introduced, leading naturally to the production of defects in the bulk and at the surface. In particular, the formation of surface adatoms and surface vacancies, the development of an extended surface morphology at higher ion fluences and the removal of a few monolayers (ML) are illustrated in detail. Sputtering of particles from the outermost surface layer(s) is outlined, first in the regime of linear collision cascades which applies to energies in the keV range. Energy dissipation in this regime proceeds, to a large extent, via elastic collisions between target species, while contributions from electronic excitations are usually small for metals and semiconductors, though not necessarily negligible. Atoms located within a few (two to three) monolayers from the surface may acquire enough energy from a recoil to be ejected from the solid. Apart from a discussion of the basic analytical theory of sputtering [1.35] and its comparison with experimental data, some specific points, such as the statistics of sputtering yields, the depth of origin of sputtered species and the possible fluence dependence of the yields are emphasized here. In the range of very low (near-threshold) ion energies and/or for light projectiles a collision cascade will usually *not* develop; rather, a few isolated collisions between the projectile or an energetic primary recoil and target atoms will dissipate the available energy and, eventually, may result in sputter ejection. The concept of a threshold energy is mentioned in this context, and total as well as emission-energy selected yields are presented for low impact energies. Furthermore, the sputtering of single-crystal specimens and the emission of clusters upon ion irradiation are highlighted. For sputtered neutral clusters mass, energy and angular distributions are specifically considered. Finally, the bombardment of surfaces with (large) cluster ions is investigated. For these conditions, very dense collision cascades can develop, with a large number of atoms in motion simultaneously. Novel mechanisms of energy dissipation and particle ejection become operative in this regime of ion–solid interaction.

Chapter 3 describes composition changes in multicomponent specimens induced by ion irradiation. The concept of preferential sputtering is introduced and several causes of a preferential ejection of individual atomic species are emphasized: First, the surface binding energies are species-dependent and the more weakly bound atom will have a higher probability of being ejected. Second, the energy sharing in the collision cascade is governed by the relevant collision cross sections and the conservation laws of energy and momentum. Thus, again, a dependence on species will enter. Third, sputtered atoms may penetrate a few monolayers before being ejected; this ability will clearly depend on the atom type. Apart from preferential sputtering, several other phenomena can give rise to composition changes in the near-surface region of an ion-bombarded solid; specifically, collisional mixing, radiation-induced segregation, radiation-enhanced diffusion and Gibbsian segregation are mentioned. The occurrence of these effects and their magnitude are closely related to the production of defects as illustrated in Chap. 2. Generally, under continuous ion bombardment, these processes result in an alteration of the surface composition. The transition to the stationary conditions is reflected also in the composition of the sputtered flux: its stoichiometry changes in a manner complementary to that of the surface and will deviate from that of the originating surface layer(s).

With these (theoretical) concepts at hand, composition changes in alloys and compounds are examined but no attempt at a comprehensive compilation of the available data is made. Rather, selected (and mostly recent) examples are presented in order to illustrate the pertinent processes. In addition to composition changes at the surface, the information conveyed by the (angular- and energy-differential) flux of sputtered atoms and clusters is also considered. Sputtering of isotopic mixtures (i.e. two or more isotopes of an element) is discussed separately. This situation appears well suited to the study of preferential sputtering isolated from the influence of other effects, as differences in binding energies are thought to be negligible and only the different isotopic masses govern the preferential ejection. The aim of such studies is to elucidate preferential isotope sputtering in the limit of low ion fluences and as a function of the ion energy and the emission angle and to allow for a direct comparison with the predictions of analytical sputtering theories. High-precision computer simulations have provided considerable insight into the relevant ejection mechanisms.

Chapter 4 is devoted to ion bombardment of crystalline semiconductor surfaces. Although the basic processes of ion–surface interactions are essentially identical to those outlined in Chaps. 2 and 3, there exist sufficient distinctions which appear to justify this separate treatment. In particular, the formation of surface defects and adatoms by single-ion impacts is studied closely using atomic-resolution-imaging data. Surface roughening and multi-layer removal upon prolonged ion bombardment are followed in the same way. The similarity to the corresponding growth processes is emphasized in this context. Several examples originating from different experimental approaches are used to illustrate the development of an amorphized near-surface layer in elemental and compound semiconductors upon low-energy ion irradiation. Comparison with the results of computer simulations is made repeatedly; defect production rates were modeled theoretically to derive (surface and bulk) displacement thresholds (energies) for amorphization. The destruction of the crystalline order in these semiconductor materials is seen to have also a tremendous influence on their electronic properties. Composition changes in compound semiconductors for different irradiation parameters are discussed and the relevance of damage and annealing rates for these materials is emphasized.

Chapter 5 characterizes some aspects of the ionization of sputtered atoms and molecules. Models of secondary-ion formation and the associated ionization probability are briefly discussed. As $Cs^+$ ions are widely used as primary ions in applications of surface mass spectrometry, the incorporation of cesium upon irradiation, the concurrent work-function changes and the influence on the ionization are described in detail. Isotopic mass effects are presented as a suitable means to study basic models of ion formation for certain classes of materials. The formation of molecular ions in sputtering is exemplified using different kinds of molecular species. Some of these illustrate the extreme sensitivity of the (mass spectrometric) techniques employed. Finally, the first results on ion emission from surfaces under ion bombardment with highly charged ions are described.

Chapters 2–5 make repeated reference to specific experimental techniques and (to a lesser extent) different methods of computer simulation: mass spectrometric techniques such as secondary-ion mass spectrometry (SIMS) and secondary-neutral mass spectrometry (SNMS) are used to monitor the flux of sputtered species in terms of mass, energy and angular distribution from elemental and multicomponent targets; composition changes at the surface are typically detected by Auger electron spectroscopy (AES), photoelectron spectroscopy (XPS) or ion scattering spectroscopy (ISS); information on the structural state of the surface can be obtained from scanning tunneling microscopy (STM) or, for larger surface areas, from low-energy electron diffraction (LEED); computer simulations employed in low-energy ion irradiation studies can be grouped into the so-called molecular dynamics (MD) simulations which solve the classical equations of motion for a suitable number of atoms, and a second type of simulation in which the binary-collision approximation is assumed to be valid. A brief outline of the experimental and computational techniques and their essential features in the context of low-energy ion irradiation of surfaces is given in the appendix, together with a selection of pertinent references.

   This book includes an extensive, although not comprehensive bibliography of work performed in the field of low-energy ion bombardment of solid surfaces as covered in Chaps. 2–5. A great wealth of information on the processes related to the interaction of energetic ions with surfaces and solids can be found in the proceedings of the two series of biennial conferences on *Atomic Collisions in Solids* [1.36–40] and *Ion Beam Modifications of Materials* [1.41–45].

# 2. Interaction of Low-Energy Ions with Solids

The irradiation of solids by energetic particles gives rise to a variety of phenomena. At the very surface, backscattering of incident particles, emission of electrons and photons, and ejection of target atoms and molecules (i.e., sputtering) may take place. In a near-surface region of the solid, extending to a depth which depends primarily on the incident particle's energy and the mass matching, the decelerated projectiles transfer energy and momentum to the target atoms, displacing them from their original positions. This process may cause rather extensive displacement cascades and point defects (vacancies and interstitials), while the concurrent accumulation of the implanted species may create other types of damage within this zone. Generally, these processes are closely correlated and often a synoptic view has to be adopted for a complete understanding of any single one of them.

Ion implantation has been widely used for material modification, in particular for the doping of semiconductor materials. Originally, the relevant energies were mostly in the tens to hundreds of keV range, but with the decreasing feature sizes of microelectronic devices energies in the low keV and even sub-keV range become relevant. The damage associated with these interaction processes has been studied, especially in terms of the required annealing steps. Sputtering, that is, the ejection of particles from the surface, has been employed for thin-film deposition, in-situ surface cleaning, microerosion, and, in combination with analytical tools such as electron and mass spectroscopies, for depth-profiling thin-film structures. Again, ion bombardment might produce damage and compositional transformations here; as for these applications the ion energies are typically in the low keV range, the modified near-surface region of the solid will be correspondingly shallower. While an introduction to the sputtering of monatomic systems is given in Sect. 2.3, the processes of surface composition changes induced during the ion bombardment of multicomponent materials are presented in Chap. 3.

This chapter first describes theoretical concepts like stopping cross sections, related to the energy loss and the slowing down of energetic particles in solids, as introduced by Bohr, Lindhard and coworkers [2.1–5] several decades ago. That approach was applied quite successfully to model the penetration and damage ranges of ions in solids, e.g. in the context of ion implantation. The dissipation of the ion's energy results in the displacement, permanent or temporary, of target atoms [2.6]. Vacancies and interstitials created during this *displacement cascade* may in part annihilate during an ensuing *cooling phase*, so that the number of Frenkel pairs surviving at the end of the event can be considerably smaller than the number of atoms originally set in motion (Sect. 2.2.1). Apart from defects in the bulk, ion irradiation typically creates also surface vacancies and surface adatoms. As will be shown, these surface defects may dominate at low ion energies (100 eV and less). A thorough discussion of these surface features is given in Sects. 2.2.2 and 2.2.3, highlighting both experimental results obtained by STM and data from computer simulations. It is this kind of surface defect that is of importance in ion-assisted deposition processes.

Closely related to the creation of defects at the surface is the ejection of particles from the surface, caused by a collision cascade in its vicinity. Sputtering is discussed in Sect. 2.3; apart from a brief description of the theoretical approach advanced by Sigmund [2.7, 8], characteristic features of sputtering [2.9, 10] in the context of linear collision cascades, such as the statistics of the ejection process, the depth of origin of sputtered species and the fluence dependence of the yield are emphasized. A more detailed discussion of sputtering at very low (near-threshold) energies (Sect. 2.3.2), sputtering of single crystal specimens (Sect. 2.3.3) and the emission of clusters (Sect. 2.3.4) is provided. This chapter is concluded by a short outline of recent results on cluster-ion bombardment of surfaces and the associated sputtering-yield variations.

## 2.1 Energy Loss and Slowing Down of Incident Particles

### 2.1.1 Stopping Cross Sections

The kinetic energy of an energetic particle is dissipated via elastic (nuclear collisions) and inelastic (electronic excitation) processes. The differential energy loss or stopping power of a layer $\mathrm{d}x$ of the target can be expressed as a sum of the nuclear (n) and electronic (e) contributions of the stopping powers [2.7, 11]:

$$\frac{\mathrm{d}E}{\mathrm{d}x} = \left(\frac{\mathrm{d}E}{\mathrm{d}x}\right)_{\mathrm{n}} + \left(\frac{\mathrm{d}E}{\mathrm{d}x}\right)_{\mathrm{e}} = -N\left[S_{\mathrm{n}}\left(E\right) + S_{\mathrm{e}}\left(E\right)\right] , \tag{2.1}$$

where $N$ is the number density of atoms in the sample, and $S_{\mathrm{n}}(E)$ and $S_{\mathrm{e}}(E)$ are the nuclear and electronic stopping cross sections, respectively. The nuclear stopping cross section is defined as [2.7]

$$S_{\mathrm{n}}\left(E\right) = \int_0^{T_{\max}} T\mathrm{d}\sigma\left(E, T\right) , \tag{2.2}$$

where $\mathrm{d}\sigma$ is the interaction cross section, $T$ the transferred (recoil) energy and $T_{\max}$ the maximum of $T$ in a head-on collision, $T_{\max} = \gamma E$, with $\gamma =$

$4M_1M_2/(M_1+M_2)^2$; $M_1$ and $M_2$ are the masses of the incoming ion and of a target atom, respectively. The interaction of two atoms is described by an interaction potential which depends only on the nuclear charges and the internuclear distance $r$. The conservation laws define a unique relationship between the scattering angle and the energy $T$ transferred by an atom hitting another atom at rest.

The probability distribution for the energy transfer $T$ is given by the cross section $d\sigma(E,T)$. For energies high enough that the scattering is determined by the Coulomb repulsion between the nuclei, this leads to the well-known Rutherford scattering cross section, which strongly favors collisions with small energy transfers. Rutherford scattering is valid only for $\varepsilon \gg 1$, where the reduced energy $\varepsilon$ is defined via [2.3, 5]

$$\varepsilon = \frac{M_2 E}{M_1 + M_2} \frac{a}{Z_1 Z_2 e^2} , \tag{2.3}$$

where $a = 0.885 a_0 \left( Z_1^{2/3} + Z_2^{2/3} \right)^{-1/2}$ is the screening length and the Bohr radius $a_0 = 0.0529\,\mathrm{nm}$. For lower energies ($\varepsilon < 1$) the screening of the Coulomb interaction is essential. Different interatomic potentials $V(r)$ have been proposed for this regime in the literature, many of them with the general form [2.12]

$$V(r) = \frac{Z_1 Z_2 e^2}{r} \Phi \left( \frac{r}{a} \right) , \tag{2.4}$$

where $\Phi(r/a)$ is the screening function that screens the repulsive forces of the nuclei because of the partial shielding by the surrounding electrons. $\Phi(r/a)$ is often approximated by [2.13]

$$\Phi \left( \frac{r}{a} \right) = \sum_{i=1}^{n} c_i \exp \left( -d_i \frac{r}{a} \right) , \tag{2.5}$$

$c_i$ and $d_i$ being constants in the screening functions for various potentials and $\sum_{i=1}^{n} c_i = 1$. (Note that $n = c_1 = d_1 = 1$ for the Bohr potential, while $n$ is 3 or 4 for other screening functions.) Another potential widely used in the past in simulations is the Born–Mayer potential

$$V(r) = A_{\mathrm{BM}} \exp \left( -r/a_{\mathrm{BM}} \right) , \tag{2.6}$$

where $A_{\mathrm{BM}}$ is an energy parameter and $a_{\mathrm{BM}}$ is a screening length. Andersen and Sigmund [2.14] proposed values of $a_{\mathrm{BM}} = 0.0219\,\mathrm{nm}$ and $A_{\mathrm{BM}} = 52 \left( Z_1 Z_2 \right)^{3/4}\,\mathrm{eV}$; other choices, however, have been employed [2.14].

The nuclear stopping cross section $S_n(E)$ can be expressed in a reduced form $s_n(\varepsilon)$, independent of the ion–target combination; this dimensionless quantity is connected with $S_n(E)$ in units of $\mathrm{eV}\,\mathrm{cm}^2$ through

$$S_n(E) = \frac{\pi a^2 \gamma}{\varepsilon/E} s_n(\varepsilon) = \frac{4\pi Z_1 Z_2 e^2 a M_1}{M_1 + M_2} s_n(\varepsilon) , \tag{2.7}$$

$$S_{\mathrm{n}}(E) = 8.46 \times 10^{-15}$$

$$\times \frac{M_1 Z_1 Z_2 s_{\mathrm{n}}(\varepsilon)}{(M_1 + M_2)\left(Z_1^{2/3} + Z_2^{2/3}\right)^{1/2}} \quad (\mathrm{eV\,cm}^2) \ . \tag{2.7a}$$

The nuclear stopping power then follows from $(\mathrm{d}E/\mathrm{d}x)_{\mathrm{n}} = -NS_{\mathrm{n}}(E)$. Rather simple analytical expressions for $s_{\mathrm{n}}(\varepsilon)$ have often been proposed. For the so-called Kr–C potential, Wilson et al. [2.15] suggested, for example, an approximation

$$s_{\mathrm{n}}(\varepsilon) = \frac{0.5 \ln(1 + \varepsilon)}{\varepsilon + 0.107\,\varepsilon^{0.375}} \ . \tag{2.8}$$

Figure 2.1 illustrates this dependence for a wide range of $\varepsilon$. Also indicated are the $\varepsilon$ values for 1 keV He, Ar and Xe ions incident on Cu.

Assuming an interatomic potential of the form $V(r) \propto r^{-1/m}$, an approximate cross section was derived from classical collision theory [2.2, 3, 5, 7]:

$$\mathrm{d}\sigma(E, T) \cong C_m E^{-m} T^{-1-m} \mathrm{d}T \ , \qquad 0 \leq T \leq T_{\max} \ , \tag{2.9}$$

with $C_m$ given by

$$C_m = \frac{\pi}{2} \lambda_m a^2 \left(\frac{M_1}{M_2}\right)^m \left(\frac{2Z_1 Z_2 e^2}{a}\right)^{2m} \ ; \tag{2.10}$$

$m$ characterizes the power potential employed to describe the interaction between atoms; it varies slowly from $m = 1$ at high energies, that is, for

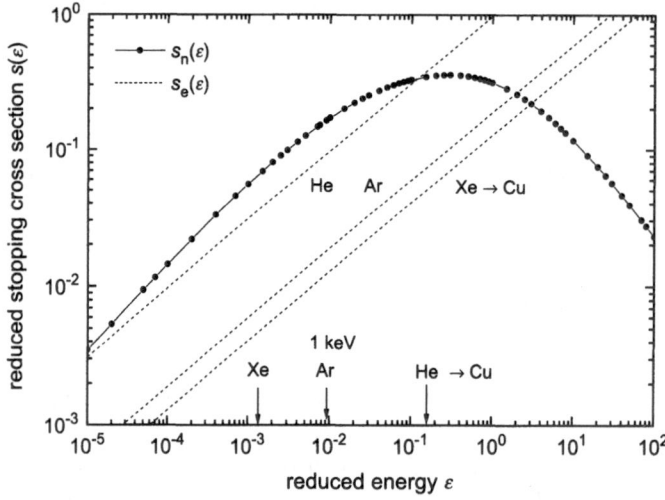

**Fig. 2.1.** Reduced nuclear and electronic stopping cross sections, $s_{\mathrm{n}}(\varepsilon)$ and $s_{\mathrm{e}}(\varepsilon)$, as a function of the reduced energy $\varepsilon$. The $s_{\mathrm{e}}(\varepsilon)$ values (*straight lines*) are computed according to (2.17) for He, Ar and Xe projectiles on Cu. The $s_{\mathrm{n}}(\varepsilon)$ data are calculated from (2.8). The reduced energies $\varepsilon$ for 1 keV He, Ar and Xe ions in Cu are marked by *arrows* on the lower axis

Rutherford scattering, down to $m \approx 0$ at very low energies. The quantity $\lambda_m$ is a dimensionless function of the parameter $m$ which increases over this range of $m$ from $\lambda_1 = 0.5$ to $\lambda_0 \approx 24$. An improved evaluation of $m$ and $\lambda_m$ was achieved [2.16] by fitting the power-law cross section to calculated cross sections for the Born–Mayer interaction. For the typical energies of sputtered atoms, $m$ was found to be about 0.1 or slightly higher for the Born–Mayer interaction. Note from (2.9) that an energy-independent cross section is approached for $m \approx 0$:

$$d\sigma\,(E,T) \cong \frac{\pi}{2}\lambda_0 a^2 T^{-1} dT \,, \qquad 0 \le T \le T_{\max} \,. \tag{2.11}$$

The nuclear stopping cross section (see (2.2)) for the power potential is thus (see (2.9))

$$S_{\mathrm{n}}\,(E) = \frac{1}{1-m} C_m \gamma^{1-m} E^{1-2m} \tag{2.12}$$

or, in reduced units,

$$s_{\mathrm{n}}\,(\varepsilon) = \frac{\lambda_m}{2\,(1-m)} \varepsilon^{1-2m} \,. \tag{2.13}$$

$S_{\mathrm{n}}$ is found to increase approximately linearly with $E$ at low energies ($m \sim 0$); it reaches a plateau at intermediate energies ($m \sim 0.5$) and decreases at high energies ($0.5 < m < 1$).

At very low energies the repulsive potentials discussed so far are often not sufficient to describe the interaction, as attractive forces may become significant. In such a case, combined potentials are used which employ a repulsive interaction at short separations and an attractive potential (e.g. a Morse potential) for large interatomic distances. Several very detailed descriptions of relevant interaction potentials, in particular for use in computer simulations [2.13], have been provided.

Unlike nuclear stopping, the interaction of the penetrating ion with target electrons does not cause appreciable scattering of the incident particle, but electronic stopping may be important for the slowing down and the energy loss. At high energies ($\varepsilon \gg 1$) electronic excitation dominates the energy loss of ions, and the electronic stopping cross section $S_{\mathrm{e}}(E)$ follows the Bethe expression [2.17]. In the low-energy range considered in this work, the electronic stopping is proportional to the ion's velocity [2.3, 4]; this is valid for ion velocities extending to

$$v \le Z_1^{2/3} v_0 = Z_1^{2/3} e^2 / \hbar \,, \tag{2.14}$$

with the Bohr velocity $v_0 = 2.18 \times 10^8 \,\mathrm{cm\,s^{-1}}$, or, in terms of energy (in keV),

$$E \le 24.88\, Z_1^{4/3} M_1 \,. \tag{2.15}$$

According to Lindhard and Scharff [2.4], the electronic stopping cross section can be expressed by

$$S_e(E) \approx 8\pi e^2 a_0 Z_1^{1/6} \frac{Z_1 Z_2}{\left(Z_1^{2/3} + Z_2^{2/3}\right)^{3/2}} \frac{v}{e^2/\hbar} = K E^{1/2} , \qquad (2.16)$$

and, in dimensionless units,

$$s_e(\varepsilon) \cong 0.0793 \, Z_1^{1/6} \frac{Z_1^{1/2} Z_2^{1/2}}{\left(Z_1^{2/3} + Z_2^{2/3}\right)^{3/4}} \frac{(M_1 + M_2)^{3/2}}{M_1^{3/2} M_2^{1/2}} \varepsilon^{1/2} = k\varepsilon^{1/2}, \quad (2.17)$$

with $k$ in the range 0.1–0.4 except for very light ions on heavy targets. The electronic stopping power is then $(dE/dx)_e = -N S_e(E)$. Note that $S_e(E)$ exhibits some nonmonotonic variation with $Z_1$ and the factor $Z_1^{1/6}$ is an average value for these $Z_1$ oscillations. Another concept of electronic energy loss is due to Oen and Robinson [2.18]; at low energies (around 10 eV) its magnitude can be up to a factor of ten smaller than the corresponding value of Lindhard and Scharff (2.16). Typically, electronic and nuclear stopping are of the same order of magnitude for $v \approx 0.1 Z_1^{2/3} v_0$ or $E \approx 0.25 Z_1^{4/3} M_1$ (keV). To give some examples: For 1 keV He on Cu ($\varepsilon = 0.16$), nuclear and electronic stopping powers are, respectively, 26 and 29 eV/nm; so electronic stopping is important for all energies. By contrast, for 1 keV Ar ions bombarding Cu ($\varepsilon = 0.0095$), the nuclear and electronic stopping values are 600 and 66 eV/nm, respectively; only beyond 100 keV do electronic effects become comparable. For even heavier projectile ions, 1 keV Xe on Cu ($\varepsilon = 0.0014$), the nuclear stopping power amounts to 1030 eV/nm while electronic stopping is about 7% of this value. Therefore, for the energies considered here, electronic stopping is about an order of magnitude smaller than nuclear stopping, except for very light incident ions.

The effects of energy dissipation on the different atoms in multicomponent systems will be treated in Chap. 3.

### 2.1.2 Ion Ranges in Solids

Owing to the energy losses described above, an incident ion has a finite range in the target. Different types of range can be defined for an ion penetrating into a solid. The average path length $R(E)$ is simple to calculate but difficult to measure. For the approximation of continuous slowing down (i.e. many small energy-loss steps), the ion's path length can be derived from a knowledge of the stopping cross sections:

$$
\begin{aligned}
R(E) &= \int_E^0 \frac{dE}{dE/dx} = -\frac{1}{N} \int_E^0 \frac{dE}{[S_n(E) + S_e(E)]} \\
&= \frac{1}{N} \int_0^E \frac{dE}{[S_n(E) + S_e(E)]} ,
\end{aligned} \qquad (2.18)
$$

or, in dimensionless units,

$$\rho(\varepsilon) = \int_0^\varepsilon \frac{d\varepsilon}{s_{\mathrm{n}}(\varepsilon) + s_{\mathrm{e}}(\varepsilon)} \; , \tag{2.19}$$

with the reduced path length defined as [2.3]

$$\rho = N\pi a^2 \gamma R \; . \tag{2.20}$$

For the power-law stopping cross section and neglecting electronic stopping, the average path length is [2.11]

$$R(E) = \frac{1-m}{2m} \gamma^{m-1} \frac{E^{2m}}{N \, C_m} \tag{2.21}$$

and, in reduced units,

$$\rho(\varepsilon) = \frac{1-m}{m\lambda_m} \varepsilon^{2m} \; . \tag{2.22}$$

It has been noted that path length values of good accuracy are obtained with these stopping cross sections with $m = 1/3$ for $\varepsilon \leq 0.2$ ($\rho \propto \varepsilon^{2/3}$) and with $m = 1/2$ for $0.01 \leq \varepsilon \leq 2$ ($\rho \propto \varepsilon$).

The average projected range $R_{\mathrm{p}}$ (the projection of $R$ on the incidence direction of the ion beam) and the penetration depth $x$ are more readily accessible by measurements and are therefore the quantities employed to characterize ion implantation. Because of the numerous deflections an incident particle may experience in its nuclear collisions with target atoms, the average path length $R$ can be considerably longer than the mean projected range $R_{\mathrm{p}}$. The ratio of these quantities depends sensitively on the mass ratio of the ion and the target atoms, $M_2/M_1$ [2.7, 11]. For $\varepsilon < 1$, the path length correction $R_{\mathrm{p}}(E)/R(E)$ is much less than 1 for $M_1 \ll M_2$ and close to unity for $M_1 \gg M_2$; the latter also holds for $\varepsilon \gg 1$. An approximate formula for the average projected range in amorphous and polycrystalline materials has been derived by Schiøtt [2.19, 20]. For the low-energy regime ($0.002 \leq \varepsilon \leq 0.1$), which is of relevance also for the present work, the average projected range $R_{\mathrm{p}}$ (in units of $\mu\mathrm{g/cm}^2$) as a function of ion energy $E$ (in keV) is given by

$$R_{\mathrm{p}} = C_l(\mu) M_2 \left[ \left( \frac{Z_1^{2/3} + Z_2^{2/3}}{Z_1 Z_2} \right) E \right]^{2/3} , \tag{2.23}$$

where $\mu = M_2/M_1$ and $C_l(\mu)$ decreases from 0.23 to 0.13 as $\mu$ increases from 0.1 to 100. Equation (2.23) indicates that for a light ion in a heavy target the range $R_{\mathrm{p}}$ is approximately proportional to $Z_1^{-2/3}$, whereas for a heavy ion in a light target the range is not very dependent on $Z_1$. The energy dependence is always proportional to $E^{2/3}$; see above ($\rho \propto \varepsilon^{2/3}$). A considerable number of compilations [2.21–26] of ion ranges for a wide variety of ion–target combinations exists. Also, computer simulation programs for range calculations are in rather extensive use, in particular the widespread TRIM code in its different versions [2.12, 13]. These simulations usually also provide the higher-order moments of the range distribution, such as the straggling (i.e., the square root of the variance), the skewness and the kurtosis. These numbers might be of importance in, for example, defining ion implantation profiles over an extended concentration range.

Figure 2.2 gives an example of ion range determinations at rather low implantation energies. Gnaser et al. [2.27, 28] have utilized sputtered neutral mass spectrometry to derive range parameters for He$^+$ ions implanted into silicon and nickel in the energy range from 250 eV to 80 keV. As the implant distributions are located close to the surface (a few nm) for the lower energies, the high depth resolution and the excellent sensitivity (detection limits of the order of $10^{18}$ cm$^{-3}$) of SNMS became advantageous. Figure 2.2 depicts range data for He$^+$ in Si [2.27]. The experimental results were compared with theoretical data evaluated using the Ziegler–Biersack–Littmark formalism [2.12]; there is good agreement, in particular, for the projected range $R_p$ at energies above 2 keV. It is noted that the theoretical calculations assume an infinite target, which, at low energies, underestimates $R_p$ and, to a lesser extent, overestimates the range straggling $\sigma$; applying a correction procedure for a real (semi-infinite) target would thus shift the theoretical curves even closer to the experimental ones. Figure 2.2 also displays values of $R_p$ and $\sigma$ obtained with the TRIM code [2.27]. These again fit the experimental values very well, reproducing the values of $R_p$ better at low implantation energies. With respect to the shape of the implantation profiles it was noted that at higher energies the distributions exhibited a distinct negative skewness (i.e. a more gradual slope towards the surface and a steeper slope on the bulk side),

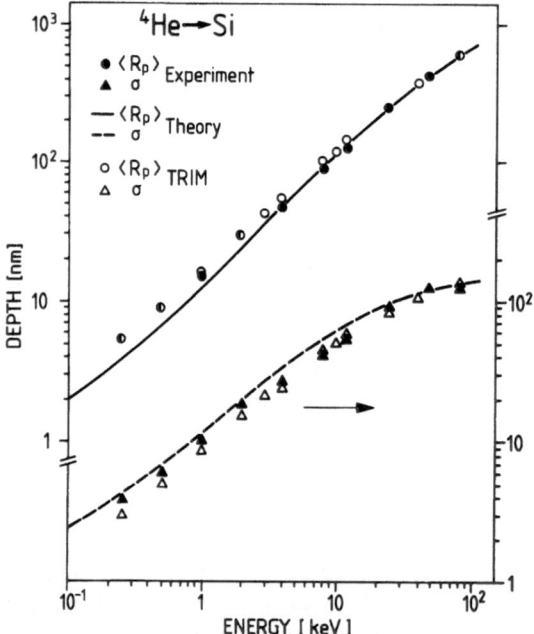

**Fig. 2.2.** Mean projected range $R_p$ and range straggling $\sigma$ of $^4$He ions in Si versus the implantation energy. The data were derived from SNMS depth profiles. Also given are theoretical data of Ziegler et al. [2.12] and values obtained with the TRIM code. Data from [2.27]

and a positive skewness at lower energies. Both of these are corroborated by the TRIM calculations. Since the authors observed no indication of ion channeling they concluded that the results refer to amorphous silicon.

If an ion enters a single crystal in the direction of a low-index crystallographic axis (or plane), the penetration can be much deeper because of the "steering" effect [2.29, 30] along open channels between regular rows (or planes) of atoms (channeling). For an entrance direction within a small angular range of a few degrees around the channel orientation, the ion is directed by these row of atoms and the number of collisions with target atoms becomes very small; hence, the range will increase drastically (as compared to an amorphous target) and energy loss is mostly due to electronic excitations. Because of the increase of nuclear stopping with decreasing energy, the ion is stopped by nuclear collisions near the end of its path. How deep the ion can actually channel into the solid depends sensitively on the initial incidence parameters. The ideal channeling conditions are related to the type and energy of the ion and the crystallographic nature of the channel; moreover, channeling can be disturbed by lattice disorder and atom vibration. The maximum range in a channel, however, may be several times the projected range in a nonchanneling direction.

Partly because of their importance in tailoring microelectronic device properties, an enormous amount of range data (for both random and channeling ion incidence) has been accumulated [2.31, 32]. An evaluation of these results is beyond the scope of the present work. Pertinent information can be found in monographs devoted to ion implantation and related topics. It is, however, stressed that ranges and range distributions at very shallow depths (some ten nm and less) are becoming increasingly important; reliable data in this regime appear to be rather rare.

### 2.1.3 Energy Transfer to the Surface

The kinetic energy of an ion approaching the surface is not necessarily transferred totally to that surface. A certain fraction may be carried away by reflected ions and/or by sputtered particles. The energy deposition coefficient $f$ or the related quantity of the energy reflection coefficient (termed also the "sputtering efficiency" [2.33]) ($\equiv 1 - f$) is expected to be a function of the incident mass, kinetic energy and angle of incidence, as well as the mass of the substrate atoms and, possibly, the surface morphology. Extensive computer simulations [2.34] of heavy-ion reflection were performed and compared with results from a theoretical description [2.35] based on the Boltzmann equation. (The ion's potential energy is in part consumed by the neutralization process which, for slow ions, occurs before it hits the surface, leading often to electron emission; some of that energy might also be carried away as electron kinetic energy.) Earlier experimental studies [2.36–40] into the energy transfer concentrated mostly on higher impact energies (above a few keV), partly because of the difficulties associated with measurements at low energies, or on light projectiles because of their importance for plasma–wall interaction in thermonuclear fusion devices. Winters and coworkers [2.41–43] have more recently extended the energy range to very low values. They measured the energy deposition coefficient $f$ for $He^+$, $Ar^+$ and $Xe^+$ ions striking surfaces of C, Si, Cu and Ag with energies in the range 100–4000 eV, and for Au surfaces with energies in the range 10–4000 eV. These studies employed a highly sensitive pyroelectric calorimeter. Figure 2.3 presents data on $f$ for Au and an ion energy range of 20–4000 eV. Generally, the energy transfer decreases with decreasing ion energy and with increasing mass ratio $M_2/M_1$. For the energy range 100–4000 eV, $Xe^+$ ions deposit $\geq 93\%$ of their kinetic energy for all samples investigated, with $f$ approaching 100% at the higher energies. For $Ar^+$ and $E \geq 1\,keV$, $f(E) \geq 0.9$ (lowest for Au) and falls for lower energy to $\sim 0.8$ for Ag and Au and $\sim 0.9$ for the other elements. Because of their small mass, $He^+$ ions exhibit a more distinct target-mass dependence; $f$ is highest for carbon (close to 100% for $E \geq 1\,keV$, decaying to $\sim 0.9$ at 100 V) and lowest for the heavy elements (Ag, Au). For these elements $f$ decreases from $\sim 0.88$ at 4 keV to $\sim 0.73$ at 100 eV. The values for Si and Cu fall roughly in between. The Au data for $E \leq 100\,eV$ indicate that for $Xe^+$ $f \sim 0.95$ down to 20 eV impact energy, while for $Ar^+$ and $He^+$ the transferred energy decreases to about 60% in that energy regime. The authors [2.42, 43] report a semiquantitative agreement of the experimental results with the corresponding data from computer simulations with the TRIM code. The latter indicate that when the sputter yield is low (e.g. $He^+$ bombardment or low energies), the reflected energy is carried away largely by reflected primary species. Conversely, in cases of moderate and high sputtering yields (and mass ratios $M_2/M_1$ not substantially larger than one), the fractions of energy carried

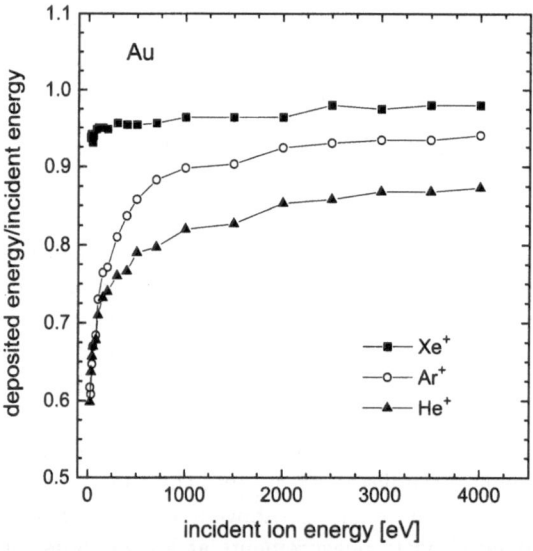

**Fig. 2.3.** Fraction $f$ of the incident energy of $Xe^+$, $Ar^+$ and $He^+$ ions deposited in gold for normal incidence. Data from [2.41, 42]

away by sputtered particles and by reflected ions may become comparable. These features are also seen in MD simulations [2.44] of the energy transfer of (5–400 eV) Ne, Ar and Xe atoms incident on Cu.

## 2.2 Displacement Cascades and Generation of Defects

The nuclear stopping of ions creates collision (displacement) cascades in the solid, in which a large number of target atoms are set in motion (recoil atoms) [2.45]. Any collision cascade intersecting the surface may cause sputtering of atoms from the surface, while in the bulk of the material the production of various types of defects proceeds. Obviously, the characterization of these damage mechanisms is of importance for the understanding of any processes related to ion irradiation of solids. In most metals and semiconductors, the threshold displacement energy (i.e. the minimum recoil energy required to produce a stable Frenkel pair after the cooling phase, when no further defect migration is expected) is between 15 and 40 eV [2.46]. Thus, a recoil with an energy a few times this value may create only isolated Frenkel pairs. Conversely, primary recoil events of hundreds of eV can result in several defects in close proximity to each other. For even higher recoil energies the number of atoms in motion within the cascade increases and an extended disordered zone in the center of the cascade may form, surrounded by interstitial atoms. As simulations [2.47] show, this disordered state may last several picoseconds. Furthermore, both simulations and experiments [2.48, 49] indicate the possible existence of individual subcascades initiated by different primary recoils.

A rather important quantity with respect both to defect production within the solid and to sputtering from the surface is the number of atoms participating in such a collision cascade [2.7, 11, 50]. If larger than unity, this number is proportional to the initially available energy. The average number of atoms $\nu(E_n, E_0)$ set in motion with an initial energy greater than some value $E_0$ in a cascade initiated by a primary ion or recoil of energy $E_n$ is

$$\nu(E_n, E_0) \approx \Gamma \frac{E_n}{E_0}, \quad \text{for} \quad \gamma E_n \gg E_0 \tag{2.24}$$

where $E_n$ is the fraction of the primary ion's energy spent in elastic collisions and $\Gamma$ is dependent on the atomic interaction ($\Gamma < 1$). For the power-law cross section, this parameter, $\Gamma_m$, depends weakly on $m$: $\Gamma_{0.5} = 0.361$ and $\Gamma_0 = 0.608$ [2.11]. The average number of atoms $F(E_n, E_0)dE_0$ recoiling in an energy interval $[E_0, dE_0]$ follows directly from (2.24):

$$F(E_n, E_0) \approx \Gamma \frac{E_n}{E_0^2}, \quad \text{for} \quad \gamma E_n \gg E_0 . \tag{2.25}$$

$F(E_n, E_0)$, the so-called recoil density [2.7, 11], is of utmost relevance for sputtering as it defines, ultimately, the flux of atoms moving towards the surface, leading eventually to particle ejection (see Sect. 2.3).

For damage creation, the value of $E_0$ must be at least the effective displacement threshold $E_d$, required to produce a Frenkel pair in an energetic collision; the mean number of Frenkel pairs was first estimated by Kinchin and Pease [2.51] to be

$$\nu\left(E_n, E_d\right) \approx \frac{E_n}{2E_d} \ . \tag{2.26}$$

This expression was later modified [2.52] and a constant of proportionality smaller than unity ($\sim$ 0.8) was introduced in (2.26). With these modifications, a damage function is defined in terms of the displacement threshold and the energy $T$ of a target atom recoil:

$$\nu\left(T, E_d\right) \cong \begin{cases} 0 & T < E_d \\ 1 & E_d \leq T < 2E_d \\ 0.8T/2E_d & T \geq 2\,E_d \end{cases} . \tag{2.27}$$

Experimental data and computer simulations revealed, however, that the actual damage function very often does not exhibit this step-like behavior [2.53–57]. In particular, the number of Frenkel pairs produced as a function of recoil energy was found to be smaller than that predicted by (2.27). This damage-production "efficiency" (i.e. the ratio of observed to predicted defects) is roughly unity for irradiation with light ions ($H^+$, $He^+$) but decreases for higher recoil energies (and heavier projectiles). Saturation values of $\sim$ 0.4 were reported for Cu and Ag. Clearly, the damage depends sensitively on the displacement threshold energy; this quantity is anisotropic in crystals and typically defect production is easiest for recoils near close-packed lattice directions.

The original concept of a displacement cascade was developed some forty years ago. Brinkman [2.58] suggested that when the energy of the ion in the target falls below a critical value, its mean free path between collisions approaches the interatomic spacing, resulting in the production of a high density of displaced atoms. In this picture, vacancies reside in the core of the cascade and interstitial atoms are located at the periphery. It was also proposed that these interstitials may be transported from the cascade region via replacement collision sequences (RCSs) along close-packed lattice directions, resulting in an efficient separation of interstitial atoms and vacancies. The concept of RCSs was further elucidated by one of the first MD computer simulations, by Vineyard and coworkers [2.59–61], which corroborated the existence of these replacement events. The occurrence of thermal spikes in collision cascades was first discussed by Seitz and Koehler [2.62]. Such a thermal spike results from the conversion of the kinetic energy of the primary recoil atom into heat in the localized region of the cascade. A thermal spike would require that the energy was partitioned among atoms in the cascade such that local equilibrium was reached before the spike dissipated. Evidence for the significance of thermally activated processes was reported later on in high-energy cascades.

Despite these early advances in the understanding of displacement cascades, it was found difficult to develop a comprehensive, analytical theory of cascade dynamics covering both the short- and long-time regimes; this is mostly due to the complex and many-body nature of collision cascades [2.63]. Traditionally, an initial "collisional" phase ($< 10^{-13}$ s) has been distinguished from the subsequent "cooling" (sometimes called "thermal spike") phase ($> 10^{-13}$ s). While analytical theories based on binary-collision approximations and linearized Boltzmann transport equations provided insight into fundamental atomic-displacement mechanisms, they could not describe the cascade evolution beyond the "collisional" regime. Conversely, the "cooling" phase was treated by applying concepts of thermodynamics, assuming that a local equilibrium was rapidly reached. With the availability of more powerful computers, extensive, albeit time-consuming molecular-dynamics simulations of energetic collision cascades became feasible; these computations provided detailed insight into the mechanisms relevant to both of the timescales mentioned above, with some simulations extending into the tens of picoseconds range.

### 2.2.1 Bulk Damage

The pioneering work of Gibson et al. [2.59] shed some light on the defect production process in low energy ($< 400$ eV) events. It was found that RCSs propagate along close-packed directions and that this mechanism effectively separates interstitials from vacancies. Using more powerful computers, Guinan and Kinney [2.54] carried out MD simulations on tungsten at energies from 25 eV to about 5 keV. Their results demonstrated that point defects created during the collisional phase of the cascade may experience extensive diffusion with the possibility of recombination during the cooling phase and that this process becomes more pronounced with increasing recoil energy. This observation provides a reasonable explanation for the experimental finding that the defect production efficiency decreases with increasing recoil energy.

King and Benedek [2.57] studied collision cascades in Cu in the energy range 25 to 600 eV by means of MD simulations, utilizing a crystallite with up to $\sim 15\,000$ movable atoms. The computations showed that during the collision phase, the number of interstitials and vacancies created, i.e. Frenkel pairs, increases quite rapidly with time and most of the energy of the lattice is kinetic. During the collision phase the kinetic-energy spectrum is far from a thermal-equilibrium Maxwellian. At a time $t \sim 2 \times 10^{-13}$ s, the instantaneous number of Frenkel pairs reaches a maximum (about 35 for a 500 eV event) and then begins to decrease as a result of extensive thermal rearrangement and annihilation of defects during the cooling phase. Equipartition of kinetic and potential energy is established during the cooling phase. After $\sim 7 \times 10^{-12}$ s, the spectrum of atomic kinetic energies is closely approximated by a Maxwellian distribution corresponding to a temperature of $\sim 17$ K, and no further defect migration is observed. The defect distribution at the end of the *collision phase* (defined as the time at which all atom kinetic energies have fallen below 5 eV) reveals a pronounced depleted zone at the center of the cascade, surrounded by an interstitial "cloud". At the end of the *cooling phase*, four Frenkel pairs remained in a 500 eV event. In this case, interstitials migrated a total of 170 times during the cooling phase and vacancies migrated 65 times. The four interstitials, oriented as [100] dumbbells, were found to surround the vacancies. The authors describe the early-stage expansion of the cascade by using the radius of gyration of the interstitial cloud and report that, initially, the cascade expands at about 12.5 times the speed of sound. This supersonic expansion rate rapidly slows to below the speed of RCSs along the ⟨110⟩ direction (about five times the sound speed) and falls below the sound speed at about $10^{-13}$ s.

The damage function $\nu(T)$, defined as the average number of Frenkel pairs created as a result of a lattice atom recoil of energy $T$, has been derived by King and Benedek [2.57] from their MD simulations in the range $25 \leq T \leq 500$ eV. The damage function determined is shown in Fig. 2.4, which also depicts the number of Frenkel pairs at the end of the collisional phase. The damage function exhibits a plateau at $\nu(T) \sim 0.5$ Frenkel pairs in the range $T \sim 30\text{--}125$ eV. The onset of multiple defect production ($\nu(T) > 1$) is slow compared to the Kinchin–Pease damage function (which is displayed

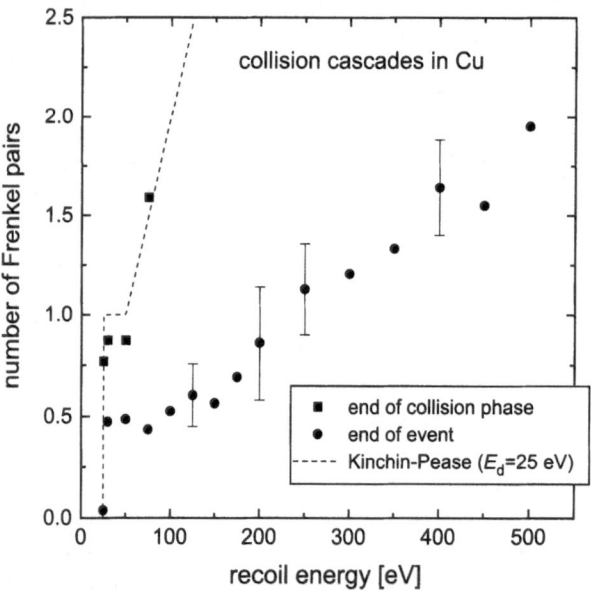

**Fig. 2.4.** Average number of Frenkel pairs (damage function, $\nu(T)$) as a function of recoil energy $T$ at the end of the collisional phase and at the end of the event as derived from MD simulations of displacement cascades in Cu. The damage function according to Kinchin and Pease with a displacement energy $E_d = 25$ eV is also depicted. Data from [2.57]

in Fig. 2.4 for a threshold energy $E_d = 25\,\text{eV}$) and only reaches $\nu(T) = 2$ at $T \sim 500\,\text{eV}$. By contrast, the number of Frenkel pairs produced during the collision phase as a function of recoil energy exhibits a reasonable agreement with the Kinchin–Pease (KP) values. The pronounced step of $\nu(T)$ in the vicinity of the minimum threshold and the plateau ($\nu(T) < 1$) have also been observed in experiments employing electron irradiation. According to King and Benedek, the pronounced deviation of the damage function from the simple KP model can be understood as follows: while at the end of the collision phase the number of Frenkel pairs corresponds closely to the KP value, at the end of the event the defects produced have been strongly reduced because of the athermal defect recombination.

King and Benedek [2.57] also investigated the importance of replacement sequences for defect production and atomic mixing in alloys. One of the major results of this study is that long linear RCSs are rare. However, short replacement sequences are often connected end to end, forming long, non-linear replacement chains. Eight percent of all replacements are of the $\langle 110 \rangle$ type. Closed chains make up about 60% of the total number of chains but, on average, have many fewer replacements than open chains. Unlike closed chains, which do not result in any defect production, open chains create a vacancy at one end and an interstitial at the other. The total number of replacements at the end of the events increases steadily with the recoil energy and amounts to about 60 at 400 eV. Roughly half of the replacements lie in closed chains; although these do not produce Frenkel pairs, their presence might entail considerable atomic mixing in alloys.

MD simulations of displacement cascades at higher impact energies were performed by several groups [2.47, 64–67]. Averback and coworkers [2.65–67] studied Cu and Ni at energies up to 5 keV. In agreement with the data at lower energies discussed above, they observe that the majority of replacements occur in the central core of the cascade, while a few of the replacements form trails leading to interstitial atoms lying outside the center. Conversely, the vacancies form a compact depleted zone. This dense cloud of replacements in the core is apparently formed by a process akin to melting. About 1 ps after the event was initiated, the simulations reveal a well-defined disordered zone embedded in a somewhat distorted crystalline matrix. Examination of the atomic distribution shows that this disordered zone initially grows to a maximum size of $\sim 2.5\,\text{nm}$ in radius (for a 5 keV recoil in Cu) and then shrinks as recrystallization occurs at its periphery. The recrystallization is complete at $t = 8\,\text{ps}$. Radial pair distribution functions for the atoms in the disordered zone were constructed and found to resemble closely that of liquid Cu; in particular, the disappearance of the (200) crystalline peak in the pair distribution function at 0.36 nm was noted. The authors stress that the lifetime of the disordered (melted) region is a few ps, i.e. some 20–50 lattice vibration periods. Hence, a quasi-equilibrium state could be established in that time interval.

Furthermore, the authors derived [2.65, 66] the temperatures and atomic densities from the computations as functions of distance from the center of the cascade at different times in the cascade evolution. (They assumed the average kinetic energy per atom to equal $3/2k_BT$.) These data are depicted in Fig. 2.5. At early times ($\sim 0.25$ ps) the average temperature in the center of the cascade ($R < 1$ nm) amounts to about 5000 K, and the temperature gradient outside this core is $\sim 3000$ K/nm. The initial cooling rate in the center of the cascade is about $10^{15}$ K/s. At 1.4 ps, the temperature begins to fall below the melting temperature of Cu at a position $R = 1.7$ nm; this corresponds roughly to the radius of the disordered zone. The atomic density variations correspond to the temperature profiles. Expansion in the hot central region gives rise to a much reduced density, about 85% at 1.14 ps. It was observed also that a high-density ridge is formed outside the hot core. As the high-temperature zone cools, the compression at the cascade rim relaxes and the density returns to its equilibrium value. The authors point out that these thermal-spike effects and the local melting may have pronounced consequences for atom relocation (e.g. atomic mixing in multicomponent specimens). According to their simulations, only a small fraction of mixing occurs

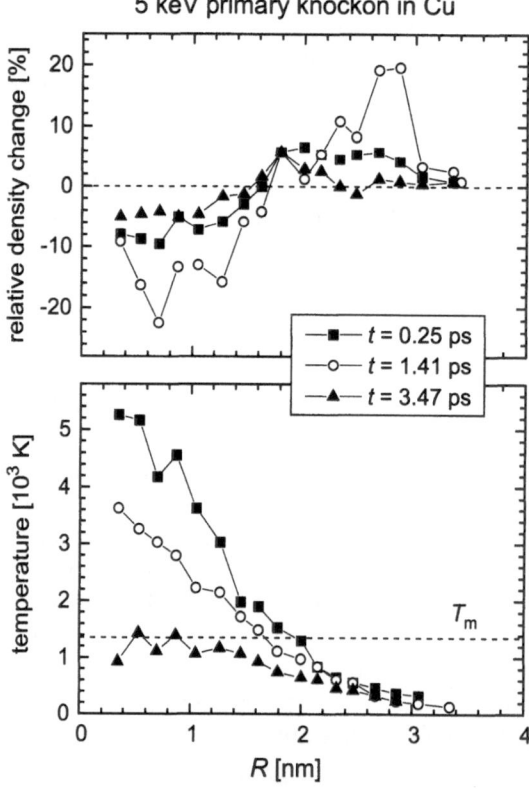

**Fig. 2.5.** MD simulation data of temperature and density profiles at three instants of time in a collision cascade in Cu initiated by a 5 keV recoil. Data from [2.65, 66]

during the collision phase and the vast majority of displacements can be attributed to the thermal-spike phase.

The local melting in the cascade as observed in the MD simulations of Averback and coworkers [2.65–67] also has important implications for defect production. Only those interstitial atoms that are transported beyond the boundaries of the melt zone survive eventual recombination. For a 5 keV event the simulations produce an average of 12 Frenkel pairs, corresponding to an efficiency of 0.2 with respect to the modified KP relation. This low efficiency is apparently due to the recombination of those interstitials that do not escape the melted region. Consequently, the defect production efficiency should decrease with increasing recoil energy since the dimensions of the cascade and of the melt zone increase with energy. At sufficiently high energies, the cascade may split into subcascades; in this regime, the efficiency is expected to become energy-independent. These anticipations are supported by experimental observations.

Surviving interstitial atoms escape the region of local melting via RCSs; these have an average length of 2.3 nm for a 5 keV recoil in Cu. Since attractive elastic interactions exist between interstitial atoms, some interstitial clustering may occur. Owing to the loss of interstitials by RCSs, the interior of the cascade contains an excess of vacancies (the depleted zone) after the melt has resolidified. The dynamic collapse of depleted zones into dislocation loops and stacking-fault tetrahedra at low temperatures has been observed in transmission electron microscopy (TEM) studies; vacancy collapse in Cu has not been found to occur, however, at energies below 10 keV.

Averback and coworkers [2.67] also studied 3 keV collision cascades in Cu at elevated temperatures (up to 700 K), monitoring the temperature in the center of the cascade (averaging the kinetic energy inside a sphere with a radius of 3.5 lattice constants) as a function of elapsed time for various ambient temperatures. At $t \sim 0.5$ ps, the effective temperature is of the order of the liquid–vapor critical temperature ($\sim 6000$ K) for all ambient temperatures. The density in this central region is lower than in the surrounding matrix but a few percent greater than the equilibrium value for the liquid. The time-dependence of the cascade temperature can be fitted by the form $T(t) - T_0 \sim t^{-1.35}$, where $T_0$ is the ambient lattice temperature. This dependence corresponds rather closely to analytical thermal spike models: Seitz and Koehler [2.62], for example, report an exponent of $-1.5$.

## 2.2.2 Formation of Surface Vacancies and Adatoms

While earlier studies into the production of defects in solids due to ion irradiation concentrated mostly on effects within the bulk of the material or on the sputtering of particles from the surface, refined experimental techniques such as the scanning tunneling microscope (STM) and more extensive MD simulation have made it feasible to investigate also defect production at the very surface and/or in the near-surface region of ion-bombarded solids.

In this and the following section, some of these recent STM results will be discussed; furthermore, data obtained by MD simulations will be presented, with a rather wide range of ion energies (30 eV to 20 keV).

The creation of adatoms on surfaces due to energetic-ion irradiation has been observed by Harrison and coworkers [2.68, 69] in MD computer simulations of 5 keV $Ar^+$ impact on Cu(100): a comparatively large number of atoms was relocated onto the outermost surface layer. Further observations of this phenomenon were reported in later simulations by Harrison and his coworkers [2.70] and in those of other groups [2.71, 72]. The effect of adatom formation was demonstrated experimentally by means of scanning tunneling microscopy by Michely and Comsa [2.73–75]. After 600 eV $Ar^+$ bombardment of Pt(111), STM topographs exhibited small monolayer-high Pt adatom islands on the original surface; these resulted from the nucleation of individual adatoms generated by ion impact. These features are exemplified in Fig. 2.6a, which represents the topography obtained upon exposing a Pt(111) surface at 150 K to a 200 eV $Ne^+$ ion fluence of $1.14 \times 10^{14}$ ions/cm$^2$, which corresponds to the removal of 4% of a monolayer (ML) by sputtering [2.73]. The topography is dominated by small, rather irregular adatom islands of monoatomic height with a typical length scale of 1 nm. At the temperature used, the adatoms created are relatively stable against recombination with vacancies present at the surface. As monovacancies are almost immobile at 150 K, no larger vacancy islands are formed and hence the isolated surface vacancies formed by ion bombardment are largely invisible in the STM topographs (a few very small, dark structures are visible in Fig. 2.6a). Annealing of the sample at 400 K results in the morphology shown in Fig. 2.6b. The adatom islands are now more compact, are bounded by $\langle 110 \rangle$-oriented steps and have a triangular or hexagonal shape. From STM topographs like this, the authors were able to determine the adatom yield $Y_a$ as a function of bombarding energy for different ion species (the values derived will be given below).

Annealing at even higher temperatures causes recombination of surface adatoms and vacancies, and bulk vacancies start to migrate to the surface. This provides the opportunity to determine sputtering yields from the STM topographs and to compare them with $Y_a$ (see below). Figure 2.6c shows the sample after further annealing at 750 K. During this annealing step, the adatom islands dissociated completely into adatoms, which then recombined with surface vacancies. The topograph exhibits only vacancy islands, which are exclusively of monatomic height with a lateral dimension of 5–10 nm.

Several other STM investigations by various groups [2.76–83] have demonstrated, for a variety of surfaces, that these and related effects of defect production are rather common phenomena under ion bombardment of surfaces. Data for semiconductors are discussed in Chap. 4.

MD simulations of low-energy ($\leq 100$ eV) bombardment of a Cu(100) surface by Cu atoms were performed by Karetta and Urbassek [2.71]. These authors studied the time-dependent vacancy and interstitial distribution in

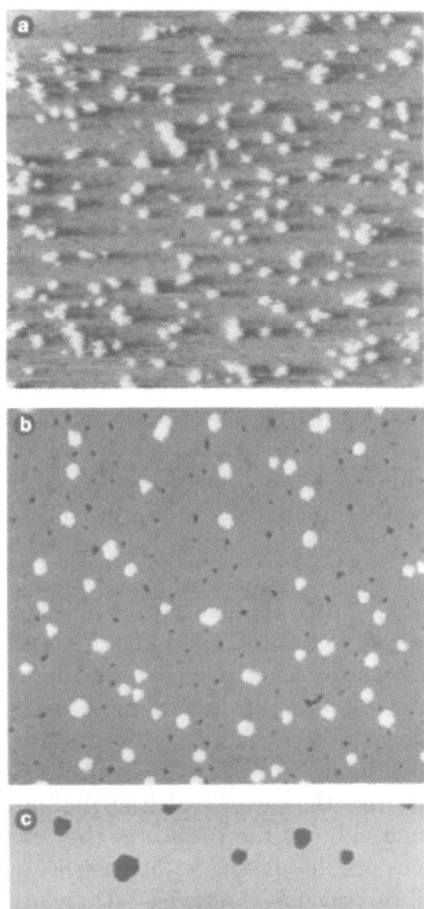

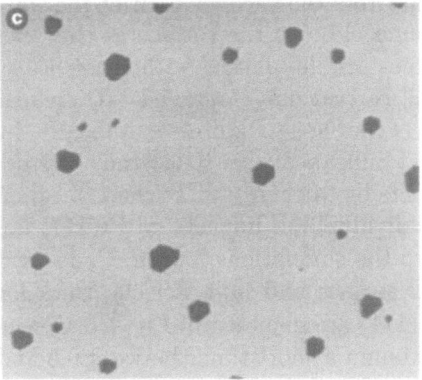

**Fig. 2.6a–c.** STM images of adatom formation on a Pt(111) surface due to 200 eV Ne$^+$ ion bombardment at a fluence of $1.14 \times 10^{14}$ ions/cm$^2$. (**a**) was imaged immediately after ion bombardment at a sample temperature of 150 K (size 55 nm × 48 nm, illumination from the left), (**b**) was recorded after annealing at 400 K (55 nm × 48 nm, gray scale) and (**c**) corresponds to a further annealing step at 750 K (110 nm × 96 nm, gray scale). Data from [2.73]

the target and the number of surface vacancies and adatoms produced; they find three categories of damage to occur: vacancies, adatoms and interstitials of the dumbbell type. They monitored the number of vacancies $N_v$, the number of interstitials $N_i$, and the number of adatoms $N_a$ as a function of time. In the early stages of the events (0.2–0.3 ps) the lattice is maximally disordered and about 6.5 vacancies and 1.5 interstitials have formed. As these are largely unstable, many of them recombine within the next half picosecond. After 0.6 ps, the number of interstitials is constant and after about 1.5 ps the number of vacancies also levels off to a constant value; only a fraction of the initially created defects survives. According to the authors, these timescales are reasonable, considering that interstitials may form within the target via replacement sequences (which travel with at least the speed of sound), while vacancies are created closer to the surface, where several channels for relaxation are available. The number of adatoms takes longest to reach its equilibrium value (1.5 to 2 ps). Slow sputtered particles that do not escape may end up as adatoms at a late stage of the event. For an impact energy of 100 eV, the final numbers of vacancies, interstitials and adatoms and the sputtering yield are 1.57, 0.51, 1.54 and 0.52, respectively; for 30 eV, the corresponding values are 0.07, 0.04, 1.03 and no atom sputtered within 210 simulation runs. Obviously, at the lower energy the number of vacancies and interstitials formed is drastically reduced, but the number of adatoms is still appreciable. Of these, 40% are projectile atoms and the remainder are surface atoms pulled on top of the surface layer, with the resulting vacancy immediately filled.

The depth distribution of defects at 4 ps after the ion impact indicates that following spontaneous recombination, the vacancies and interstitials are spatially separated. The latter are only formed deeper than the third layer, while vacancies assemble close to the target surface, in the first three layers. The majority of all vacancies (94%) are located in the first layer. This spatial separation is a sign of the production mechanism of interstitials via RCSs which separate the remaining interstitials from the vacancies left at the start of the sequence. No interstitials are found close to the surface as they can easily decay there, either forming adatoms or annihilating with vacancies.

Recently, Gades and Urbassek [2.72] carried out a detailed MD simulation of rare-gas bombardment of a Pt(111) surface to address specifically the formation of adatoms and of surface and bulk vacancies. This study was devised to complement the STM experiments by Michely and Teichert [2.74] on adatom formation and sputtering by individual ion impacts on Pt(111) described above (see Fig. 2.6). Similarly to the simulations for Cu, Gades and Urbassek studied the time evolution of surface and bulk defects. Data for 100 eV and 600 eV Xe irradiation of Pt(111) are depicted in Fig. 2.7. Generally, the number of defects shows a maximum a short time (between 0.5 and 1.5 ps, depending on the bombarding energy) after ion impact; with progressively longer times the number of defects is reduced because of annealing and eventually reaches a final value. (Note that the simulations refer to a temperature of 0 K.) This annealing rate increases with increasing bombarding energy. For adatoms, for example, only 50% are annealed at 100 eV, whereas up to 80% are annealed out at 1 and 3 keV. The authors stressed that the time needed for the equilibration of adatoms and other defects is considerably

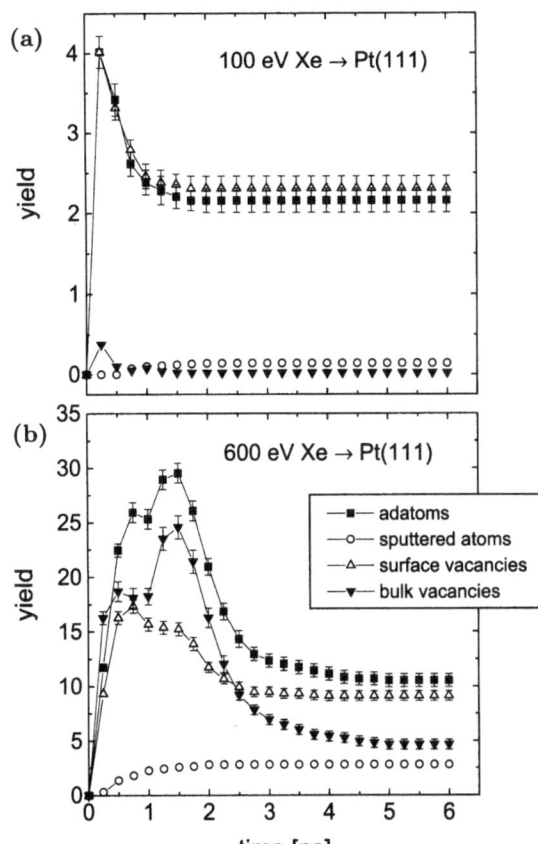

**Fig. 2.7a,b.** MD data on the time evolution of defects for (**a**) 100 eV and (**b**) 600 eV Xe bombardment of Pt(111) at perpendicular incidence; the yields of adatoms, sputtered atoms, surface vacancies and bulk vacancies are given. From [2.72]

longer than the time during which sputtering takes place; this effect is more pronounced at higher irradiation energies.

As shown in Fig. 2.7, for 100 eV Xe bombardment essentially only adatoms and surface vacancies are formed; at higher energies bulk vacancies are also created up to the fifth or ninth layer for 600 eV or 3 keV bombardment, respectively. Considering the depth of origin of the adatoms and sputtered atoms, Gades and Urbassek note that adatoms come from increasingly larger depths for higher impact energies: while at 100 eV they all originate from the first layer, for 600 eV bombardment more than 10% come from the second layer; at 3 keV the surface becomes very rough and even atoms from the fifth layer may end up as adatoms, with about 60% coming from the topmost layer. A similar trend was observed for sputtered atoms, albeit not in such a drastic form; even for 3 keV ion irradiation 90% of the ejected species originate from the first monolayer, with contributions coming from up to the third layer. The evaluation of the yields of adatoms $Y_a$, surface vacancies $Y_{sv}$ and sputtered atoms $Y_s$ provides a means to assess the transition from surface defect formation at low energies to bulk defect formation at higher energies. At 100 eV, $Y_{sv} = Y_a + Y_s$ since all adatoms and sputtered atoms originate from the first layer and the number of bulk vacancies is very small, as the bombarding energy is not sufficient to generate defects in the bulk that are stable after the spontaneous annealing phase about 1 ps after ion impact (see Fig. 2.7). At higher energies (600 eV), the simulations indicate $Y_a > Y_{sv}$ and $Y_a$ is about four times higher than $Y_s$. The number of bulk vacancies ($Y_{bv}$) also starts to increase. At even higher energies (3 keV), $Y_a$ exceeds $Y_{sv}$ by about 50% and $Y_{bv}$ is considerably larger than $Y_{sv}$, i.e. defect formation becomes a bulk rather than a surface effect.

In the following, the experimental STM results of Michely and Teichert (see above) and the computer simulations for the Pt(111) surface will be presented in quantitative terms. Specifically, adatom yields and the adatom/sputter yield ratios from both approaches will be compared. For further details the reader is referred to the original publications [2.72, 74].

Figure 2.8a shows the adatom yields $Y_a$ (= number of adatoms per ion) from the experiments and the simulations for three projectiles (Ne, Ar, Xe) as a function of the impact energy. In the STM experiments [2.72] the sample was held at a temperature of 150 K during the bombardment and then annealed at 400 K to obtain the best conditions for STM imaging and adatom determinations. The total fluences ranged from $2.8 \times 10^{15}$ cm$^{-2}$ for 40 eV Ne$^+$ to $2.4 \times 10^{12}$ cm$^{-2}$ for 5 keV Xe$^+$. It is seen that $Y_a$ steadily increases with increasing energy, both for the experiment and for the simulations, but the latter predict considerably higher yields than the experiment (the authors state, however, that the $Y_a$ values derived from the STM images may constitute lower limits of the true adatom yields). The experimental values of $Y_a$ are found to span a large range, from $1.1 \times 10^{-2}$ for 40 eV Ne$^+$ to 66 for 5 keV Xe$^+$ impact. One may suspect that for the latter bombardment conditions "large"-yield events can drastically enhance the adatom yields, as demonstrated in a related study [2.84]: impact events have been observed in which as many as 500 atoms have been pushed onto the original surface. Such "spike" effects and even the evolution of a molten area were seen also in MD simulations of collision cascades initiated by 10 keV Au$^+$ impacts on Au performed by Ghaly and Averback [2.85, 86]. In the latter work, thermal expansion of a liquid droplet was found to cause material to flow onto the surface and then to crystallize. Those investigations are discussed in more detail in Sect. 2.2.3.

Figure 2.8b shows the adatom-to-sputter yield ratios $Y_a/Y_s$ for the computer simulations [2.72] and for the STM experiments [2.74] for the three projectile species. Since the sputtering yields agree very well, the divergence of $Y_a$ noted in Fig. 2.8a translates into a similar discrepancy of $Y_a/Y_s$ between the two approaches, with the simulations generally producing the higher ratios. In both cases a strong increase of $Y_a/Y_s$ is reported with decreasing impact energy for all projectiles. This appears rather plausible if one assumes

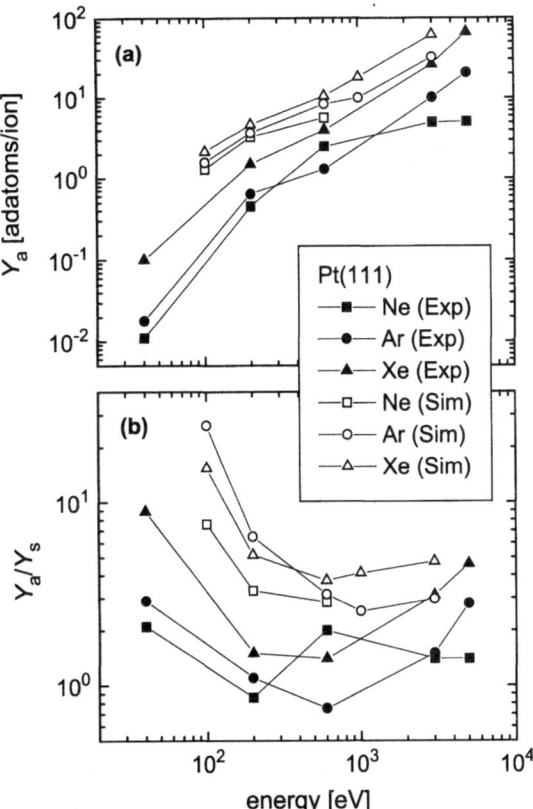

**Fig. 2.8a,b.** (a) Adatom yields and (b) adatom/sputter yield ratios for Pt(111) as a function of bombarding energy for three projectiles. The data are from the STM experiments of Michely and Teichert [2.74] and from the computer simulations of Gades and Urbassek [2.72] (*closed* and *open symbols*, respectively)

that the energy required to form an adatom, $U_a$, is smaller than that necessary to sputter-eject an atom, $U_s$: at low energies atoms may form adatoms but may not be able to escape from the surface, and thus $Y_a/Y_s$ will increase. Gades and Urbassek [2.72] give an estimate of $U_a/U_s$ based largely on a nearest-neighbor model and derive, for the (111) surface of an fcc metal, $U_a/U_s \cong 0.4$. For the conditions of a linear collision cascade (i.e. above a few hundred eV bombarding energy), they give a yield ratio $Y_a/Y_s \sim 4$ for atoms sputtered with an energy in excess of $U_s$, in agreement with their simulation data. The experiments, on the other hand, produce in this energy regime ($\geq 500\,\text{eV}$) a value $Y_a/Y_s \sim 1$. While for Ne$^+$ bombardment this number stays roughly constant with increasing energy, the $Y_a/Y_s$ ratio increases for Ar$^+$ and Xe$^+$ (to about 2.8 and 4.6, respectively, for 5 keV impact). Thus, the experimentally determined constant yield ratio ($Y_a/Y_s \sim 1$) at intermediate energies and its increase with decreasing energy appear to be in reasonable agreement with the simulations and the aforementioned concept of binding energies. The increase of $Y_a/Y_s$ with the irradiation energy, on the other hand, is tentatively ascribed by the authors to the possible occurrence of spike effects as discussed above. The necessary dense collision cascades are not expected to develop under $Ne^+$ impact; in fact, $Y_a/Y_s$ remains roughly constant for $Ne^+$ above some 100 eV bombarding energy [2.74].

The STM experiments definitely confirm the creation of adatoms in low-energy ion bombardment of surfaces. MD computer simulations [2.72] support the outcome of the experiments but produce consistently higher adatom yields. Both approaches indicate that, at low impact energies ($\leq 100$ eV), adatom yields are considerably higher than sputtering yields (by an order of magnitude or possibly more). At energies around 1 keV the yield ratio $Y_a/Y_s$ appears to have a minimum (in the experiment) or to level off to a constant value (in the simulations). As is discussed in Sect. 2.2.3, the occurrence of "large" emission events [2.84–86] might be important at high collision densities (e.g., $Xe^+$ at 5 keV or more on Pt or other heavy targets).

Snapshots from an MD simulation [2.87, 88] of the evolution of a displacement cascade for 1 keV Cu irradiation of Cu(100) are shown in Fig. 2.9. The panels show cross sections through the crystal at different times (given at the upper left) after the ion impact. The color coding indicates the "temperature" normalized to the melting temperature of Cu ($T_m = 1358$ K). The authors [2.87] define the temperature (and similarly the density) of an atom as the average kinetic energy (in the center-of-mass frame) of all atoms in a sphere of radius $R$ around the atom; $R$ was chosen equal to the cutoff radius of the potential used (0.47 nm). The simulations indicate that the core of the collision cascade is at temperatures above the melting point and at densities below those of solid or liquid Cu for an extended time interval. Resolidification is completed only after 1.8 ps. Conversely, particle ejection is terminated after about 0.3–0.4 ps. The number of atoms in the molten core is time-dependent; it reaches a maximum (about 850 atoms) around 0.13 ps and falls to $\sim 100$ atoms at 0.4 ps, staying fairly constant at this value for another picosecond [2.87]. Similarly, the kinetic energy of the atoms in the molten core remains for quite a long time just above the melting temperature; it decreases only slowly between 0.4 and 1.5 ps because of the lattice conductivity of solid copper. During this time interval, the potential energy is about 0.15 eV above the kinetic energy. The authors [2.87] noted that this value is close to the latent heat of melting of Cu (0.135 eV). At the end of the simulation after 3.6 ps (see the lowest panel in the right-hand column of Fig. 2.9), the atoms in the region of the cascade have almost cooled to the temperature of the surrounding atoms. Still, considerable bulk damage is observed which, to some extent, might anneal out during much longer time periods not accessible by the simulation. A fairly large number of adatoms remaining on top of the original surface are also clearly visible.

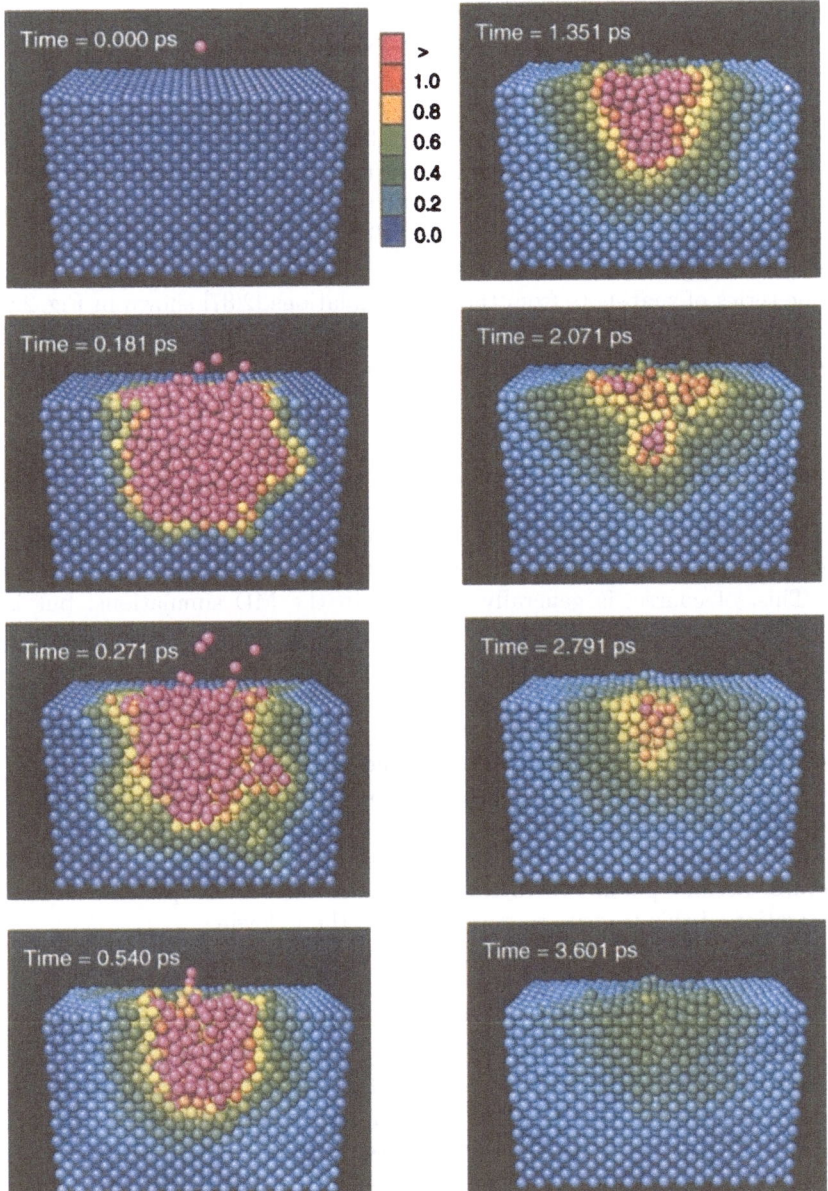

**Fig. 2.9.** Snapshots from an MD simulation [2.87] of the evolution of a displacement cascade for 1 keV Cu irradiation of Cu(100). The panels show cross sections through the crystal at different times (given at the *upper left*) after the ion impact. The color coding indicates the "temperature" normalized to the melting temperature of Cu ($T_m = 1358$ K)

Colla and Urbassek [2.87] also define in their simulations a pressure via the virial theorem and utilize this parameter to investigate the possible occurrence of shock waves in the irradiated material during the time evolution of the cascade. They observe, roughly 0.3 ps after ion impact, a compression wave moving outwards from the core, confined to only 1 or 2 monolayers. The atoms constituting this shock wave form distinct (111) layers that have detached from the central region. The authors stress that the longitudinal velocity of sound in Cu is largest in (111) directions, which may explain this preferential crystallographic orientation.

The series of snapshots from the MD simulations [2.87] shown in Fig. 2.9 thus illustrates both the occurrence of bulk defects and melting as discussed in Sect. 2.2.1, and the formation of surface defects (adatoms and vacancies) considered in this section.

### 2.2.3 Extended Surface Damage

The results presented in the preceding section referred to single-ion impact events, i.e. to a situation where the impinging ions encounter a pristine surface. This, of course, is generally the case in the MD simulations, but is also valid for those STM investigations which were performed at ion fluences that largely satisfied this condition (removal of much less than a monolayer). With progressive surface erosion, the development of surface (and bulk) damage more extensive than single adatoms and surface vacancies is expected to occur. In particular, pronounced topographic features may start to evolve, as are well documented for very-high-fluence ion-bombarded samples. Monitoring the transition from the generation of isolated surface defects (adatoms and vacancies) to the initial stages of the formation of gross surface defect structures became possible with scanning tunneling microscopy. A thorough investigation of this transition is presented in the following.

Michely and Comsa [2.75, 79, 80, 89] have investigated, in a series of papers, ion bombardment effects on metal surfaces using STM. Specifically, they studied the surface morphology of Pt(111) after 600 eV Ar$^+$ bombardment as a function of ion fluence and specimen temperature. Without doubt, a very important outcome of their work was the finding that the sample temperature during sputtering has a decisive influence on the surface topography, at least for the Pt(111) surface: above a transition temperature, the presence of surface diffusion during irradiation tends to favor the formation of large, smooth, well-developed surface structures. The STM micrographs in Fig. 2.10 [2.89]

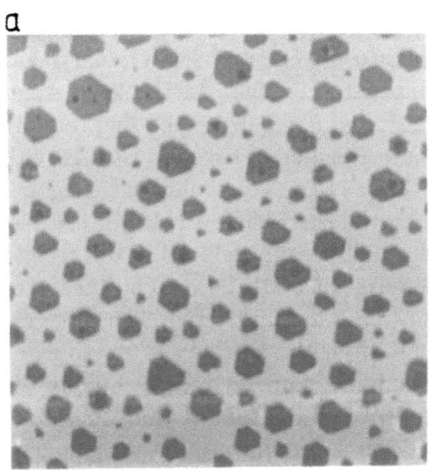

**Fig. 2.10a–e.** Comparison of the surface morphologies (**a**), (**b**) and (**c**), obtained by STM at a sample temperature $T_s = 625\,\mathrm{K}$ with the surface morphologies (**d**) and (**e**), obtained at $T_s = 910\,\mathrm{K}$ under $600\,\mathrm{eV}$ $\mathrm{Ar}^+$ bombardment of Pt(111). The image sizes are $82.5\,\mathrm{nm} \times 82.5\,\mathrm{nm}$ in (a), (b) and (c), and $330\,\mathrm{nm} \times 330\,\mathrm{nm}$ in (d) and (e). The ion fluence is $3.05 \times 10^{14}\,\mathrm{ions/cm^2}$ in (a) (0.45 ML removed), $3.4 \times 10^{15}\,\mathrm{ions/cm^2}$ in (b) and (d) (5 ML removed) and $3.4 \times 10^{16}\,\mathrm{ions/cm^2}$ in (c) and (e) (50 ML removed). (b) and (c) are presented as images with illumination at glancing incidence from the left, all other images are gray scale topographs. From Michely and Comsa [2.89]

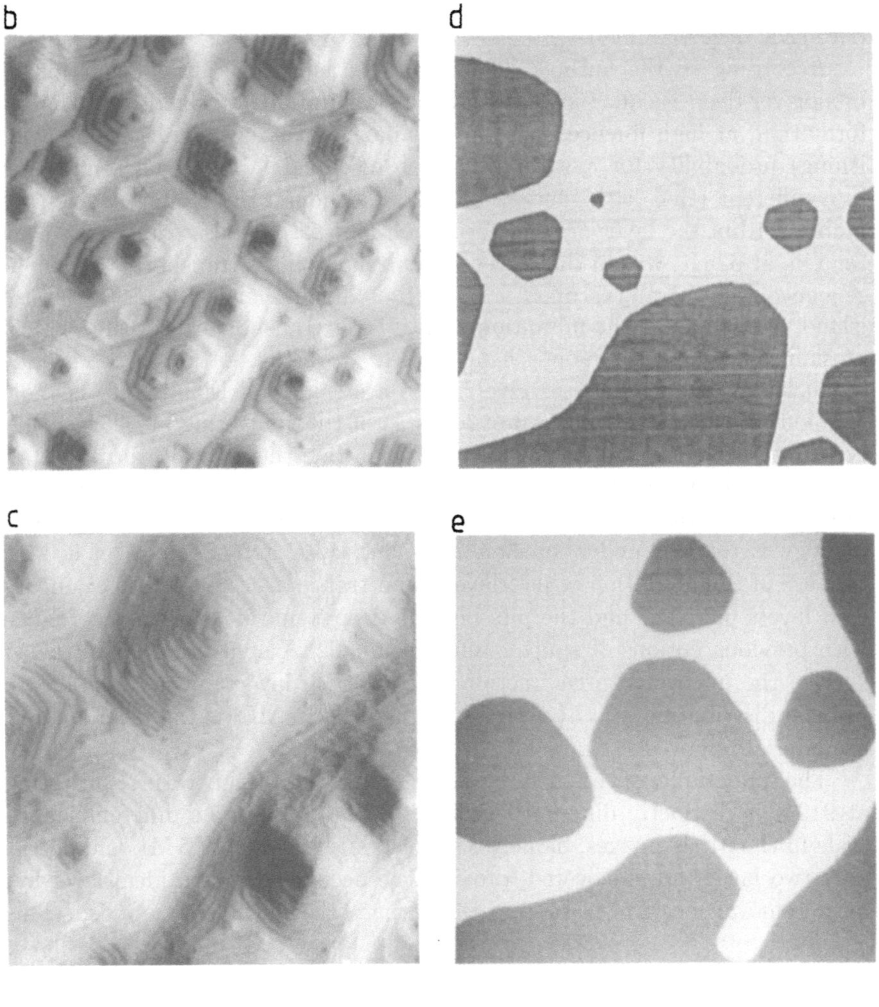

demonstrate the dramatic differences in the surface morphological evolution with ion fluence at two sputtering temperatures (625 and 910 K), one below and the other above the transition temperature range. At the lower temperature, after an ion fluence of $3.05 \times 10^{14}$ ions/cm$^2$, corresponding roughly to the removal of 0.45 monolayers, a large number of vacancy islands with diameters of 0.5 to 8 nm are observed. They are of triangular to hexagonal shape, bounded by monatomic steps along the close-packed $\langle 110 \rangle$ directions. In the bottom of the larger islands, new small vacancy islands in the second layer can already be observed. After removal of 5 ML by an ion fluence of $3.4 \times 10^{15}$ ions/cm$^2$ (Fig. 2.10b), twelve layers were uncovered. The topography is now dominated by pits that result from the stacking of up to eight vacancy islands. Again a tendency for a sixfold symmetry is seen. Finally, after removal of 50 ML at a fluence of $3.4 \times 10^{16}$ ions/cm$^2$, more than 30 atomic layers are exposed, with a few deep, wide pits dominating the morphology. The regularity of the pit shape is now very pronounced; also, the remainders of almost completely removed layers tend to form adatom islands stacked onto each other.

According to the authors [2.89], the characteristic initial formation of monolayer-deep vacancy islands and the stacking of the islands, i.e. the pit formation, at high fluences can be explained by the low (and perhaps vanishing) probability for a vacancy to become filled by an atom originating in a different (that is, higher) layer. Thus, although the monovacancies are mobile within the layer of their formation (*intra*layer mass transport), they remain, at 625 K, within that layer (no *inter*layer mass transport). As a consequence, the intralayer mass transport of monovacancies leads to lateral island growth, while the inhibition of interlayer transport causes the vertical erosion, by forcing the monovacancies created on the bottom of the islands to nucleate. These processes give rise to a stacking of islands into islands. The authors stress two important features of the sputtering at 625 K: (i) A constant pit slope is preserved during erosion (i.e. the pit walls form facets), so that the diameter grows correspondingly; therefore, the number density of pits decreases with increasing fluence. (ii) No steady-state (saturation) morphology is reached under ion bombardment, even upon removal of 50 ML; because of the inhibition of interlayer atom transport, the number of uncovered layers increases and the pits become deeper and deeper. In accordance with previous notions of sputter removal, for 625 K sputtering of a Pt(111) surface the uncovered area fraction $x_i$ of the $i$th layer appears to follow a Poisson distribution, $x_i = (A^i/i!)e^{-A}$, where $A$ (in ML) is the total amount of material removed.

The topography of the surface observed after high-temperature sputtering at 910 K is distinctly different (see Fig. 2.10d,e). (Note the different length scale in the 910 K images, namely 330 nm.) After erosion of 5 ML and 50 ML only two layers are uncovered, proving the occurrence of an ideal layer-by-layer removal mechanism despite the rather high ion fluence applied. This finding is in agreement with earlier results [2.90, 91] of thermal-energy atom scattering investigations, which revealed the transition to layer-by-layer erosion to occur in a temperature range between 650 and 700 K. For the high-temperature sputtering conditions, monovacancies created on the bottom of a vacancy island are effectively filled by a nearby adatom, so that the nucleation of vacancies to form stable vacancy clusters (or islands) is unlikely. This filling is an efficient *inter*layer mass transport from a higher to a lower layer. The suppression of vacancy island nucleation inhibits the formation of pits. Michely et al. [2.75] have shown in a related study that on Pt(111) surfaces above 700 K the 2D evaporation of step atoms from vacancy island step edges onto the island terrace becomes very efficient; very probably, this mechanism is responsible for the transition from pit formation to the layer-by-layer removal morphology. The STM images taken at the higher sputtering temperature also demonstrate that, in contrast to 625 K sputtering, at 910 K a steady-state topography develops under ion bombardment, characterized by only two uncovered layers. This saturation morphology is reached upon the removal of less than 1 ML.

Related results [2.92] have been found for *room-temperature* bombardment of an Ag(111) surface by 1 keV Ar$^+$ ions. Figure 2.11 depicts STM images taken at different ion fluences: although no layer-by-layer removal was observed, it is seen that even under these low-temperature conditions Ag atoms exhibit a sufficiently high mobility that no more than five atomic layers are exposed. This is evident also from two micrographs taken a few minutes apart (Fig. 2.11e,f), without further ion bombardment in this time interval. Some of the surface structures have distinctly changed. These examples demonstrate convincingly that the evolution of surface topography may vary considerably for different materials and will depend strongly on parameters such as sample temperature and possibly others (e.g. ion energy, flux density).

At higher impact energies, more extended defect structures can be created at or near the surface. The formation of craters (pits) and dislocation loops was observed by Jäger and Merkle [2.93–96] on Au surfaces upon irradiation with 10–20 keV monatomic and diatomic heavy ions (Bi$^+$ and Bi$_2^+$) by means of transmission electron microscopy. The defect production efficiency was found to be on the order of unity and vacancy loops were observed, a large fraction containing more than 300 defects. This very pronounced defect production was ascribed to the high energy density within the collision cascades (nonlinear effects) due to the heavy projectiles and to the presence of the nearby surface. More recently, evidence for such drastic events was provided by MD simulations carried out for similar irradiation conditions [2.85, 86].

In conjunction with their investigations of defect production in the bulk, Averback and coworkers [2.85, 86, 97, 98] studied displacement cascades near the surface. For example, for 10 keV Au bombardment of an Au surface, interstitials are ejected from the central core of the cascade in the initial

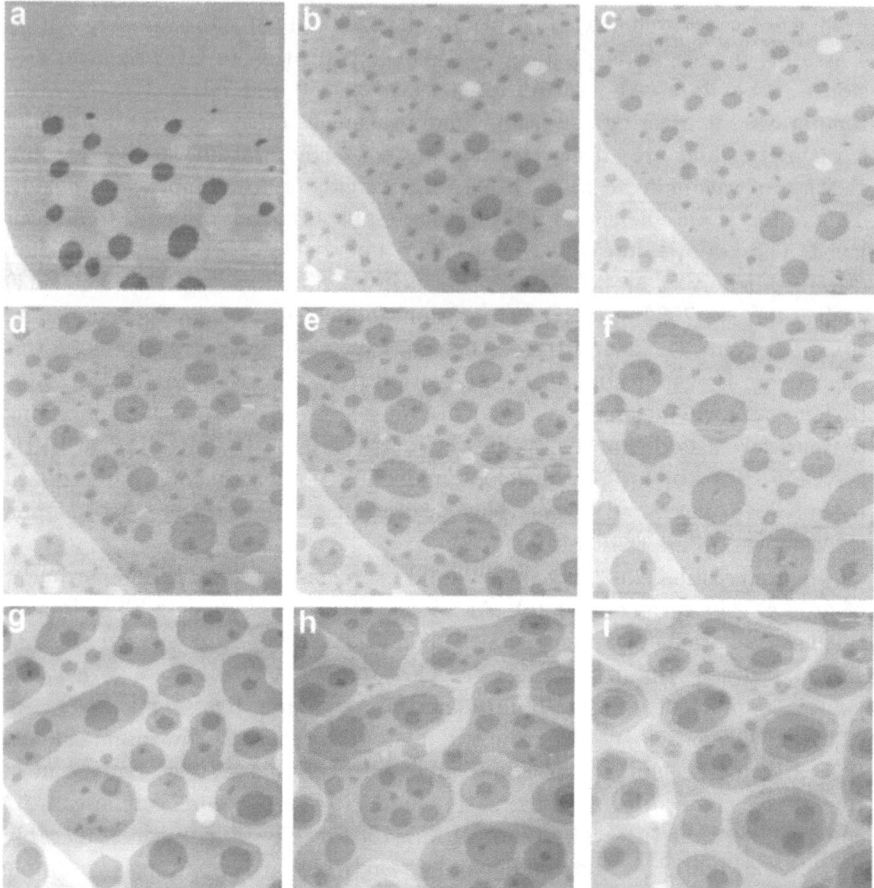

**Fig. 2.11a–i.** STM images of an Ag(111) surface irradiated at room temperature with 1 keV $Ar^+$ ions. The micrograph (**a**) was taken on a pristine surface whereas the others refer to ion fluences of 0.3 (**b**), 0.5 (**c**), 1.5 (**d**), 4.5 (**e**) and (**f**), 9 (**g**), 18 (**h**) and $36 \times 10^{14}$ $Ar^+$ ions/cm$^2$ (**i**). Image (f) was recorded at the same fluence as (e) but 8 minutes later without further ion bombardment during this time interval. The individual images have a size of 190 nm × 190 nm and were recorded with a positive bias of 1.7 V. The dark regions seen in (a) were produced *before* ion bombardment during scanning with a negative bias voltage. From [2.92]

stage of the event (some 0.1 ps), creating a depleted zone; RCSs propagating along ⟨110⟩ directions are observed, with a shock wave spreading outward with a slower speed behind it. Similarly to the observation for the bulk, local melting occurs after a few picoseconds in the core region of the cascade, and interstitials initially knocked to the periphery are now reabsorbed in the melted zone. Only those deposited at the ends of long RCSs survive. At this stage, strong cavitation may occur below the surface, caused by the high local temperatures and internal pressures; the pressure in the interior of such a cavity and in the surrounding liquid can be as high as a few gigapascals. At the same time, the temperatures are several thousand kelvin. Because of these high pressures and since the pressure at the surface is zero, a driving force for liquid flow can exist. The simulations clearly illustrate the flow of Au atoms onto the surface. (Some 500 Au atoms have been observed in individual events.) The final stages of the cascade evolution involve cooling and resolidification. The latter occurs by the inward motion of the solid–liquid interface, while many of the atoms that have flowed to the surface remain there. Consequently, too few atoms are available to fill the original lattice sites and the inward motion may give rise, eventually, to complex dislocation structures with vacancy character. The formation of dislocation loops by surface melting found in these MD simulations is probably closely related to the corresponding defect structures observed by Jäger and Merkle [2.95, 96] via transmission electron microscopy.

The MD simulations of 10 keV Au bombardment of Au demonstrate that extended defects and a large number of adatoms may be formed in such high-energy-density (thermal spike) events. According to the authors, this does not necessarily have an influence on sputtering: all sputtered particles are ejected in the first 0.2 ps of the cascade evolution, and little modification of the surface has occurred in that time period. The authors stress, however, that other irradiation conditions can produce displacement cascades that strongly modify sputtering. Typically, in such events the pressure that develops in the cavity overcomes the strength of the thin layer of material above it and ruptures the surface. Sputtering then derives from the exfoliating material from the interior of the cavity; atoms may leave the surface over times much longer (some ps) than the usual collisional phase. Not surprisingly, the ejection of some 100 sputtered atoms in single events has been found in such cases [2.99]. While they are by no means typical, their occurrence is very probably important in the context of the observation of very large clusters in sputtering (see Sect. 2.3.4).

## 2.3 Sputtering of Particles from Elemental Targets

Recoil atoms created in the collision cascade contribute to the ejection of atoms from the surface [2.100]. Those atoms that move towards the surface with sufficient energy to overcome the surface binding forces can be sputtered. This indicates that the sputtering yield depends on the energy deposition in the cascade at or near the surface. Frequently, three different regimes with respect to the type of displacement cascades occurring are defined [2.11, 101, 102]: near-threshold, linear-cascade and spike (or nonlinear-cascade). The first is operative when the energy transferred from the incident particle to the target atoms is only sufficient to produce a few isolated recoils (knockons); it dominates sputtering for low (near-threshold) energies and/or light ions (such as $H^+$, $D^+$ and $He^+$ at $E < 10$ keV). Generally, these sputtering conditions are difficult to describe by analytical theories of sputtering, but considerable insight has been provided by computer simulations (see Sect. 2.3.2). The second regime refers to interactions between the incident particle and target atoms which result in collisions cascades of the latter, but with a limited fraction of atoms set in motion. Such a cascade may consist of a series of binary collisions between moving and stationary atoms. Sputtering in this linear-cascade regime was theoretically described by Sigmund [2.8, 11] using a linearized Boltzmann transport equation. For bombardment with high-energy, heavy ions or with molecular ions, the density of recoil atoms within the cascade is sufficiently high that encounters between moving atoms become frequent; then, the linearity assumption breaks down and the third regime of high-density (spike or nonlinear) cascades is reached. This will be alluded to in Sect. 2.4 in the context of cluster-ion bombardment of surfaces.

Collision cascades leading to sputtering, and the ejection events proper have been modeled theoretically by many different concepts [2.8, 11, 101–110]; apparently, the approach of Sigmund [2.8, 11, 101] has the widest applicability to describe sputtering in the linear-collision-cascade regime. It will be briefly outlined, therefore, in the following section. Computer simulations of sputtering have contributed greatly to the elucidation of the pertinent processes. Several reviews [2.13, 70, 111–113] summarize these data.

### 2.3.1 Linear Collision Cascade

Sputtering in the linear cascade regime has been described analytically by Sigmund [2.7, 8, 11, 101]. This approach assumes that only a small fraction of the target atoms within the cascade volume is in motion and that the low-energy recoil flux is distributed isotropically; the latter assumption breaks down at low bombarding energies. Another assumption concerns the fact that the target surface does not exert a decisive influence on the development of the collision cascade. Under these conditions the sputtering yield $Y$ was predicted to scale linearly with the energy deposited in elastic collisions at the surface, $F_D(E, \theta, x = 0)$:

$$Y(E, \theta) = \Lambda F_D(E, \theta, 0) , \tag{2.28}$$

where $\Lambda$ is a material-specific constant (see below). $F_D(E, \theta, x)$ is the energy deposited by the bombarding ion (energy $E$, incidence angle $\theta$ relative to the surface normal) in low-energy recoils in the depth interval $(x, x + dx)$. The depth-integrated deposited-energy density $\int F_D(E, \theta, x)dx$ corresponds to the total energy available in the cascade for creating recoil atoms, $E_n$; cf. (2.24). Hence, $F_D$ is the space-resolved version of the recoil density $F(E, E_i)$ as given in (2.25), with $E_i$ being the energy of the recoiling atoms. (Sigmund has utilized a *linearized* form of Boltzmann's equation to describe the motion of recoil atoms and to derive the asymptotic solution for $\nu(E_n, E_i) \sim \Gamma E_n/E_i$, see (2.24). This linearization is valid for a *dilute* cascade and the term *linear collision cascade* refers therefore to that situation.)

To evaluate $\Lambda$, information on the surface potential barrier and the stopping cross section of (slowly moving) recoil atoms is required. The simplest but not necessarily the only model for the binding of atoms at the surface is a planar surface barrier $U$ [2.11, 113] (commonly the cohesive energy of the solid is used for $U$, see below). Then, the probability $P(E_i, \theta_i)$ for an atom to escape from the surface reads

$$P(E_i, \theta_i) = \begin{cases} 1, & E_i \cos^2 \theta_i \geq U \\ 0, & E_i \cos^2 \theta_i \leq U \end{cases} . \tag{2.29}$$

$E_i$ and $\theta_i$ are the energy and the angle (relative to the normal) with which the atom approaches the surface from *within* the target. The stopping power is needed as the number density of atoms moving with energy $(E_i, dE_i)$ in the cascade and, hence, also the current density depend inversely on $(dE_i/dx)$; for large values of the latter, fewer atoms are moving in the energy interval $[E_i + dE_i, E_i]$. Using (2.29), the recoil density (2.25) and the stopping cross section in the power-law form (2.12), Sigmund derives [2.8, 11]

$$\Lambda = \frac{\Gamma_m}{8(1 - 2m)} \frac{1}{NC_m U^{1-2m}} , \tag{2.30}$$

where $N$ is the atomic density of the sample and $C_m$ is given by (2.10). Owing to the uncertainties associated with the stopping power of an atom moving with very low energy $(E_i \sim U)$, he originally proposed $m = 0$ and therefore

$$\Lambda = \frac{3}{4\pi^2} \frac{1}{NC_0 U} . \tag{2.31}$$

In his original work [2.8], Sigmund used $\Gamma_m (m = 0) = 6/\pi^2$, $C_0 = \pi\lambda_0 a_{BM}^2/2$, $\lambda_0 \sim 24$ and the Born–Mayer screening length $a_{BM} \doteq 0.0219$ nm. This yields $C_0 = 0.0181$ nm$^2$. Later work [2.16] established, however, that with this choice the power-law cross sections (2.9) underestimate stopping at the energies relevant for atom ejection by roughly a *factor of two*. Note that $\xi_0 = 3/4NC_0$ is a characteristic depth of origin of the sputtered atoms (for a discussion of this quantity, see below).

The deposited energy for incoming ions of energy $E$ and incidence angle $\theta$ with respect to the surface normal can be expressed according to [2.8, 11] via

$$F_D(E, \theta, 0) = \alpha N S_n(E) . \tag{2.32}$$

Here $\alpha$ is a dimensionless function of the incidence angle $\theta$, the mass ratio $M_2/M_1$ (and the energy $E$, if electronic stopping is important). Absolute values of $\alpha$ have been given by Sigmund and were later interpolated from experimental sputtering yields. For $M_2/M_1 \leq 0.5$, $\alpha$ is nearly constant ($\sim 0.2$), but rises strongly with an increase of $M_2/M_1$ above 0.5. Approximations for $\alpha$ in the range $0.5 \leq M_2/M_1 \leq 10$ have been proposed, e.g. $\alpha = 0.3 (M_2/M_1)^{2/3}$.

(The increase with $M_2/M_1$ is caused by the rising importance of large-angle scattering events.) $\alpha$ also increases rapidly with the ion's incidence angle $\theta$. Combining (2.28), (2.31) and (2.32), the total sputtering yield is given by

$$Y(E, \theta) = 0.042 \frac{\alpha S_n(E)}{U} .$$
(2.33)

Equation (2.33) is based on $m = 0$ and $C_0(\lambda_0)$ as given above. It was found [2.16] that a more accurate evaluation of $\lambda_m$ and $C_m$ is feasible. An excellent agreement of experimental yield values with the theoretical predictions of (2.33) was noted in Sigmund's original work [2.8] and was later corroborated by many studies using a large variety of ion energies and ion–target combinations [2.114, 115]. It appears in fact that (2.33) is one of the most widely used (and quoted) single formulas of ion–solid interactions.

While (2.33) represents the integrated yield, the differential yields with respect to the emission energy $E_0$ and the emission angle $\theta_0$ of the sputtered atom are frequently of importance. The planar surface potential effects a refraction of the atom's path upon passage through the surface. The perpendicular momentum component is reduced by an amount equivalent to the binding force. Sigmund [2.8, 11] has established the differential yield of atoms sputtered with an emission energy $E_0$ into the solid angle $\Omega_0$ around the emission angle $\theta_0$ as

$$\frac{\mathrm{d}^3 Y}{\mathrm{d}E_0 \mathrm{d}^2 \Omega_0} = F_D(E, \theta, 0) \frac{\Gamma_m}{4\pi} \frac{1-m}{N C_m} \frac{E_0}{(E_0 + U)^{3-2m}} \cos\theta_0 .$$
(2.34)

Thus, the continuously falling recoil spectrum within the target is transformed into an energy spectrum peaking at an energy $(E_0)_{\max}$ which depends only on the specific sample (via $U$), but not on the incident ion and energy:

$$(E_0)_{\max} = \frac{U}{2(1-m)} ;$$
(2.35)

towards high emission energies it falls off as $1/E_0^{2-2m}$ for $E_0 \gg U$. A considerable number of experiments have been carried out with the aim of assessing the predictions of (2.34) with respect to the energy spectra of sputtered atoms, in particular the high-energy fall-off proportional to $\sim E^{-2}$ and the peak position at about $U/2$. Most of these data have been summarized in several reviews [2.106, 116, 117], including a very recent one due to Betz and Wien [2.117]. Typically, measured energy spectra were fitted to a distribution $dY/dE_0 \propto E_0/(E_0 + U)^n$, using $U$ and $n$ as fitting parameters. Despite some ambiguities involved in such a procedure, it appears that, in agreement with (2.34), both the high-energy asymptotic behavior ($n \sim 3$) and the peak at roughly half the cohesive energy have been verified by experimental results at sufficiently high impact energies (a few keV and higher). Deviations have been observed in measurements in which the impact energy was reduced below about 1 keV [2.118–123]: then, the peak of the energy spectrum tends to shift to lower energies and the width of the distribution becomes narrower (see Sect. 2.3.2). In this energy range the collision cascade does not exhibit the large number of high-generation recoils required to yield a recoil density $\propto 1/E_i^2$. Rather, isolated collisions dominate sputter ejection; thus, for too low bombarding energies the asymptotic high-energy slope of the emission spectrum becomes steeper, and fitting with an $E^{-n}$ power dependence is not feasible [2.124, 125]. Quite pronounced deviations in the surface binding from the cohesive energy were also reported in some cases of metal sputtering, possibly due to the fact that these surfaces were covered with adsorbate atoms (e.g. oxygen); these conditions often tend to broaden and to shift the energy spectra of sputtered species (see Sect. 3.1.2 for a discussion of such effects).

There have been many discussions in the literature [2.126–130] as to the appropriate choice for the surface binding energy $U$. As mentioned, most often the cohesive energy (or the heat of sublimation [2.131]) has been used. Different authors have argued, however, that the energy required to remove an atom from the surface should be greater by some 30–40% than that value: at an unperturbed surface an in-surface atom is bound by $U = (2Z_S/Z)\,E_{coh}$, where $Z_S$ and $Z$ are the surface and bulk coordination numbers and $E_{coh}$ is the cohesive energy. For the (100) surface of an fcc crystal this yields $U = 1.33E_{coh}$. This reasoning is based on the application of pairwise interaction potentials. Gades and Urbassek [2.129] have emphasized that pair potentials describe the binding properties of metals only poorly. Owing to the delocalized nature of metallic bonding, attractive binding forces in metals have to be expressed by many-body potentials. Using such potentials to describe atomic emission processes [2.129], they obtain for metals surface binding energies that are smaller than the values derived for pair potentials, that is, $U = (2Z_S/Z)\,E_{coh}$. (For a more thorough discussion of these features, see Sect. 3.1.2.) Gades and Urbassek noted also that the use of many-body potentials leads to a stronger refraction of the emitted atoms away from the surface normal. Consequently, the maximum in the energy distribution of sputtered particles is shifted by about 50% to higher energies. It appears that the accuracy of the available experimental data does not allow for an unambiguous verification of this proposition. (The surface binding energy in multicomponent samples is discussed in Sect. 3.1.2.) In computer simulations of the BCA type [2.130], the surface binding energy can be chosen as an input parameter. Data for Ne bombardment of Ni obtained in this way indicate that the sputtering yield varies as $U^{-1}$ for binding energies between 1.5 and 10 eV at impact energies $\geq 1$ keV.

The differential yield expression (2.34) predicts a cosine law for the angular distribution of the sputtered flux from an amorphous or polycrystalline sample. (For single-crystal sputtering, see Sect. 2.3.3.) Such a dependence is characteristic of an isotropic flux in the target. Experimentally, this simple cosine distribution is often not observed. At low energies the required isotropy of the collision cascade is not achieved and the emission distributions tend to be under-cosine (heart-shaped) [2.132]: fewer atoms are ejected normal to the surface and a larger fraction at oblique angles, as for the latter the reversal of the momentum from the incoming ion is more easily obtained. At higher energies the measured distributions are often found [2.133–140] to be over-cosine and can be described by $\cos^\beta \theta_0$, where $\beta$ is a fitting parameter. Values of $\beta$ varying between 1.0 and 2.0 have been reported, but in exceptional cases much larger numbers have been found. Possible explanations for the over-cosine emission have been based on surface-induced anisotropy of the recoil flux below the surface and/or anisotropic surface scattering of the flux passing through the surface. Robinson [2.141] suggested that an atom leaving the surface at an oblique angle experiences a net deflection toward the normal because of an asymmetric distribution of scattering atoms. A rough estimation showed that this effect might increase $\beta$ from 1 to 1.55 [2.102].

Cosine-type distributions are not observed under oblique ion incidence. Then, the emission distribution is peaked at or near the specular direction [2.142, 143]. Furthermore, it has been pointed out that the polar angular distribution can be different for atoms sputtered from the first and second atomic layers. As the angle of ejection is correlated with the depth of origin of the sputtered species, that is, grazing ejection prefers shallow depths, the angular spectrum for first-layer-emitted atoms is broad, whereas the distribution of atoms sputtered from the second layer is sharper and directed towards the normal. This notion is supported by data from molecular dynamics simulations and experimental measurements on binary alloys (see Sect. 3.3.2). The transition from under-cosine (at low energies) to over-cosine distributions is visible also in simulations by Biersack and Eckstein [2.130] of Ne$^+$ bombardment of Ni. On increasing the impact energy from 50 eV to 5 keV, the change occurred at about 300 eV. On the other hand, the mass of the projectile (H, He, Ne, Ar and Xe) appears to have little influence on the distribution for 1 keV bombardment.

With respect to experimental determinations of angular distributions, it should be noted that they are usually very sensitive to surface contamination (a continuous surface layer forces atoms from the topmost target layer into pronounced forward emission) [2.144, 145] and ion-bombardment-induced surface topography [2.146]. The influence of surface roughness on the angular ejection was simulated by Yamamura et al. [2.146]. The occurrence of such effects results in over-cosine distributions. At oblique ion incidence a tendency for preferred emission towards the specular direction is observed. A correlation between the dominant angle of ejection and the total sputtering yield was reported by Betz et al. [2.147]: the lower $Y$ is, the more the direction is peaked toward the specular direction. In the single-knockon regime, this preferred emission is even more pronounced, as demonstrated both experimentally and by simulations for light-ion sputtering.

Equation (2.33) gives the sputtering yield for normal incidence of the projectile. With increasing impact angle, the yield increases because of the higher energy deposition in the vicinity of the surface. The analytical sputtering theory of Sigmund provides predictions concerning the dependence of the sputtering yield on the incidence angle of the ions, $\theta$. For not-too-oblique incidence,

$$\frac{Y(E, \theta)}{Y(E, 0)} = (\cos \theta)^{-b} , \tag{2.36}$$

where $b$ is a function of $M_2/M_1$. For $M_2/M_1 > 5$, $b \sim 1$, and thus the dependence is roughly $1/\cos \theta$. For $M_2/M_1 < 3$, $b \sim 5/3$. While the yield rises with increasing incidence angle, for large values of $\theta$ (approaching $90°$), scattering of the incident beam increases, however, and the yield decreases rapidly. Generally, the maximum of the yield occurs between $60°$ and $80°$, as confirmed by experiments [2.114]. At low energies, the linear cascade cannot describe the angular dependence of the yield and a pronounced influence of the target material (via the differences in threshold energies) has been observed [2.148].

Detailed investigations [2.149] into the incidence-angle dependence of the sputtering yield for light ions ($H^+$, $D^+$, $He^+$) in the low keV range reported a yield increase distinctly more pronounced than $1/\cos \theta$, in particular for $H^+$ and $D^+$ ions. Furthermore, the maximum is observed at angles of $\theta \sim 80°$ or even larger. The maximum of the normalized yield was found to increase with increasing projectile energy and increasing surface binding energy of the target material. The authors [2.149] discuss their data in terms of possible models for light-ion sputtering (see Sect. 2.3.2).

**Statistics of Sputtering.** The number of atoms ejected for any single incoming ion appears to be a strongly fluctuating quantity. This notion was inferred first from theoretical arguments [2.150] and later from different types of computer simulation [2.151–155]. (Unfortunately, no setup has been devised as yet which might allow for an experimental verification.) Eckstein [2.151] has carried out an extensive investigation into the probability distribution of single-ion yields for a large variety of ion–target combinations, impact energies and incidence angles. His data in general support the previous inferences of large fluctuations and show, furthermore, that these are strongly dependent on the specific bombardment conditions. The data indicate that the distribution of the probability of sputtering $N$ atoms can be reasonably well described by a two-parameter negative binomial distribution. For certain bombardment conditions (e.g. high energies, heavy ions and/or oblique incidence) these distributions tend to become very broad and a number of atoms much larger than the average yield can be ejected. For 50 keV Xe bombardment (at an incidence angle $\theta = 60°$) the mean yield is 22.2 atoms/ion, but the probability of sputtering 70 or 55 atoms per single projectile is only two or one orders of magnitude, respectively, lower than the maximum probability. Such high-yield events are probably similar to those observed by Averback and coworkers in MD simulations (cf. Sect. 2.2.2). When the single-knockon regime (low energies, light ions) is approached, the distributions become more Poisson-like and the widths decrease.

In a similar investigation, Conrad and Urbassek [2.152] have expressed the conjecture that the observed fluctuations of the sputtering yield are mainly caused by the fluctuations of the energy deposited near the surface. In fact, they show that both quantities are almost identically distributed. They argue convincingly that the proportionality of the yield to the deposited energy (see (2.28)) holds not only in the average over many cascades but also for every individual event. This would be due to the fact that for a fixed amount of deposited energy, the number of recoils generated above an energy $U$ does not fluctuate much around its mean value. For crystalline targets Robinson [2.154] has examined the sputtering yield as a function of the ion's impact point on the surface. Apart from some distinct variations which can be ascribed to axial and planar channeling effects, he observes that the sputtering yield is very sensitive to the impact point and small changes in position often produce large changes in yield. He infers that the sputtering yield is a random (or chaotic) function of the point of impact.

Related conclusions can be can be drawn from simulations of Betz et al. [2.155] for Ar bombardment of a Cu(111) surface. The single-event yield is decisively influenced by the specific location of the ion's impact point, particularly for energies below 500 eV. In agreement with Eckstein's data [2.151], this work also reports rather broad yield distributions: at 1 keV, events may

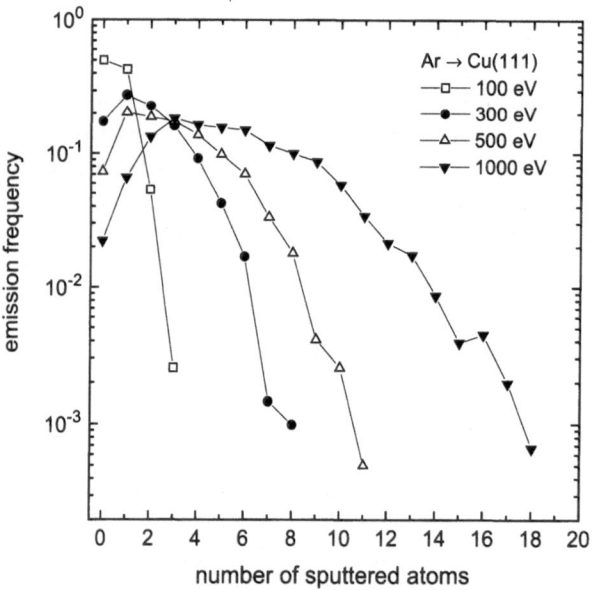

**Fig. 2.12.** Emission statistics derived from MD computations for Ar projectiles impinging on Cu(111) at the indicated energies. The frequency for the sputtering of a given number of atoms in an individual event is plotted versus this number. Data from [2.155]

produce up to 18 atoms at a frequency about 100 times smaller than for the most probable yield of $\sim 5$ atoms/ion. These data are depicted in Fig. 2.12.

**Depth of Origin of Sputtered Species.** In the context of (2.31), the characteristic depth of origin $\xi_0$ was assumed constant (i.e. independent of the emission energy) by using $m = 0$ in the power-law cross section. (As mentioned there, a value of the stopping cross section [2.16] a factor of two larger is applicable in the energy range of sputtered particles.) Some dependence on the emission energy appears plausible, however [2.156]. For the general case $(m \neq 0)$,

$$\xi_0 \cong \frac{1-m}{1-2m} \frac{E_0^{2m}}{NC_m} \, . \tag{2.37}$$

Hence, the higher the energy of a sputtered atom, the more likely it comes from deeper inside the target. This expectation is corroborated by computer simulation using the TRIM code [2.130]. It is in disagreement, on the other hand, with the results of molecular-dynamics simulations [2.70], which show generally that virtually all atoms are ejected from the *top two* surface layers. Considerable effort [2.156, 157] has been invested to clarify these discrepancies and to compare the results with the predictions of the analytical theory. Vicanek et al. [2.16] performed a theoretical analysis of the depth of origin of sputtered atoms. The results confirmed the previous assertion that the depth of origin is determined primarily by the stopping of low-energy recoil atoms (i.e. via $C_m$ in (2.37)), while angular scattering is only of minor importance. The authors established, furthermore, that the standard power-law cross section *underestimates* stopping in the energy regime pertinent to atom ejection $(\varepsilon \approx 10^{-4}\text{--}10^{-3})$ by about a factor of two. Correction of this underestimation by increasing the stopping cross section would result in $\xi_0 \approx 0.25$ nm and thus bring $\xi_0$ very close to the MD data [2.70].

Experimental determinations [2.158–166] of $\xi_0$ have been carried out on different alloy systems. Although some discrepancies emerged again, the data generally indicate that a large fraction of the sputtered flux originates from the topmost surface layer. Specifically, for a Ga–In alloy [2.159, 160], $\sim 87\%$ of the sputtered atoms come from the first layer, with essentially no dependence on bombarding energy from 25 to 250 keV and a slight increase (to 94%) at 3 keV. For a Ru(0001) specimen covered with thin overlayers of Cu [2.161, 162], the first-layer contribution to sputtering was determined to be 67% for 3.5 keV Ar$^+$ bombardment. Lam and coworkers [2.163–165] monitored the temperature-dependent steady-state surface composition in segregating alloys (cf. Chap. 3). From a detailed evaluation of the resulting surface layer profiles, they were able to determine first-layer contributions to sputtering (for 3 keV Ne$^+$ impact) for various Ni-based alloys. These fractions amount to $\sim 50\%$ (for Ni–Ge), $\sim 65\%$ (Ni–Cu), $\sim 70\%$ (Ni–Pd) and almost 100% in Ni–Si.

In a recent experiment, Wittmaack [2.166] succeeded in determining the depth of origin $\xi_0$ of sputtered atoms, while avoiding possible ambiguities related to the different sputtering conditions in multicomponent systems. He combined in situ secondary-ion mass spectrometry and (mass-resolved) low-energy ion-scattering spectrometry to measure sputter profiles through the interface between isotopically pure layers of $^{30}$Si and $^{28}$Si. Thereby, he was able to monitor, in parallel, the compositon of the ionized fraction of the sputtered flux and the composition of the *outermost* layer of the ion-bombarded surface as a function of the eroded depth. From the spacing between the two profiles in the interface region of the layered structure, the mean depth of origin of sputtered $^{28}$Si ions ejected with an energy of $\sim 40$ eV at an angle of 48° to the surface normal was found, giving $\xi_0 = (0.2 \pm 0.04)$ nm.

The MD simulations of Betz et al. [2.155] for Ar$^+$ on Cu(111) showed that for impact energies below 200 eV all atoms are sputtered from the first layer. With increasing energy, deeper-lying layers start to contribute to the emitted flux; at 1 keV about 15% of all atoms originate from below the topmost layer.

A different approach to deriving information about the escape range of low-energy species from solids was taken by Madey and coworkers [2.167]. They studied the transmission of slow oxygen ions ($< 10$ eV) through ultra-thin films of rare gases (Ar, Kr, Xe) and ascribed the observed attenuation mainly to elastic scattering of the ions by the rare gas atoms. They found that 10% of O$^+$ ions can be transmitted through 1.6 atomic layers of Ar, 2.9 ML of Kr and 4.0 ML of Xe; the associated attenuation cross sections were $6.0 \times 10^{-15}$ cm$^2$ for Ar, $2.2 \times 10^{-15}$ cm$^2$ for Kr and $1.5 \times 10^{-15}$ cm$^2$ for Xe. For Xe, the authors also found indications that the angular distribution of the ions changes because of large-angle scattering. These data are in good agreement with associated MD simulations [2.168].

**Fluence Dependence of Sputtering Yields.** Conventional techniques [2.114] used in the determination of sputtering yields typically employ large ion fluences ($10^{17}$ to $10^{19}$ ions/cm$^2$). Changes of the sputtering yield under these experimental conditions have been reported, but they might be largely due to the development of macroscopic surface morphology, bulk impurities or implantation of the primary ions [2, 169, 170]. Under moderate vacuum conditions, the presence and removal of surface contaminants also play an important role. Standard sputtering theories and computer simulations have usually treated the sputtering yield as a fluence-*independent* quantity.

On the other hand, Burnett et al. [2.171] have determined the change of the sputtering yield of Ru for very low fluences ($10^{13}$–$10^{16}$ Ar$^+$ ions/cm$^2$). Bombarding a well-characterized Ru(0001) surface with 3.6 keV Ar$^+$ ions, they observed a decrease of $Y$ with increasing ion fluence by about a factor of two and a stationary value for fluences $> 2 \times 10^{15}$ Ar$^+$/cm$^2$. These data are depicted in Fig. 2.13. (The experiments employed nonresonant laser ionization of sputtered Ru atoms followed by mass-spectrometric detection of the ions created.) This fluence range is low enough to rule out extended surface topographical changes and significant implantation of primary ions as the cause of the yield reduction. Nevertheless, this fluence will create single vacancies

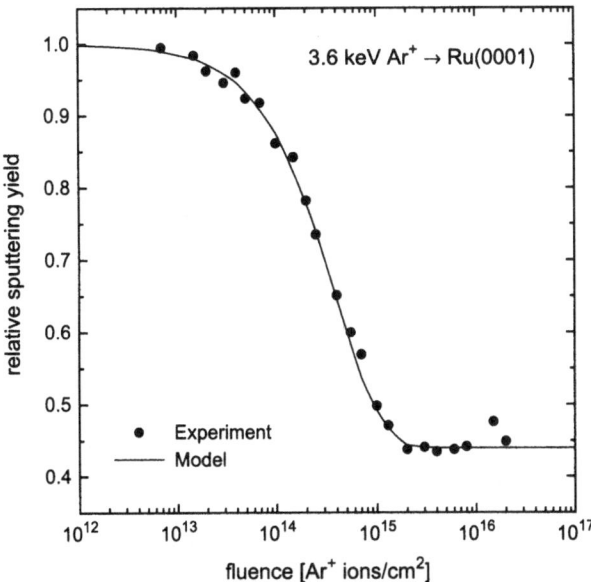

**Fig. 2.13.** The intensity of sputtered Ru atoms as a function of 3.6 keV Ar$^+$ ion fluence; the data are normalized to the low-fluence value and indicate a lowering of the sputtering yield with increasing ion fluence. The *solid line* is a fit to the experimental data (see text). Data from [2.171]

or vacancy islands on the surface, as seen from the STM data on Pt(111) discussed in Sect. 2.2.2. The authors [2.171] split the detected ion signal into independent contributions originating from virgin portions of the surface and from previously bombarded areas, and were able to fit the yield-versus-fluence data with an exponentially decreasing function, $\propto \exp(-\sigma\Phi)$, where $\Phi$ is the ion fluence and $\sigma$ is a damage cross section for a single-ion impact. From this fit (the solid line in Fig. 2.13) they derive $\sigma = (2.7 \pm 1.0) \times 10^{-15}$ cm$^2$, which corresponds to $4.3 \pm 1.6$ surface Ru atoms; the ratio of the sputtering yields for the previously impacted and the virgin surface was found [2.171] to amount to $0.49 \pm 0.08$. Figure 2.13 indicates a very good agreement between the experimental results and the model. Burnett et al. tentatively ascribe the observed factor-of-two reduction of the yield to a lowering of the deposited energy $F_D$ at the surface: they argue that surface vacancies may allow an increased penetration of the primary ions into the target before the first collision and, hence, $F_D(x)$ would shift deeper into the target.

In order to obtain a better understanding of the processes leading to this yield reduction, the same authors [2.172] carried out MD simulations on different single-crystal surfaces; specifically, they tried to find out if changes in the surface structure (e.g. the presence of a surface vacancy) can explain the experimental results. To this end, the authors compared yields for rare-gas (Ne, Ar, Xe) ion bombardment of Cu(100) at energies from 100 to 500 eV for complete surfaces (no atoms removed), in the neighborhood of a surface atom which has been removed and for a surface where 0.5 ML has been removed. The simulations showed that the sputtering yield typically decreases for impact points within half of the lattice constant of a missing atom position: for example, for 500 eV Ar the yield drops from 3.3 to 2.3; outside this area the yield reaches the value of the undamaged surface within a lattice constant. The effect appears most pronounced for lower energy and higher mass, but is largely absent for Ne projectiles. When half of a monolayer is removed, the yield near a surface atom which sits on top of the surface (as its immediate neighbors have been removed) increases quite remarkably (for 500 keV Ar, from 3.3 to 4.2), but is depressed further away (3.3 to 2.7). The authors [2.172] thus conclude that the yield of a virgin surface may indeed decrease with fluence (as seen in the experiment) until an equilibrium surface topography has developed. A caveat, however, remains: while in the experiment the effect was most pronounced for Ne and least for Xe, in the computations the opposite tendency was found.

Sputtering yields for very low fluences were determined by Michely and Comsa [2.89], utilizing STM micrographs (see Sect. 2.2.2). For 600 eV $Ar^+$ bombardment of Pt(111) surfaces they established values of $Y$ for fluences between $4.6 \times 10^{13}$ cm$^{-2}$ and $1.2 \times 10^{15}$ cm$^{-2}$ and sample temperatures in the range 170–760 K. Within the scatter of all data, of a few percent, the yield (average 2.1) is found to depend neither on the fluence nor on the temperature. Thus, no indication of a surface-morphology dependence of the sputtering yield for 600 eV $Ar^+$ irradiation of Pt(111) is found.

Colla and Urbassek [2.173] investigated, via MD simulations, the dependence of the sputtering yield on the random coverage of the surface with adatoms. For 1 keV Ar impact on a Pt(111) surface, they report an increase of $Y$ with increasing coverage, with $Y$ reaching a maximum (with a value about 10% higher than for the free surface) at a coverage of $\sim 0.25$ and falling for high adatom coverages. For a more-than-half-filled layer of atoms the yield is essentially identical to that from the pristine surface. The authors ascribe their findings to two effects. (i) Sputtering is easiest for isolated adatoms, which are bound most weakly to the surface; these occur most frequently at small coverage. (ii) For large coverage, the energetic benefit of sputtering atoms from the adatom layer is largely balanced by the geometric restriction on ejecting atoms from the original surface layer due to blocking by adatoms. They also stress that the effect of surface topography should be considerably stronger in the case of covalent materials, with their directional bonding. In fact, an almost 40% decrease of the sputtering yield of 1 keV Ar on Si(100) has been found in MD computations for a reconstructed surface as compared to an unreconstructed one; this is due to the higher surface binding energy of the former [2.174]. By contrast, an increase of $Y$ by 6% upon reconstruction was found [2.173] for 1 keV Ar irradiation of Pt(111).

### 2.3.2 Single-Knockon (Near-Threshold) Regime

At low bombarding energies and for light incident ions, sputtering yields generally decrease rapidly with decreasing impact energy. In addition, the energy and angular distributions of emitted species may change drastically as compared to the linear-cascade case. A quantitative theoretical description of sputtering in this regime was found difficult to establish [2.11, 175], however. The concept of the linear collision cascade employed in the previous section to describe sputtering faces limitations for bombardment with very light ions (e.g. H, D, He) and, for heavier ions, at impact energies approaching the threshold for sputter ejection. (The occurrence of such a threshold energy $E_{th}$ was inferred from the early sputtering experiments [2.176, 177] but is still not a well-defined quantity despite experimental efforts over a long period of time.) For light ions (such as $H^+$ and $He^+$) a large amount of sputtering yield data from both experiments and computer simulations exists [2.130, 178, 179], because of their importance for thermonuclear fusion research. Conversely, theoretical descriptions are rather limited for light-ion sputtering and for energies close to threshold. Although the collision cascade under these circumstances is still linear in the sense defined in Sect. 2.3.1 (i.e. only a small number of atoms are in motion and binary collisions prevail), typical cascades are often too small, so that the assumption of an isotropic recoil spectrum is not fulfilled. Instead, rather specific collision sequences may lead to atom ejection [2.180–185]. Possible processes are depicted schematically in Fig. 2.14; they all illustrate that these collisions may involve only

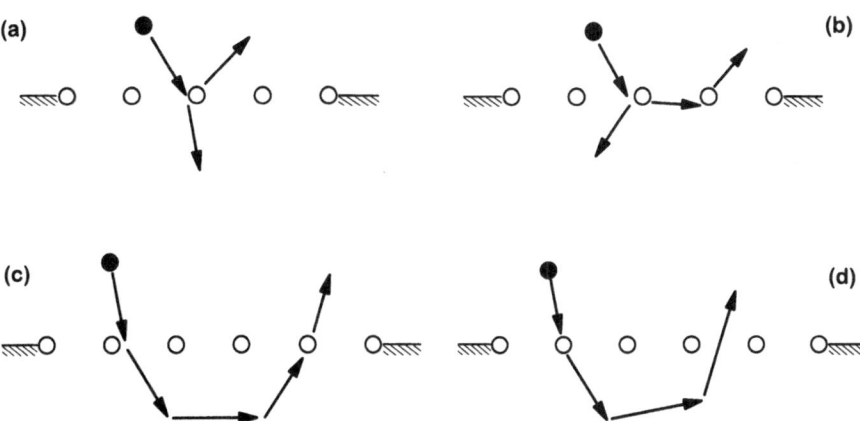

**Fig. 2.14a–d.** Illustration of some possible ejection mechanisms in the single-knockon regime: (**a**) a primary recoil is produced in the first collision and ejected directly (or after a further deflection by a target atom); (**b**) a higher-order (secondary) recoil is ejected; (**c**) the projectile undergoes multiple collisions, is backscattered and ejects a surface atom in a near-head-on collision; (**d**) a secondary recoil undergoes several collisions (at small scattering angles) that effect reversal of momentum and is ejected directly (or collides with another atom which is ejected)

a small number of atoms (i.e. they are short) and a collision cascade does not develop. (Note, however, that many small-angle-scattering collisions can also be very efficient for momentum reversal and sputtering at near-threshold energies.) As particle reflection coefficients increase with decreasing energy (and are quite high for light ions), reflective scattering collisions near the target surface contribute increasingly to sputtering. Their importance was noted by Winters and Sigmund [2.184] for the sputtering of nitrogen atoms from a tungsten surface. The proposed extension [2.11] of the sputter-yield formula (2.28) to heavy-ion bombardment in the near-threshold range would yield, for $m = 0$ (see (2.32)), $F_D(E, 0, 0) = \alpha N S_n(E) = \alpha N C_0 \gamma E$ and, combined with (2.31),

$$Y(E) \cong \frac{3}{4\pi^2} \alpha \frac{\gamma E}{U} \quad \text{for} \quad E \gg U. \tag{2.38}$$

Sigmund [2.11] has, however, expressed some reservations regarding the theoretical soundness of this approach. A general concern in the threshold regime is the importance of the ion–target mass ratio for sputtering; different processes of particle ejection are expected for $M_1 \ll M_2$ and for $M_1 \gg M_2$. In the first case, ions can be reflected with little loss in energy and knock out a surface atom for $T_{\max} = \gamma E \geq U$; hence, the threshold would be $E_{th} \approx U/\gamma$. The situation is less clear in the second case.

The relevance of single-knockon collisions to sputtering has been illustrated by computer simulations. Eckstein et al. [2.185–187] simulated near-threshold sputtering and identified distinct collision types leading to sputtering. For light ions (e.g. D on Cu) at normal incidence, the only process of importance is due to a primary knockon atom generated directly by the projectile on its way back out of the sample after having undergone one (or more) collisions with one (or several) target atoms (see Fig. 2.14c) For oblique incidence and higher energies, the mechansim depicted in Fig. 2.14a may contribute significantly. For heavier ions (Ar on Cu) two mechanisms are operative: the first involves a primary knockon with the ion moving into the target (Fig. 2.14a), while in the second a secondary knockon atom created by the projectile effects sputtering, possibly after further collisions with other target atoms (Fig. 2.14d). For the situations $M_1 = M_2$ and $M_1 \gg M_2$, only the latter kind of collision sequence contributes significantly to sputtering at normal incidence. These mechanisms may change drastically for oblique ion incidence: then, processes such as those in Fig. 2.14a or b can become relevant. Rather detailed accounts of these mechansims and their importance for near-threshold sputtering are given in [2.182, 185].

The concept of a threshold energy $E_{th}$ for sputtering, defining the projectile energy at which the sputtering yield becomes zero, was already employed by Wehner and coworkers in 1960 [2.176]. Many early studies in sputtering were aimed at establishing such threshold values. As the sputtering yields are very low when approaching this limit and are, furthermore, subject to fluctuations (see Sect. 2.3.1), experimentally determined threshold energies may often reflect also the sensitivity of a specific setup. Notwithstanding these problems, some coherent picture has emerged. Stuart and Wehner [2.176] argued that the originally proposed threshold energy $E_{\text{th}} = U/\gamma$ is too low for heavy-ion sputtering, as the momentum of the projectile has to be reversed into the backward direction and more than one collision is needed to effect this. They proposed $E_{\text{th}}/U = 4$ for $1 < \mu < 13$, where $\mu = M_2/M_1$. For light-ion sputtering, Behrisch et al. [2.181] used $E_{\text{th}}/U = 1/\gamma\,(1-\gamma)$ for $\mu > 5$. This follows from the observation that light ions sputter upon reflection at a target atom and their maximum energy after one collision would be $(1-\gamma)E$. Other relations were proposed, one by Bohdansky et al. [2.188, 189], $E_{\text{th}}/U = 8\mu^{-2/5}$ for $\mu < 3$, and another by Eckstein et al. [2.185],

$$E_{\text{th}}/U = 7.0\mu^{-0.54} + 0.15\,\mu^{1.12}\,, \tag{2.39}$$

for a wider range of mass ratios, obtained by fitting a large amount of experimental and calculated data. Figure 2.15 displays these results [2.187] and the fit according to (2.39). Several other forms for the dependence of the ratio $E_{\text{th}}/U$ on $\mu$ have been proposed [2.185]. Generally, the determination of the threshold energy requires the extrapolation of the yield data with an (assumed) energy dependence towards zero yield.

The lack of a comprehensive description of the sputtering yield at low and near-threshold energies and the need for sputtering data for light ions in fusion research, as well as for low-energy bombardment in thin-film deposition, triggered attempts to develop empirical formulas for the sputtering yield under these conditions, based on simple scaling laws. The yield was character-

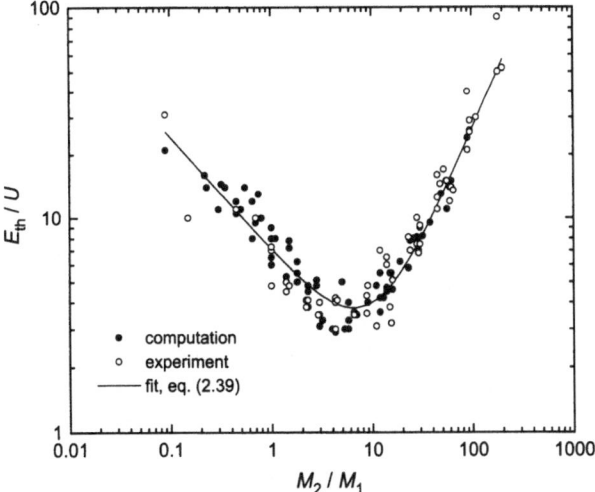

**Fig. 2.15.** The relative threshold energy, $E_{\text{th}}/U$, versus the ratio of target-to-projectile mass. The *solid symbols* represent calculated data, the *open symbols* experimental data. The *solid line* corresponds to (2.39), and is a fit to the data depicted. From [2.187]

ized by a normalized function which depends only on the surface binding energy and the masses of the ion and the target [2.188–194]. The surface binding energy enters these empirical expressions through the definition of the threshold energy for sputtering using one of the above-mentioned expressions. An empirical formula originally proposed by Bohdansky [2.188, 189] was found suitable to estimate sputtering yields for most elements bombarded with light ions at normal incidence with energies up to a few keV. He also demonstrated that by modifying Sigmund's analytical expression for the linear cascade, a rather universal relation for the sputtering yield could be derived for normal-incidence ion impact. The first modification concerned the evaluation of the effective deposited energy because at low energies a considerable number of recoils carry insufficient energy to overcome the surface barrier. Secondly, as for light ions the deposited energy is overestimated (through the dependence on $\alpha$), a correction for this was made by dividing $\alpha S_n(E)$ by $R/R_p$. This ratio represents the average number of surface crossings of the primaries. With these modifications to (2.33), the following expression for the sputtering yield was proposed [2.188]:

$$Y(E) = \frac{0.042}{U} (R_p/R) \, \alpha S_n(E) \left[ 1 - (E_{th}/E)^{2/3} \right] \left[ 1 - (E_{th}/E) \right]^2$$

(2.40)

Owing to the lack of knowledge of the above-mentioned parameters $\alpha$ and $R_p/R$ entering (2.40), these often are subsumed in a second fitting parameter (besides $E_{th}$) [2.186, 187]:

$$Y(E) = Q s_n(\varepsilon) \left[ 1 - \left( \frac{E_{th}}{E} \right)^{2/3} \right] \left( 1 - \frac{E_{th}}{E} \right)^2 .$$

(2.40a)

This formula is considered valid for both light and heavy ions, extending to the threshold regime of sputter ejection. It was fitted to a large number of experimental and simulation data, and values of $Q$ and $E_{th}$ were derived by this procedure for many ion–target combinations. Matsunami et al. [2.190] derived a similar empirical expression by adding a factor $\left( 1 - (E/E_{th})^{1/2} \right)$ to (2.33). The appropriateness of such a correction term has been suggested before. Using this modified yield formula, Matsunami et al. [2.193] performed an extensive comparison with experimental yield data. Similar empirical approximations were provided by other authors. While originally intended to describe the sputtering yield at near-threshold energies, some of those formulas were extended to higher energies, even going beyond the maximum of the nuclear stopping power. The extensive compilation of Matsunami et al. [2.193] was widely publicized.

Littmark and Fedder [2.195] developed a theoretical formalism for the sputtering of heavy targets with low-energy light ions, assuming that only primary recoils are candidates for sputtering in this case. A comparison of their results with experimental yield data for $H^+$ and $He^+$ on heavy targets

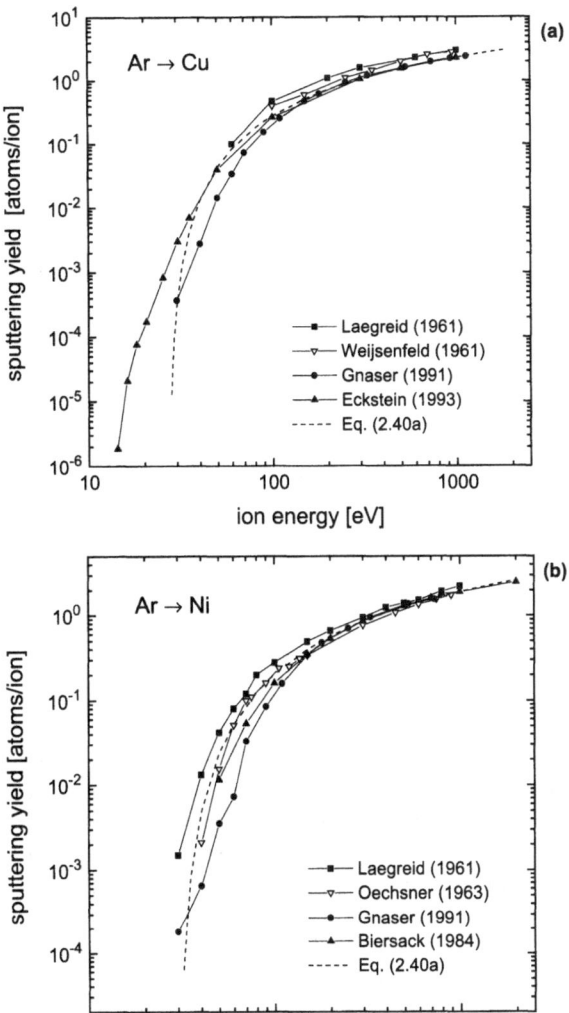

**Fig. 2.16a,b.** Sputtering yield of Cu (**a**) and Ni (**b**) under perpendicular $Ar^+$ ion bombardment versus the ion energy. Comparison is made between various experimental data sets, results obtained with the computer code TRIM and the empirical sputter yield formula given in (2.40a). Data from [2.186, 196–201]

(e.g. Ni and Au) showed excellent agreement for bombarding energies from 0.1 to 10 keV. In contrast to the empirical formulas, this theoretical approach is free of fitting parameters.

(Relative) sputtering yields for several elemental and alloy specimens in the energy range $30 \leq E \leq 1000\,eV$ have been measured recently by secondary-neutral mass spectrometry [2.196]. Sputtered neutral species were ionized during their passage through a low-pressure plasma that also provided the low-energy bombarding ions. Data for $Ar^+$ ion irradiation of Ni and Cu are shown in Fig. 2.16 together with other results from experiments

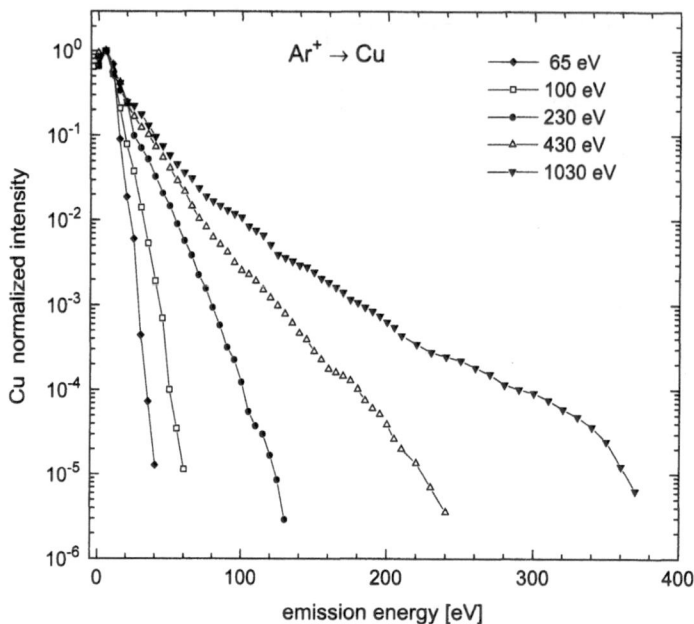

**Fig. 2.17.** Normalized energy spectra of neutral Cu atoms sputtered from a Cu specimen by $Ar^+$ ions of the given energies. The data were recorded using secondary-neutral mass spectrometry with plasma post-ionization

[2.197–201] and computations with the TRIM code [2.185, 186]. Also, a yield dependence as given by (2.40a) is included in Fig. 2.16 using values of $E_{th}$ and $Q$ from [2.186]. For both elements the various data sets show reasonable agreement in the energy interval 200–1000 eV, but pronounced differences towards lower impact energies. Owing to these divergences, the empirical expression is of little help as it can be adjusted to the various data sets by means of the two fitting parameters. Figure 2.16 clearly demonstrates that an unambiguous determination of sputtering thresholds is a difficult task.

Energy spectra of sputtered atoms at low bombarding energies are shown in Fig. 2.17. Neutral Cu atoms sputtered by $Ar^+$ ions from elemental copper were monitored, again using secondary-neutral mass spectrometry. With decreasing $Ar^+$ impact energy the emission energy distributions exhibit a steeper fall-off, which is related to the maximum energy a sputtered Cu atom can have for a given $Ar^+$ energy. Owing to the large dynamic range of these measurements (about five orders of magnitude), a rough estimate of this maximum energy can be derived from the spectra in Fig. 2.17. These values are surprisingly high compared to the respective $Ar^+$ ion bombarding energies and provide distinct evidence that only a very small number of distinct collision sequences can lead to the ejection of Cu atoms, since otherwise too much energy would be lost. Furthermore, the close mass matching of Ar and Cu atoms ($\gamma = 0.948$) imposes additional restrictions in this respect. Such

special collision sequences are discussed by Yamamura et al. [2.182, 183] and
are also inferred from computer simulations [2.185–187, 202, 203]. Some computations [2.185] indicate, furthermore, that an optimum number of collisions
in these sequences may exist which result in the highest emission energies.
Not surprisingly, the energy spectra of Cu atoms at these low bombarding
energies (Fig. 2.17) cannot be fitted by a power law $1/E_0^2$ as predicted by
(2.34). Urbassek [2.124] has developed a theoretical concept that illustrates
the deviations from the power dependence at low impact energies.

### 2.3.3 Sputtering from Single Crystals

In contrast to the sputtering from random targets discussed so far, sputtering processes in crystalline materials are strongly influenced by the crystallographic orientation of the surface relative to the incident beam direction
[2.204–206] and, for emission-angle-selective experiments, on the position of
the detector relative to the crystal axes. Such observations had already been
reported by several groups around 1960. Theoretical concepts and experimental data for monocrystal sputtering have been outlined in reviews by
Robinson [2.207] and Roosendaal [2.208], respectively. Angular distributions
of particles sputtered from single crystals have been summarized by Hofer
[2.116].

The *total* sputtering yield generally exhibits a distinct dependence on
the orientation of the crystal surface sputtered: the yield $Y_{(uvw)}$ from a
surface $(uvw)$ increases with increasing interatomic spacing $d_{uvw}$ along the
$[uvw]$ direction in the crystal. For normal incidence on fcc targets this means
$Y_{(111)} > Y_{(100)} > Y_{(110)}$, in agreement with experimental findings, although
at low energies ($\sim 500\,\text{eV}$ and less) a reversal was observed for Cu, i.e.
$Y_{(100)} > Y_{(111)}$. It was realized that crystallographic variations in the surface
binding energy cannot account for the large yield changes observed (with
the possible exception of those at very low energies). The yield variations
resembled, both in angular and energy dependence, other observations that
are known to be affected by ion channeling (e.g. ion ranges). Channeling
[2.29, 30] is understood as the movement of a penetrating ion nearly parallel
to densely packed atomic rows or planes via a series of glancing collisions so
that close encounters with lattice atoms have a low probability. As the nuclear
energy loss is greatly reduced, the ion's range increases drastically; the reduced energy deposition near the surface will lower the sputtering yield. This
qualitative concept was cast into a more thorough theoretical description by
several authors [2.209–213].

The theoretical model of Onderdelinden [2.213] is based on the influence
of channeling on energy deposition and, ultimately, on the sputtering process;
it was utilized frequently to explain this crystallographic variation of $Y_{(uvw)}$
at incidence energies above a few keV. According to this approach, channeling
will effectively reduce the sputtering yield for ion energies $E$ larger than an
energy $E_{uvw}^c$ which is essentially determined by the distance $d_{uvw}$ between

the atoms measured along the $[uvw]$ direction; $E^c_{uvw} \propto (d_{uvw})^3$. As the ratio of nonchanneled fractions of the ion beam for two crystallographic directions $[uvw]$ and $[rst]$ is roughly constant and equal to $(d_{uvw}/d_{rst})^{3/2}$, the sputtering yield ratio for these two surfaces is given by

$$Y_{uvw}(E) / Y_{rst}(E) \propto (d_{uvw}/d_{rst})^{3/2} . \tag{2.41}$$

This model provided a good agreement with experimental data obtained from low-index surfaces in an energy range from a few hundred eV to some 25 keV [2.207, 208]. It also predicts, for Ar$^+$ bombardment of crystalline Cu, the order and the energy values of the experimentally found maxima in the yield-versus-energy curves, namely $E^{\max}_{(111)} > E^{\max}_{(100)} > E^{\max}_{(110)}$; these all lie at much lower energies ($\sim 5$–8 keV) than that for a polycrystalline (random) sample ($\sim 50$ keV) (cf. Fig. 2.1).

Conversely, at low energies channeling is of minor importance and this model appears to fail. At near-threshold energies $Y_{(uvw)}$ is dominated by the surface binding energies $U_{(uvw)}$. Jackson [2.126] has calculated the dependence of the surface binding energy on the crystallographic orientation of the surface plane: for fcc crystals the typical order is $U_{(111)} \sim U_{(100)} > U_{(110)}$. Gades and Urbassek [2.129] have improved these computations for metallic systems by employing many-body potentials. A detailed comparison of experimental data and different models for the sputtering yields of monocrystalline samples can be found in [2.207, 208]. Also, a considerable number of molecular-dynamics computer simulations have been carried out to elucidate the sputtering effect for single crystals, highlighting, in particular, ejection mechanisms of atoms and molecules.

The sputtering of crystalline targets is characterized also by the preferential ejection of sputtered atoms (and molecules) in the direction of certain preferred crystal axes (e.g. close-packed lattice rows). The observation of this effect was first reported by Wehner [2.214, 215] in the 1950s for low energies (some 100 eV and below), but was also verified later for high energies [2.216–220]; it is now established over five orders of magnitude of projectile energy and has been observed both in conventional backsputtering and in transmission. This preferential emission (often called "Wehner spots") occurs for metals, semiconductors and insulators and appears to be a general irradiation effect in crystalline solids [2.116]. While the most prominent preferential ejection directions usually correspond to close-packed lattice rows (e.g. [110] in the fcc and [111] in the bcc structure), some preferentiality was also observed in other lattice directions (e.g. [100] in the fcc and [111] in the diamond structure). Silsbee [2.221] pointed out the possibility of a lattice influence on the energy dissipation of energetic recoils. He demonstrated, applying the hard-sphere interaction, that momentum focusing along [110] in fcc lattices can be accomplished. These *focusing collision sequences* (also termed "focusons") were widely employed to interpret the observed preferential ejection along close-packed lattice directions. The attractiveness of this picture [2.221, 222] might in part be due to the simple criterion which can be derived for conditions under which focusing would occur: In the hard-sphere approximation focusing is expected if $2R > d/2$, where $2R$ is the distance of closest approach of two atoms in a string and $d$ is the atom spacing along this row. This relation also makes it plausible that focusing collisions are only expected along close-packed lattice directions (small $d$). (As mentioned in Sect. 2.2.1, such focusing or replacement collision sequences have been observed frequently in MD simulations of damage generation [2.59–61].) However, because of some inconsistencies in the focuson model and the need to explain anisotropic emission distributions observed at low impact energies when extended focusing collision sequences do not occur, Lehmann and Sigmund [2.223] have proposed a quite different mechanism to explain the observed preferential particle emission. They stress the importance of the low-energy fraction of the recoil spectrum and of the regularly ordered *surface* lattice [2.223, 224]. In this view, the existence of a positive surface binding energy results in a selection of small impact parameters when the energy of the subsurface atom is on the order of the binding energy $U$. This selection becomes the more stringent the closer $E_0$ is to $U$. A distinct feature of this approach is the possibility to explain preferential ejection down to very low bombarding energies; for the latter, extended focusing collision sequences are not expected to occur as the ion range is very shallow.

An appreciable number of investigations into the preferential particle ejection from single crystals have been performed (for a review see [2.116]). More recently, Szymczak and Wittmaack [2.225, 226] have carried out a very detailed study covering a wide range of irradiation parameters. They investigated the angular distributions of gold atoms sputtered from an Au(111) crystal as a function of target temperature (15–550 K), ion energy (0.1–270 keV) and ion mass (He, Ne, Xe) using a collector technique in combination with backscattering analysis of the deposits. The distributions produced the well-known [110] and [100] spots superimposed on an apparently random background. By careful analyses of the spot shape and the background intensity, the authors [2.226] were able to separate the sputtering yields contributing to the spots and to the background. Surprisingly, they found no bombardment conditions for which anisotropic emission prevails. Rather, the yield due to the background component always dominates, in particular for high energies. At the lowest primary ion energies that they employed, the relative contribution of the sputter emission into the preferred directions to the total sputtering yield amounts to as much as 50%, but decreases with increasing energy to about 25% for He and Ne and to 15% for Xe impact. According to the authors, this finding implies that the angular distributions of particles sputtered from single crystals are sensitive to the energy and momentum distribution in the collision cascade; they suggest that the range of [110] focusing collision sequences may amount to a few nm. While the target temperature has little influence on the total and partial sputtering yields, the spot width increases with temperature. At 15 K, the half-width of the [110] spot is only 5°, but twice as large at 550 K.

Owing to their crystalline target arrangement, MD simulations of sputtering [2.70] inherently produce information on the preferential ejection from crystalline surfaces, provided this angle-selective emission is looked for. In fact, this has been done quite frequently. As an example, Fig. 2.18 depicts results from computations that were performed by Betz et al. [2.117, 155] for 1 keV Ar bombardment of a Cu(111) single-crystal surface. Figure 2.18a

## ANGULAR DISTRIBUTION OF ATOMS
## 1000 eV AR - CU (111)

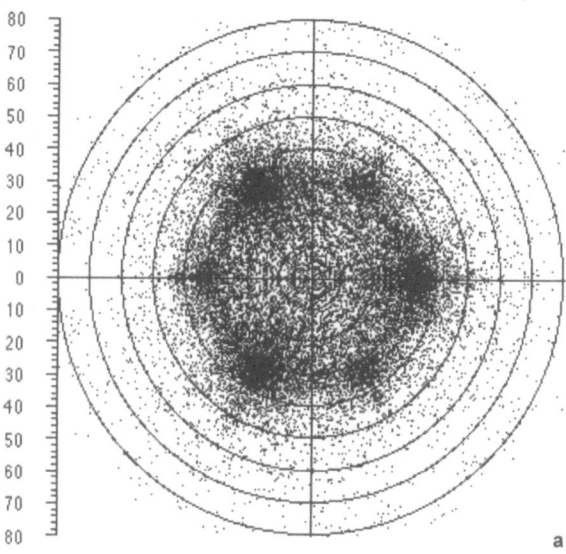

a

1 KEV AR->CU(111) ATOM POLAR ANGLE SCAN

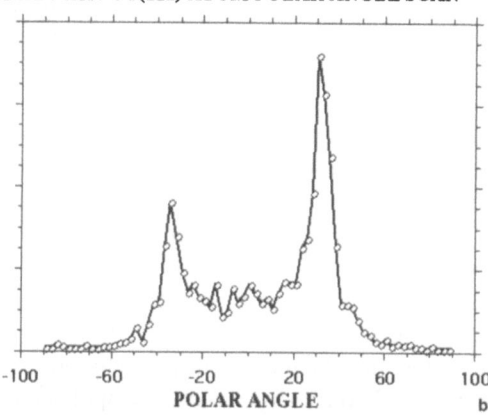

POLAR ANGLE

b

**Fig. 2.18a,b.** Calculated angular distributions of emitted atoms for 1 keV Ar bombardment of Cu(111). (**a**) Polar plot of the angular distribution showing six preferential emission directions (Wehner spots), three strong ones corresponding to the [110] lattice directions (i.e. direct focusing) and three weaker ones in the [100] lattice directions (assisted focusing). (**b**) Polar scan which includes both the [110] and the [100] direction. From [2.155]

shows a polar plot of the angular distribution of ejected atoms: six preferential emission directions are observed, three prominent ones corresponding to the [110] lattice directions, and three weak ones in the [100] directions (assisted focusing), the latter being shifted to lower emission angles ($\sim 35°$ instead of $54°$ in the crystal). The polar scan (Fig. 2.18b) illustrates the anisotropic distribution more clearly. The authors monitored also the distribution of atoms coming from the *second* layer in the target and observe only a narrow peak in the direction of the surface normal. Apparently, the passage of these atoms through the top layer confines their emission to a narrow cone around the normal. The authors in addition studied the ejection of $Cu_2$ dimers. These species exhibit again a pronounced emission along the [110] lattice direction, in agreement with experimental findings (for mass-selective measurements, see below). Contrary to atoms, however, no preferential ejection in the [100] direction was found in these simulations.

Interestingly, simulations by Eckstein and Hou [2.227–230] indicate that the position of the spots seems to be strongly dependent on the interaction potential, especially on its value at large internuclear separations. In the specific case of sputtering of an Au(111) crystal by 600 eV Xe, they observe six preferential emission directions and noted that a Moliere potential with a screening length $a = 0.00752$ nm gives satisfactory agreement with experimental data. Shorter screening lengths lead to polar emission angles that are too large, whereas for larger values of $a$ the spots are too close to the surface normal.

Other simulations [2.231] indicate that both of the aforementioned emission mechanisms contribute about equally to the total yield for a Cu(111) surface. By contrast, other computations [2.232, 233] demonstrate that just two random layers on a monocrystalline target (or vice versa) nearly destroy (or create) the angular emission characteristics of the underlying structure. These data are therefore a clear indication of the sensitive dependence of the angular distribution on the near-surface structure.

Summarizing the observations for preferential atom emission, Hofer [2.116] has noted these pertinent points: (i) Anisotropic ejection is observed down to ion energies of less than 100 eV. (ii) In the threshold regime, preferential emission may also be due to specific ion/surface atom collisions (e.g. direct recoil by a scattered projectile ion). (iii) With a variation of the irradiation energy, some shifting of the polar emission direction may occur, possibly indicating some influence of surface-atom scattering effects. While most experimental investigations have been carried out for close-packed directions, preferential ejection was observed also along other low-index lattice rows. For this case other focusing effects have been invoked (e.g. assisted focusing through symmetric arrangements of atoms).

Apart from experiments and computations which determined the preferential ejection of the *total* sputtered flux, investigations have been performed that analyzed the *mass-selected* flux from single crystal surfaces [2.234–

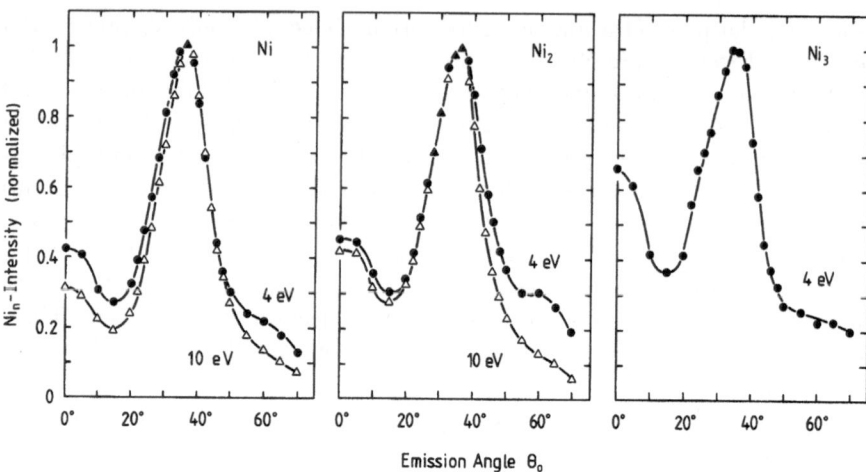

**Fig. 2.19.** Emission angle dependence of neutral Ni atoms and $Ni_n$ clusters emitted from a Ni(111) surface under 5 keV $Ar^+$ irradiation. The emission angle $\theta_0$ was varied by rotating the crystal around an axis in the surface and parallel to $[1\bar{1}0]$; the angle of $\theta_0 = 35°$ corresponds to the [110] lattice direction. The parameter (4 eV and 10 eV) is the emission energy of the ejected species. From [2.235]

239]. Gnaser and Hofer [2.234, 235] have done this using sputtered-neutral mass spectrometry. This experiment was carried out in a quadrupole-based sputtered-neutral mass spectrometer and the ejected neutral species were post-ionized by electron-impact ionization. Not surprisingly, preferential ejection along low-index directions was found, but distinct differences were noted depending on particle (cluster) size and emission energy. Figure 2.19 [2.235] shows, as an example for this kind of data, the intensities of neutral Ni, $Ni_2$ and $Ni_3$ emitted from a Ni(111) crystal surface as a function of the emission angle $\theta_0$ under 5 keV $Ar^+$ bombardment. The emission angle was varied by rotating the sample around an axis in the surface and parallel to $[1\bar{1}0]$; in this way the close-packed [110] direction was swept over the entrance of the mass spectrometer. Distinct maxima in the yields of all species occur at $\theta_0 \cong 35°$, which corresponds to the [110] direction of this fcc crystal.

As a parameter (see left and center panels of Fig. 2.19), the emission energy accepted by the mass spectrometer was varied; the angular anisotropy is seen to increase at higher energies. Measurements made at even higher emission energies on a Cu(111) surface [2.234] corroborate this observation. It should be noted, furthermore, that the measured angular anisotropy in these mass-selective experiments is usually as pronounced as that of the total yield. (The relevance of a preferential Ni dimer and Ni trimer ejection will be evaluated in the context of cluster emission, see Sect. 2.3.4). In these experiments energy spectra were also recorded for atoms and small clusters sputtered from the (111) surfaces of fcc crystals. For monomers and dimers, ejection in the close-packed direction ($\theta_0 \cong 35°$) shows a broadening of the energy spectra and a distinct shift towards higher energies as compared to the emission at near-normal directions. This finding is in agreement with data of Thompson and coworkers [2.105, 106] who found, for the total flux sputtered from Au crystals, that spectra taken along the [110] peak at higher energies as compared to those recorded 15° outside this preferred ejection direction. Time-of-flight spectra of the sputtered flux recorded by this group also provided evidence for direct- or reflected-recoil ejection along lattice directions which enhance preferred-momentum-transfer processes.

Using a rather sophisticated experimental setup, Winograd and coworkers [2.236–238, 240–242] studied the angular- and energy-selective emission of ground-state and excited-state Rh atoms from Rh(111) and Rh(100) surfaces by keV ion bombardment. Both clean and oxygen-covered surfaces were bombarded. Because of the sensitivity of the employed technique, the analyses were largely nondestructive, that is, essentially no modifications of the atomic arrangement at the surface were induced by ion irradiation. These experiments were accompanied by MD simulations [2.243–245] that were, over the years, considerably refined, in particular in terms of the atomic interaction potentials applied. Generally, the experiments were well reproduced by the computations, indicating that a detailed understanding of the angular and energy distributions can be derived solely from the regularity of the crystal lattice and an adequate atom interaction mechanism. The data displayed strong preferential ejection for an emission angle that would correspond to the [110] lattice direction from the (111) surface and another emission, albeit weaker, that might tentatively be ascribed to the [100] directions, but somewhat shifted to smaller angles. The authors [2, 237, 238, 240, 241, 246] stress, however, that the geometrical structure of the very-near-surface region controls the angular anisotropy. The registry of the second-layer atoms with respect to the top layer induces a highly directional momentum transfer in the last collision, leading, in turn, to preferential particle emission in certain crystallographic directions. Similar conclusions had already been drawn from early MD simulations of Harrison and coworkers [2.247]. Such mechanisms are akin to those of blocking and shadowing, which are well established in studies of surface structure determination by means of low- and medium-energy ion scattering. Evidence comes also from the observation that atoms emitted in preferred directions carry more kinetic energy than those ejected in other directions.

Garrison, Winograd and coworkers have recently studied microscopic mechanisms of particle ejection, on the basis of information derived from MD simulations [2.240, 241, 245], using a graphical utility for the visualization of the space–time evolution of the collision events. The authors succeeded in categorizing the collision processes leading to atom ejection during ion bombardment. For fcc (100) surfaces subjected to 5 keV Ar impact three major emission mechanisms were identified. Mechanism $\Delta_0$ (in the nomenclature of [2.240]) refers to the ejection of an atom in the same layer as the atom that energizes it. In mechanism $\Delta_1$ the atom that ejects is one layer above the atom that recoils. Finally, mechanism $\Delta_2$ represents the situation where the ejected atom is two layers displaced from the collision partner. From a quantitative analysis of their data [2.240] the authors conclude that ejection processes $\Delta_n$ rarely occur when $n$ is greater than two. By contrast, on Si(100) the $\Delta_3$ mechanism is quite common due to the openness of the Si lattice [2.245]. On the other hand, for Ni and Ru specimens the $\Delta_1$ process is dominant, contributing to 75%±8% of the ejection yield for Ni and 84%±8% for Ru targets. In comparison to the number of $\Delta_1$ interactions, the relative numbers of $\Delta_0$ and $\Delta_2$ events are small and of similar magnitude. These findings corroborate the authors' frequently expressed notion that the inherent registry of the atoms in the crystal lattice near the surface is crucial in determining the dominant microscopic sequences of events leading to ejection of atoms in sputtering and, ultimately, the experimentally observable quantities such as the emission-angle differential yields of atoms and molecules.

### 2.3.4 Cluster Emission in Sputtering

The sputtered flux from an ion-bombarded solid surface is composed not only of atoms but also of polyatomic molecules and clusters [2.116, 248, 249]. Following Urbassek and Hofer [2.249], the term *cluster* is used here in a broad sense designating any aggregation of atoms, while *molecules* is reserved for those atomic aggregations which exhibit strong bonding and may exist as preformed entities in the solid or at the surface. These authors stress, furthermore, that the emission of these species is a rather ubiquitous observation and is found during ion bombardment of metals, semiconductors and insulators, and bioorganic and polymeric materials; for both elemental targets and multicomponent specimens. Apart from its inherent importance for understanding energy-sharing processes in the solid and the possible transfer of (a part of) this energy into the gas phase (via the ejected particles), cluster emission has some practical applications: Ion-bombardment induced desorption of (large) organic molecules and biomolecules [2.248] is utilized in surface mass spectrometry for the characterization of the solid irradiated. Furthermore, molecular and volatile reaction products may be emitted from a surface bombarded with reactive ions, thereby increasing the erosion rate (reactive-ion etching [2.250], as applied, for example, in semiconductor device fabrication).

Compared to the emission of atomic species in sputtering, the understanding of the formation and emission of clusters under ion irradiation is much more incomplete, despite a large number of investigations over the last four decades [2.248, 249, 251–260]. Some of the unresolved issues concern the size (mass) distribution of the cluster flux and how this depends on the binding energy, the internal energy and the ionization in the case of charged clusters. Many computer simulations [2.261–270] have been dedicated to the investigation of cluster emission in sputtering. Both computations and experiments

[2.271–275] have indicated the importance of fragmentation processes. Also, the description of the detailed atomistic mechanism which imparts energy to an aggregation of atoms at the surface and causes its ejection as a bound entity (i.e. the number and type of recoils from within the solid involved) is largely fragmentary. An important means of investigating these processes is the study of the energy and angular emission spectra of sputtered clusters; these have been studied using mass-spectrometric techniques. (The possible distortions introduced by the necessity to post-ionize *neutral* clusters by interaction with electrons or photons to render them accessible to mass spectrometry have to be taken into account here.) Some of this work has been reviewed in recent publications. The present exposition is intended to describe new results and to focus on data pertinent to low-energy bombardment conditions.

**Mass Distributions.** The emission of clusters in sputtering was first reported some four decades ago: Honig [2.276] observed positively charged $Ag_2^+$ dimers upon bombardment of an Ag surface. Later studies found clusters with an increasing number in constituents of both positive and negative charge states [2.277, 278]. Interestingly, most of these early and also some more recent studies were performed on silver. Katakuse et al. [2.278] managed to detect $Ag_n^+$ clusters up to $n \sim 200$ sputtered from Ag by 10 keV $Xe^+$. A pronounced oscillatory behavior in the abundance distributions of charged clusters was seen for certain elements and was originally interpreted in terms of binding-electron parity. Early attempts to detect neutral clusters were largely unsuccessful, mostly because of the inefficient methods of post-ionization then available and the mass interferences caused by high residual-gas contributions. Gerhard and Oechsner [2.279, 280] provided the first dimer-to-atom and trimer-to-atom ratios for neutral species. Gnaser and Hofer [2.235] have greatly extended that data set by measuring, for several transition elements and two semiconductors (Si and Ge), the abundance distributions of neutral and positively charged clusters in the same experimental setup. This direct comparison of neutral and ionized clusters is depicted in Fig. 2.20; the data refer to 5 keV $Ar^+$ bombardment at an incidence angle of 70°. Whereas the yields of most ionic clusters decrease monotonically with increasing cluster size, $Cr_n^+$ and $Cu_n^+$ clusters show oscillations such that odd-numbered clusters exhibit higher yields than the preceding even-numbered ones. As mentioned above, this agrees with previous findings [2.277] and was ascribed to the electronic configurations of these elements (Cr and Cu have incompletely filled 4s orbitals). Compared to the ionized clusters, the relative yields of neutral clusters typically decrease more rapidly with an increasing number of constituents. This (apparent) disparity needs some qualification, however. Owing to the different ionization processes, the absolute intensities of charged and neutral species are not directly comparable: as discussed below, the ionization potentials of clusters generally decrease with cluster size. (According to a quite simple model, the ionization potential exhibits a monotonic decay

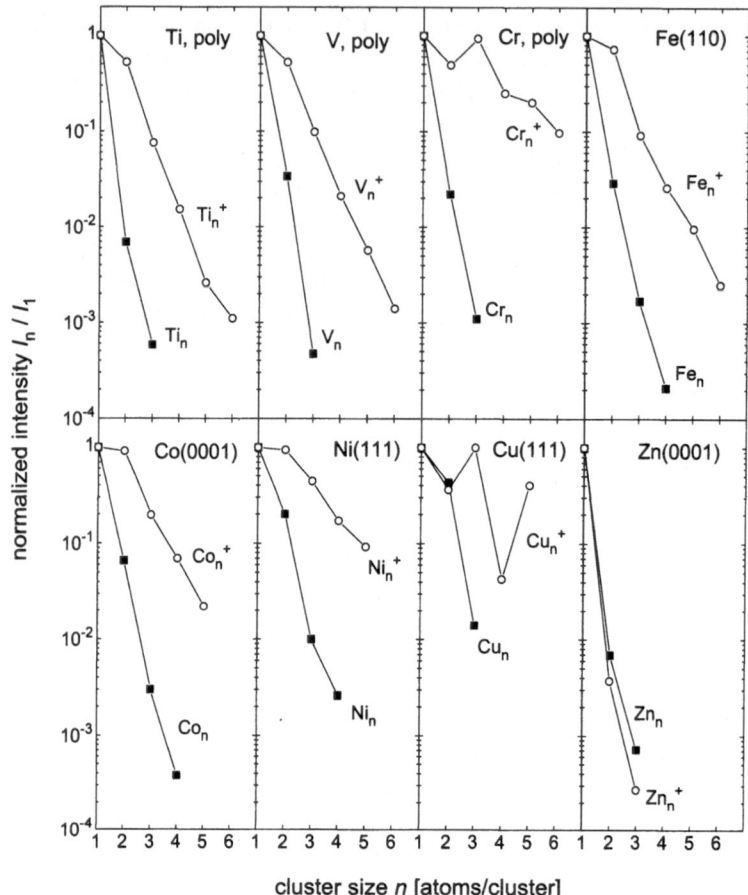

**Fig. 2.20.** Abundance distributions of clusters sputtered from various transition metals by 5 keV $Ar^+$ ions at an incidence angle of 70° from the surface normal. The *open circles* and *closed squares* represent secondary ions and sputtered neutral species, respectively. From [2.235]

and approaches asymptotically the bulk work function with increasing cluster size. See also Sect. 5.4.) Hence, larger clusters have a higher probability of ionization and subsequent detection than smaller ones and single atoms. Furthermore, even for the same cluster size, this efficiency of ionization might be different for the gas-phase ionization that neutral species are subjected to and for the ionization of sputtered species that occur at the surface.

A direct influence of cluster stability (i.e. binding strength) on the cluster distribution is evident for the sputtering of Zn. As Fig. 2.20 indicates, the Zn dimer (and trimer) yields, for both ions and neutrals, are considerably smaller than for most other elements, despite the fact that the sputtering yield of Zn under these bombardment conditions is much higher than for the other specimens. (As shown below, high total yields usually facilitate the formation of clusters.) Apparently, the much lower dimer binding energy of $Zn_2$ (about an order of magnitude smaller than that of the other elements) prevents the formation of stable clusters and/or impedes their survival, that is, it enhances the probability of fragmentation. Hence, a direct comparison of the abundance distributions of (small) ionized and neutral clusters sputtered under identical conditions from the same element reveals two distinct differences: (i) neutral clusters exhibit a stronger decay with atom number $n$ than charged clusters; (ii) the pronounced even–odd alternations of charged clusters of some elements are absent in the distribution of the neutrals. Alternating cluster-ion intensities are a common feature of clusters composed of atoms with an odd number of valence electrons. This effect is best documented for monovalent elements such as the alkali and noble metals, but is evident also for others, such as Al and C; it is present for both negative and positive cluster ions (see Chap. 5).

Data for the variation of the binding energy of Ag clusters with the number of atoms [2.281] indicate that for charged clusters this is definitely higher than for the neutral ones. This applies to a comparison of both iso-nucleonic and iso-electronic clusters ($Ag_4$ and $Ag_5^+$) and can explain the above-mentioned finding that ionized clusters show a higher abundance, in terms of binding strength. These computations do not produce, however, the strong even–odd alternations for the charged species. These appear to be due to oscillations of the ionization potentials (for positive ions) or the electron affinities (for negative clusters) with cluster size $n$ [2.282]. Hence, the oscillating abundance distribution of charged clusters can be ascribed to alternating *ionization thresholds* and not to cluster stability. As stated here, these arguments may be directly valid only for cluster emission from metallic and covalent targets. For other materials the stability may be dominated (also) by cluster geometry, producing so-called magic numbers in the abundance distribution of sputtered clusters (e.g. from alkali halides).

With the first detection of small neutral clusters in the sputtered flux, the question of their formation became an intensively discussed topic; it appears, in fact, largely unresolved to this day, despite considerable research efforts [2.251–257]. Originally, there existed some consensus that sputtered clusters leave the surface as an entity, i.e. the constituents were nearest neighbors in the solid. Ejection was envisaged to occur via a single recoil imparting to the cluster-to-be, in the last collision, enough energy to surmount the surface binding energy. This so-called *single-collision* model encounters difficulties in explaining the emission of clusters with constituents of similar mass and weak bonding. (This applies, for example, to the clusters shown in Fig. 2.20, the dissociation energy of $Cu_2$ and $Ni_2$ amounts to $\sim 2\,eV$ and is even smaller for the other dimers.) When energy is transferred only to one of the constituents, this may result in a high degree of rotational and vibrational excitation, leading ultimately to dissociation. This concept may be applicable, however, to (small) clusters with greatly different masses of the partners and strong interatomic binding. If the energy is transferred to the heavy atom, the energy in the center-of-mass can be sufficiently low that the cluster can survive the energetic ejection event. Data on the emission of neutral TaO and similar species provide supporting evidence for this scenario of single-collision emission.

Specifically for metal dimers (and trimers) such as those shown in Fig. 2.20, the so-called *double-collision* mechanism was proposed [2.280, 283]: the two constituent atoms of the dimer are emitted by two (or more) largely independent collisions with target atoms. Obviously, this formation scheme imposes severe constraints on the energies and momenta involved; these aspects are discussed thoroughly in [2.283]. While this model provides, in principle, the possibility of cluster formation from atoms which initially were not nearest neighbors, simple energy arguments restrict the range to nearest and next-nearest neighbors. As noted by several authors, this is due to the fact that at large separations the mutual interaction potential is close to zero, whereas the constituents' kinetic energy is still positive; hence, the distinction between the two models may come down to a semantic one [2.249].

For the double-collision model and not too large clusters, the cluster abundance distribution is predicted to follow roughly a power dependence on the sputtering yield [2.280]

$$Y_n \propto Y^n \tag{2.42}$$

where $Y_n$ is the yield of an $n$-atom cluster; this relation is based on statistical arguments and supposes that the cluster is ejected through multiple recoils hitting each of the individual constituents independently. Such a scenario imposes of course a severe restriction with respect to the temporal and momentum overlap of these recoil events. Several authors [2.124, 196, 279, 284–287] have studied the emission of small neutral clusters in an attempt to check the validity of (2.42). Gnaser and Oechsner [2.196] have verified (2.42) for dimers and trimers sputtered from different metals. This was accomplished by varying the ion incidence energy (and thus the yield $Y$) in the range from 30 eV to 1000 eV; to avoid the influence of changes in the angular emission distribution, the samples had a hemispherical shape and were immersed in the plasma to ensure normal-incidence ion impact. (Such an arrangement results in the detection of the emission-angle-integrated flux of sputtered species [2.196].) The plasma ions effected sputtering, while the electrons were employed for ionization of the sputtered neutral species. The post-ionized atoms and clusters were detected in a mass spectrometer. Figure 2.21 shows the dependence of the yield ratio $Y(Cu_2)/Y(Cu)$ on $Y(Cu)$. The linear correlation is in excellent agreement with (2.42). Similar data have been reported for dimers and trimers emitted from Ni and some binary alloys (see Sect. 3.3.4). Despite these good correlations of the data with (2.42) for ion energies below about 1 keV, the strict validity of (2.42) for much higher energies has been doubted. Furthermore, such a simple mechanism based on individual recoils

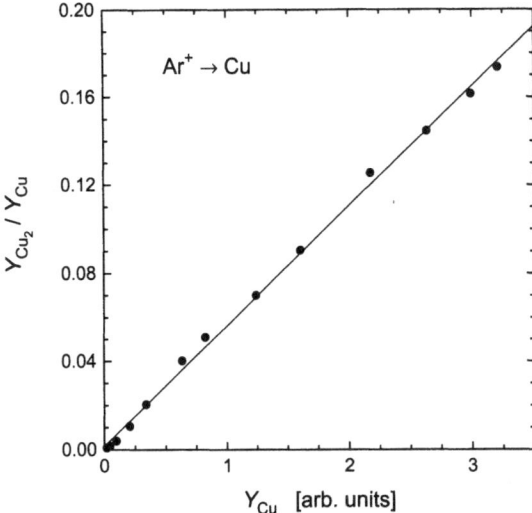

**Fig. 2.21.** Relative yields of neutral Cu dimers $Cu_2$ (normalized to the atom yield $Y_{Cu}$) sputtered from a hemispherical Cu specimen by $Ar^+$ ions as a function of the atom yield. The *solid line* is a linear least-squares fits to the data. Yield variation was effected by changing the $Ar^+$ impact energy. Data from [2.196]

(and purely statistical formation) appears less plausible for larger clusters with $n \geq 4$ and is very unrealistic for clusters with $n > 10$. As noted before, the statistical fluctuations in the collision cascade are expected to have a major influence on the formation of these clusters. This is also exemplified by the fact that even for the situation where the average sputtering yield is much smaller than 1 (see Fig. 2.21), a non-vanishing probability for the emission of two atoms from the same collision cascade and for dimer formation does exist. (The lowest dimer yield depicted in Fig. 2.21 corresponds to an $Ar^+$ bombarding energy of 50 eV.)

In an MD simulation of cluster emission in sputtering, Gades and Urbassek [2.288] introduced the concept of a *clustering probability*, which is meant to describe the probability that two sputtered atoms actually cluster, forming a dimer. Basically, it constitutes the coefficient of proportionality in (2.42). Specifically, they define the clustering probability $p$ as the ratio of the dimer yield to the average number of sputtered atom pairs:

$$\langle Y_2 \rangle = p \langle N_{\text{pairs}} \rangle , \tag{2.43}$$

where $\langle Y_2 \rangle$ is the average yield of homonuclear dimers. If exactly $Y$ atoms are sputtered, the number of pairs equals

$$N_{\text{pairs}} = \frac{1}{2} Y (Y - 1) \tag{2.44}$$

and

$$\langle Y_2 \rangle = p \left\langle \frac{1}{2} Y (Y - 1) \right\rangle = p \frac{1}{2} \left[ \langle Y \rangle^2 + (\Delta Y)^2 - \langle Y \rangle \right] , \tag{2.45}$$

where $\Delta Y$ is the standard deviation of $Y$. The authors investigated values of $p$ both for pure Cu and for an artificial target composed of a mixture of natural Cu and a material with twice the mass of Cu. While $p$ is essentially independent of the surface binding energy and the mass ($p \sim 0.035$), it increases roughly linearly with the cluster's dissociation energy $D$. For the pure Cu target, the authors do not find the parabolic dependence of the dimer yield on $Y$ predicted by (2.45); the dependence is, rather, linear. Gades and Urbassek [2.288] argue that high-yield cascades would have a high probability of creating clusters larger than dimers. Taking this effect into account, they infer slightly larger values for the clustering probability, $p \approx$ 0.04–0.05. These authors determined, furthermore, the original positions of the atoms that form dimers; almost 60% of the sputtered dimers had been nearest neighbors and were bound in the solid. The probability to form one stable dimer, $Y_2 = 1$, when exactly *two* atoms are ejected, $Y = 2$, amounts to $\sim 0.04$ and is thus comparable in magnitude with the value of $p$ mentioned before. This implies that the clustering probability is the same in high- and low-yield events, an observation that is far from trivial.

Applying the concept of a clustering probability to the $Cu_2$ dimer data given in Fig. 2.21, (2.45) may be used to approximate $p$ for these experimental data. For 1 keV $Ar^+$ bombardment of Cu, $\langle Y \rangle \approx 3$ atoms/ion, whereas according to [2.152] the standard deviation might be estimated as $\Delta Y \sim \langle Y \rangle$. With the dimer yield $\langle Y_2 \rangle \approx 0.45$ dimers/ion, it follows that $p \sim 0.06$, a value comparable to those obtained in the simulations. Computations at irradiation energies below $E/U \sim 300$ show that $\langle Y \rangle \approx (\Delta Y)^2$; for this condition (2.45) would be identical to (2.42).

*Ionized* clusters carrying an impressively large number of atoms have been found in the sputtering of metals [2.278]. The detection of *neutral* clusters by employing an electron beam or a plasma for post-ionization has generally been limited to cluster sizes $n \leq 5$ (cf. Fig. 2.20). More recently, with the application of high-power lasers for the photoionization of neutral species, the accessible size range could be extended, but still falls short of that of the detected ionized clusters. The metallic *neutral* clusters studied most intensively were $Al_n$ ($n \leq 12$), $Ag_n$ ($n \leq 19$), $Cu_n$ ($n \leq 20$), $Ga_n$ ($n \leq 13$), $In_n$ ($n \leq 32$), $Nb_n$ and $Ta_n$ [2.289–98]. Generally, a strong decay of cluster abundance with the number $n$ of atoms contained in the cluster is observed. For homonuclear clusters a fall-off according to a power dependence on $n$ has been reported,

$$I_n \propto n^{-\delta} , \tag{2.46}$$

first for $Cu_n$ and $Ag_n$ and later for most of the other above-quoted clusters. The exponent $\delta$ turned out to depend on the bombardment conditions; specifically, an empirical relation between $\delta$ and the sputtering yield $Y$ of the target was established: an increasing yield causes a weaker decay of cluster abundances, i.e. results in a larger fraction of larger clusters. Variations of $\delta$

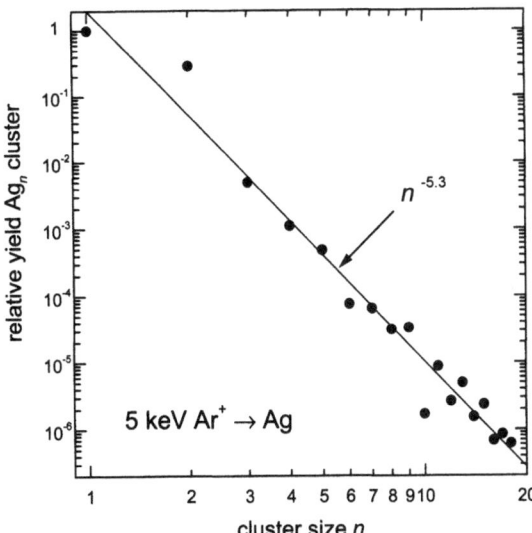

**Fig. 2.22.** Relative yields of neutral silver clusters $Ag_n$ sputtered from polycrystalline Ag by 5 keV $Ar^+$ ions (incidence angle 45°) versus the cluster size $n$. The *straight line* is a power-law fit with $n^{-5.3}$. Detection of the neutral species was accomplished by photoionization and time-of-flight mass spectrometry. From [2.297]

and $Y$ were induced either by changing the ion energy and ion species for the same sample, or by keeping the bombardment conditions constant and using different target materials. Figure 2.22 depicts the dependence of the $Ag_n$ yield on cluster size $n$ for 5 keV $Ar^+$ bombardment of Ag at an incidence angle of 45° [2.297]. The fit according to (2.46) yields $\delta = 5.3$. For 1 keV $Ar^+$ the experiments result in $\delta = 7.4$, whereas for 5 keV $Xe^+$ ions $\delta = 4.3$. Qualitatively, this is related to the broadening of the probability distribution of sputtered atoms discussed above, which appears to be the prerequisite for the formation of (large) clusters.

While the cluster yield data available up to now fit the inverse correlation with size proposed in (2.46), there exists apparently no simple theoretical model that describes this experimental finding. Wucher and Wahl [2.299] have shown that the multiple-collision model outlined above would produce an *exponential* yield decay with increasing cluster size rather than the *power-law* dependence observed for large clusters (Fig. 2.22). This model is therefore not suitable for clusters larger than $n \sim 3$. The only way to reproduce the experimentally observed yields of sputtered metal clusters is, apparently, a combination of MD and Monte Carlo (MC) computer simulations [2.265, 267]. The cluster sputtering mechanism is divided into two major steps. First, the formation of so-called *nascent* clusters, i.e. clusters which are identified immediately above the sample surface, is computed by MD. These clusters carry a great amount of internal energy and are mostly unstable on the timescales

needed for experimental detection. In a second step, therefore, the unimolecular decomposition of these clusters is followed by an MC scheme. The *final*
fragmentation products resulting from that step can be compared to the experimental data. (Actually, in the MD simulations clusters up to $n \sim$ 8–10
are observed, whereas for larger ones an extrapolation according to (2.46)
is applied.) Employing a realistic interaction potential due to DePristo and
coworkers [2.300], Wucher and Garrison [2.267] were able to determine yield
distributions for neutral $Ag_n$ clusters in close agreement with the experimental results, in terms of both the power-law decay with the size $n$ and the value
of the exponent $\delta$ in (2.46). In view of these consistencies it can be expected
that the existing (and possible additional) simulations can provide detailed
insight into the formation process of larger clusters.

Cluster mass distributions from a thermalized medium have been discussed by Urbassek [2.301]. He applied the assumption that local thermodynamic equilibrium is established in part of the ion-irradiated sample volume.
When cooling down, this region may pass through the liquid–gas coexistence
curve and undergo a (nonequilibrium) phase transition. At the critical point
$(T = T_c, p = p_c)$, the liquid and gas phases are in equilibrium and the model
yields, in this case, a mass distribution of clusters identical to (2.46), albeit
with an exponent $(\delta = 7/3)$ much smaller than the value typically derived for
sputtering of metal targets [2.292–297]. Urbassek [2.301] noted on the other
hand that experimentally measured mass distributions of clusters sputtered
from condensed gas specimens follow the power law with the above exponent
of 7/3.

From the experimental cluster yield distributions, the total fraction of
sputtered atoms which are emitted in a bound state can be determined
[2.297, 298]. For the available cluster data obtained for Ag, this bound-state
fraction was found to correlate with the sputtering yield in such a way that
it increases linearly up to $Y \sim$ 8 atoms/ion (then the fraction amounts to
$\sim 0.35$), and tends to level off for higher yields at a value as high as 0.46. This
number is mostly due to dimers ($\sim 96\%$) and trimers ($\sim 3\%$). This surprisingly high numbers of atoms emitted in clusters underscores the importance
of (neutral) clusters with respect to particle emission in sputtering.

Both experimental data [2.272–275] and molecular-dynamics simulations
[2.265, 267] have proved convincingly that cluster abundance distributions
may be subject to modification due to fragmentation processes following the
ejection from the surface. This is not surprising in view of the observation
[2.302–304] that clusters (and indeed molecules) are sputtered *hot*, i.e. with
high internal energies. Detailed decomposition experiments were performed
for ionized clusters and the results elucidated the fragmentation kinetics and
thermodynamic properties of the clusters [2.305]. (An example of such a
fragmentation process will be presented in Chap. 5 for sputtered negative $C_n^-$
clusters: a comparison of the ratio of stable to fragmented clusters of a given
size $n$ indicates that $C_n^-$ clusters with $n > 5$ have a strong tendency to de-

compose on time scales of some picoseconds.) These findings are corroborated
by MD simulations [2.265, 267], which found that for keV Ar bombardment
of Ag the majority of the emitted trimers and virtually all the larger clusters
fragment spontaneously in the first few nanoseconds after emission. This behavior could be traced back to the high internal energies the clusters carried
upon ejection.

Data on the internal energy of sputtered clusters and molecules are comparatively rare. The first experiments were performed on diatomics sputtered
from silicon containing various impurities [2.252, 253], $S_2$ dimers ejected from
elemental sulfur and $CS_2$ specimens [2.303], and a few alkali dimers ($Na_2$,
$K_2$, $Cs_2$) [2.302]. More recently, Wucher [2.304] investigated the internal excitation of sputtered $Ag_n$ clusters by laser spectroscopic methods. For dimers
he derived from the spectra vibrational and rotational temperatures around
2700 K and 6700 K, respectively, in very good agreement with corresponding
MD simulations [2.266]. For larger clusters, a large amount of internal energy
was inferred from ionization experiments using several different laser wave-
lengths; around 50% of the sputtered $Ag_6$ clusters are formed with internal
energies in excess of 0.75 eV.

The aforementioned MD simulations by Wucher and Garrison [2.267] provided further support for these observations. Sputtered $Ag_n$ clusters in their
nascent state carry a high internal energy which scales roughly linearly with
cluster size $n$, $E_{int}(n) \sim (1.4n - 1.86)$ eV. For dimers, the population of both
vibrational and rotational states can be accurately approximated by quasi-
thermal, i.e. Boltzmann-like distributions albeit with different rotational and
vibrational temperatures, $T_{rot} = 5900$ K and $T_{vib} = 3100$ K. The high degree
of excitation results in a decomposition of the nascent clusters on their way
away from the surface: the survival probability amounts to 21% for trimers,
5% for tetramers and is essentially zero for all clusters with $n \geq 5$. These dissociation reactions occur on a subpicosecond or at most picosecond timescale,
while the total decomposition time is of the order of some ten picoseconds.
After that time interval the nascent clusters have decomposed into the final
clusters and atoms; these are the ones that can be detected experimentally.

**Energy Spectra.** With regard to the emission-energy distributions of sputtered clusters and molecules, one may wish to make a distinction between the
different cluster formation mechanisms that might be operative. Considering
first a solid where preformed molecules exist, for these the dissociation energy $D$ of a diatomic molecule might be large compared to the binding energy
$U$ to the surrounding atoms. Such a situation can be realized for molecular
solids or under reactive-ion-bombardment conditions. The experimental observation of a high-energy decay in the energy spectra of the form $E_0^{-2}$ is
supportive evidence for the occurrence of the so-called single-collision emission mechanism [2.257]: molecule ejection is triggered by a single collision
with a recoil atom in the cascade (as opposed to the double- or multiple-
collision process discussed above). According to Sigmund et al. [2.256] the

single-collision picture would result in the following features: (i) both rotational and vibrational excitation obey an $E_0^{-2}$ law; (ii) the internal energy is positively correlated with the kinetic energy; and (iii) the rotational and vibrational energies are anticorrelated with each other. The experimental data of De Jonge et al. [2.303] do not fully support these expectations and require a further refinement of the model. For a more detailed discussion see [2.249].

On the other hand, in most elemental metals and semiconductors the dissociation energy $D$ of a dimer is smaller than the sublimation energy (i.e. the surface binding energy $U$). As discussed before, a single collision might impart a sufficiently high internal energy to destroy the dimer. Conversely, a double-collision process might work provided the momenta of the two atoms hit are sufficiently aligned and of comparable magnitude; if the trajectories are close to each other, the atoms are bound and form a dimer. As mentioned above, this implies rather stringent conditions on the phase space available for dimer formation and the kinetic energy is expected [2.283] to decay quite rapidly towards higher values, as $E_0^{-5}$ for dimers $(E_0 \gg U, D)$. Extension of this model to larger clusters of size $n$ indicates that the energy spectra of clusters should exhibit an asymptotic behavior at high energies proportional to $E_0^{-3n+1}$. Furthermore, the average emission energy should become smaller, implying a shift of the most probable energy to lower values. Experimental support for this notion is far from clear-cut.

Experiments [2.306] using post-ionization by electron impact produced energy spectra of small neutral dimers and trimers that qualitatively agree with these predictions, although the high-energy decay could often not be established unambiguously since rather low bombarding energies were used. Energy spectra of neutral $Cu_2$ and $Cu_3$ were recorded over a more extended emission-energy range using plasma post-ionization in conjunction with high-sensitivity mass spectrometry [2.307]; these data indicate an essentially identical high-energy fall-off ($\propto E_0^{-5}$) for both the dimers and the trimers. Larger neutral clusters became accessible only with the utilization of photoionization techniques for the detection of neutral species. The energy distributions of sputtered neutral clusters recorded by these means produced an (initially) quite puzzling result: contrary to expectations (and the predictions of the multiple-collision model), for these large clusters (detailed studies were done for $Al_n$, $n \leq 6$, $Cu_n$, $n \leq 6$, $Ag_n$, $n \leq 7$, and $In_n$, $n \leq 8$ [2.289–292, 296, 298]), typically, the most probable energy differs little for atoms and clusters, while the exponent $p$ in the high-energy decay, $E_0^{-p}$, is slightly higher for clusters than for the respective atoms, but is largely *independent* of the cluster size. For example, 3.9 keV $Ar^+$ irradiation of Cu produces $p \approx 2.8$ for Cu atoms and $p \sim 3.5$ for $Cu_n$ with $2 \leq n \leq 6$. Similar results were reported for $Al_n$ clusters. The data for $Ag_n$ clusters ($n \leq 7$) are depicted in Fig. 2.23. The asymptotic exponents derived are, for monomers, $p \sim 1.7$, for dimers, $p \sim 2.9$, and $p \sim 4$ for all larger clusters. For large clusters these and similarly derived values are somewhat uncertain as the limited energy range restricts an accu-

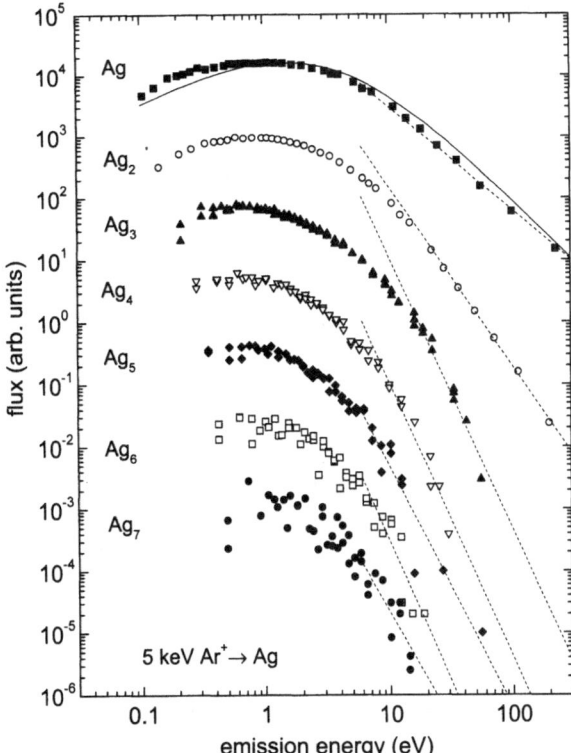

**Fig. 2.23.** Kinetic energy distributions of neutral Ag atoms and Ag$_n$ clusters sputtered from a polycrystalline silver sample by $5\,\text{keV}$ Ar$^+$ ion bombardment. (The relative scaling of the different curves is arbitrary.) The *solid line* represents the theoretical distribution calculated for Ag atoms, (2.34). The *dotted lines* are power-law fits of the asymptotic high-energy dependence of the atom and cluster yields. From [2.297]

rate determination of $p$. Despite this caveat, it is obvious that the measured energy spectra do not show the steep decay at high energies anticipated from an extension of the multiple collision model. In fact, this is hardly surprising in view of the difficulties one faces in envisioning the production of a (large) cluster by means of many independent (i.e. uncorrelated) recoil collisions in the ejection event.

MD simulations of the energy spectra of Ag$_2$ dimers are in very good agreement with the corresponding experimental data [2.265]. Such simulations also show that for keV irradiation sputtered dimers originate with the highest probability from nearest-neighbor sites and that a true double-collision mechanism accounts for the majority of emitted dimers. For larger clusters, however, simulations [2.268] provide rather convincing evidence for an emission mechanism in which, owing to correlated motion in the collision cascade, a group of neighboring atoms at the surface receives, simultaneously, nearly parallel momenta, effecting the emission of a group of bonded atoms. These computations also produced energy spectra of $Cu_n$ clusters ($n \leq 5$) which closely match corresponding experimental data. They indicate, furthermore, that (large) clusters consist predominantly of atoms from the surface belonging to a connected group and that their emission occurs at a late stage of the particle emission process; the larger the cluster, the later it is ejected. Very probably, cluster emission is correlated with single high-yield events.

**Angular Emission Distributions.** Information on the angular distributions of sputtered clusters is comparatively rare. For bombardment of monocrystalline targets, the experiments by Gnaser and Hofer [2.235] showed enhanced emission of dimers and trimers along the close-packed lattice directions (see Fig. 2.19). This can be considered a consequence of momentum alignment; any such alignment of the recoils' momenta should result in a reduction of their relative kinetic energies and, thus, an enhanced possibility of satisfying the binding condition. Apart from this preferred emission direction, there is little influence of the lattice structure on the abundance distributions: for Si and Ge clusters sputtered from the respective elemental specimens, the yields are essentially identical for a crystalline or an amorphous state of the sample.

A difference in the angular emission distributions for atoms and clusters might be expected in view of the predictions of (2.42) and (2.45) for the double-collision formation process [2.283]. The (angle-selective) variation of the total yield $Y$ would produce a relatively stronger variation of the dimer yield $Y_2$. An example of such a behavior was reported for neutral $Si_2$ dimers sputtered from $WSi_2$ by 500 eV $Ar^+$ ions. As shown in Chap. 3, under these conditions an altered surface layer develops, resulting in a depletion of Si; this, in turn, causes a forward-peaked angular emission of Si atoms compared to W atoms (the Si atoms are ejected, on average, from a greater depth, which restricts their emission angle). This preferentially forward-directed Si flux results in an even more pronounced $Si_2$ emission along the target normal. The $Si_2/Si$ yield ratio is about a factor of three higher at $\theta_0 = 0°$ than for $\theta_0 \geq 40°$. This observation implies a high sensitivity of the cluster flux to surface modifications.

## 2.4 Cluster-Ion Bombardment

The irradiation of surfaces by projectiles composed of two or more atoms has received considerable attention over the last decade or so. On one hand, this interest in cluster bombardment was motivated by possible applications, e.g. in thin-film deposition [2.308–310], enhanced electron and ion yields [2.311–313], mass spectrometry [2.313–316] and surface modifications [2.317, 318]. On the other hand, the more basic mechanisms of ion–solid interactions could be studied for a wider range of parameters by using cluster-ion impact. For example, the possibility of achieving a high energy-deposition density in the collision cascade was of interest and was utilized to explain the enhanced sputtering yields observed in bombardment with small clusters long ago [2.319–325] and the even more drastic enhancement found later with larger clusters [2.326, 327]. Another feature of interest is the unusual multiple-collision kinematics which might occur in cascades initiated by cluster impact. A large number of computer simulations have been carried out; they have provided quantitative information on many important aspects, such as the penetration depth and the energy loss of clusters in solids, the energy spectra of transmitted and reflected particles and of target recoils, and the occurrence of cascades within the cluster upon its hitting the surface [2.328–331]. The importance assigned to cluster-ion irradiation has triggered, within the last couple of years, the appearance of several reviews [2.308, 310, 332, 333] that cover both possible applications and fundamental issues. Hence, in this section some recent results on effects associated with (enhanced) particle emission under cluster-ion bombardment are highlighted.

For bombardment with high-energy, heavy ions, sputtering yields are generally observed to be higher than predicted by the analytical expression given by Sigmund [2.8]. The same finding apparently holds for irradiation with molecular (or cluster) ions. The first clear evidence for an enhanced sputtering yield under cluster bombardment was provided by the experiments of Andersen and Bay [2.319]. By comparing yields for atomic and dimer ion irradiation, they observed that in the latter case the yield was more than twice as high as the value for monomers of the same velocity. Similar findings were reported by Thompson and coworkers [2.320–322]. These deviations from a linear superposition of atomic yields were interpreted as being caused by *nonlinear* effects in the collision cascade. However, changes in the effective surface binding energy have also been invoked as a possible cause. For dimer and trimer ions, the observed enhancement factors (defined as the ratio of $Y$ for cluster impact to $n$ times the yield for monomers, with $n$ the number of atoms in the cluster) were in the range from $\sim 1$ to $\leq 10$, for ion energies of some $10\,\mathrm{keV/atom}$. For lower energies or lighter ions the enhancement tends to become very small. Apart from enhanced sputtering yields, further evidence for nonlinearities in the collision cascade came from low-energy peaks in the energy spectra of sputtered atoms [2.325, 334, 335] and the formation of extensive craters upon bombardment with heavy ions or molecules [2.94, 95, 336] (cf. the discussion in Sect. 2.2.3).

Several distinct theoretical models have been proposed to describe these observations of nonlinear collision cascades: they range from shock-wave propagation [2.337, 338] to crater formation and some type of thermal evaporation from high-temperature spikes [2.339, 340]. These approaches were aimed at quantifying this nonlinear (spike) regime of sputtering in terms of the temporal evolution [2.341], the energy density [2.341, 342] and the yield enhancement [2.342–344]. In this context, the temperature dependence of the sputtering yield was also investigated [2.345] in order to interpret experimental yield data [2.346, 347]. A large number of investigations have been devoted to elucidating these effects. Related to these observations are attempts to picture sputtering events with respect to the timescale: Kelly [2.107] distinguished four different time steps in sputtering, namely prompt collisional ($10^{-15} \leq t \leq 10^{-14}$ s), slow collisional ($10^{-14} \leq t \leq 10^{-13}$ to $10^{-12}$ s), prompt thermal ($10^{-13}$ to $10^{-12} \leq t \leq 10^{-11}$ to $10^{-10}$ s) and slow thermal ($t \geq 10^{-11}$ to $10^{-10}$ s). A similar concept was invoked by Sigmund and Szymonski [2.345]. In the elastic collision spike (corresponding to the cooling phase discussed in Sect. 2.2.1), the energy has to be shared among many atoms within a limited volume, and the resulting spike may attain a very high temperature, up to some $10^4$ K. This high initial density of kinetic energy is a necessary condition for the observation of large sputtering-yield enhancements in that stage. Although these effects are not expected to play a role in the general low-energy irradiation conditions discussed in this book, they are very probably of great importance for cluster-ion bombardment. Unfortunately, the understanding of these processes appears still quite fragmentary. Since there exist several more detailed reviews [2.101, 102, 332] of this topic, it will not be discussed here further.

Several groups [2.311–317] have used cluster ions composed of a larger number of atoms, such as $SF_6$, $ReO_4$, $Au_n$, $(CsI)_nI$, $(CsI)_nCs$, $Ar_n$, $(CO_2)_n$ and $C_{60}$, to study impact phenomena at the surface. Large water clusters ($n \sim 150$) at high energies (a few keV/atom) were employed by Beuhler and Friedman [2.326, 327] to investigate bombardment-induced surface modifications, and the occurrence of large craters and extended defect structures was reported. Typically, the total kinetic energies for these projectile ions covered a large range, from less than 1 keV to several MeV. Essentially all of these studies monitored the flux of sputtered *ions* from the surface but did not provide information on the total sputtering yield. It appears not to be possible, therefore, to separate yield enhancement effects in the sense discussed above for dimer and trimer clusters from effects that are due to an enhanced ion formation probability under cluster-ion bombardment. Generally, ion yields under cluster impact are much higher than under atomic-ion irradiation at the same energy/mass value. This tendency is clearly demonstrated by data of Benguerba et al. [2.315]; Fig. 2.24 depicts the yield of $Au^-$ ions emitted from an Au surface as a function of the energy-to-mass ratio of $Au_n^+$ bombarding ions. (These authors produced singly and doubly charged positive $Au_n$ clusters in a liquid metal source and detected the sputtered ions by time-of-flight mass spectrometry.) Comparable results were found by monitoring other secondary-ion species and by irradiating other targets. These authors also note that, at the same velocity, the secondary-ion yield induced by clusters is proportional to the square of the number of constituents. The nonlinear effects of cluster impact appear to be more significant for complex secondary ions.

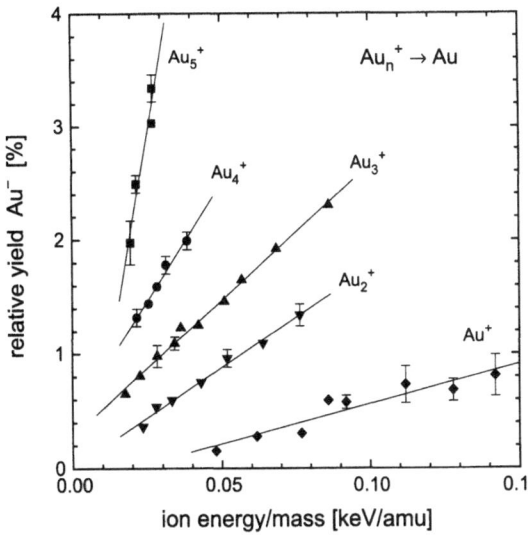

**Fig. 2.24.** Yield of $Au^-$ ions sputtered from a gold specimen as a function of primary-ion energy per unit mass (in keV/amu) for bombardment with $Au_n^+$ cluster ions with $n = 1$–$5$. Data from [2.315]

Yamada and his coworkers [2.310, 317, 318] investigated very intensively the interaction of large cluster ions (composed of several thousand atoms or molecules) with surfaces. Varying the mean cluster size and/or the impact energy, they were able to cover a wide range of values of the energy per cluster atom. These authors observed pronounced smoothing of surfaces upon the impact of these large clusters and ascribed this finding to a sputtering process that ejects atoms mostly sideways, thereby removing protruding surface features. In terms of sputtering yields, Yamada et al. [2.318] reported dramatic yield enhancements for *cluster* impact as compared to *atomic* ions. Figure 2.25 depicts some of these results, showing the values of $Y$ for several elemental samples under irradiation by 20 keV $Ar_n^+$ cluster ions ($n \sim 3000$) and by 20 keV $Ar^+$ ions. (Results very similar to those for $Ar_n^+$ were reported [2.318] for $(CO_2)_n^+$ cluster ions.) The sputtering yield is enhanced drastically (e.g. up to a factor of 30 for Si) for cluster ions as compared to atomic ions although the energy per atom of the cluster only amounts to about 6.7 eV (for $n \sim 3000$). This value is far below the threshold for single-ion impact sputtering (see Fig. 2.15); some of the processes identified in computer simulations of cluster impacts on surfaces (see below) may be responsible for the enormously high sputtering yields observed by Yamada et al. [2.318]; see also Sect. 2.2.3.

Extensive computer modeling of cluster–solid interaction was performed by several research groups [2.328–331, 348–354]. Sigmund and coworkers [2.328–331] have elucidated pertinent features of cluster-induced collision cascades that clearly distinguish them from those initiated by atomic-ion bombardment. First, with respect to cluster stopping, a so-called "clearing-the-way" effect was identified: Front-runners in a cluster may kick atoms

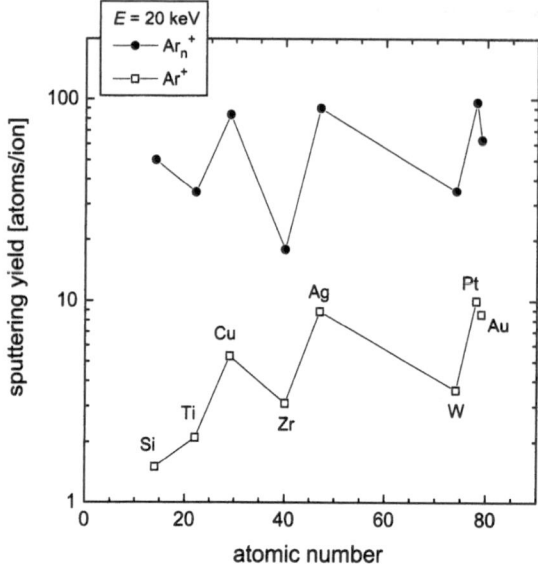

**Fig. 2.25.** Sputtering yields of various elemental targets under 20 keV $Ar_n^+$ cluster-ion irradiation (average cluster size 3000) compared to yield data for 20 keV $Ar^+$ ion bombardment. The results are depicted as a function of the elements' atomic numbers. Data from [2.318]

away from the trajectories of subsequent projectile atoms, which see a target that has been depleted of potential collision partners. This effect reduces the average energy loss of the cluster; it is significant for heavy projectiles in light targets. Second, collision cascades within the incoming cluster were found, leading to a possible acceleration of cluster atoms. Recoiling target atoms may achieve kinetic energies much larger than the maximum recoil energy in a single elastic collision. Third, a completely different mechanism may become operative when particles participate simultaneously in a target cascade and in a cluster cascade. According to Sigmund, this process is most efficient when at least two different species are involved, one light and one heavy, with the latter in both target and cluster. A light species might bounce back and forth between a heavy species in the cluster and in the target (hence the term "shuttle effect"), leading to a substantial acceleration for several reflective collisions. While the influence of the aforementioned processes on the energy dissipation and the slowing down of the clusters is rather apparent from the simulations of Sigmund and coworkers [2.328–331,350], the effect on sputtering is not as obvious.

The wide spectrum of processes that can occur on the surface upon irradiation with large clusters is illustrated in Fig. 2.26. These are snapshots from MD simulations [2.354] taken at different times after cluster impact ($Al_n$ with $n = 504$ atoms) on a Cu(111) surface for different bombarding energies

(0.1 eV, 1 eV, 10 eV and 30 eV per cluster atom). Figure 2.26A displays cluster impact at almost thermal energies of the constituent atoms (0.1 eV/atom). Under these conditions, essentially no penetration into or mixing with the target occurs as shown in panel (d). Also, no sputtering is observed. For sufficiently low temperatures of the sample (300 K in the simulations), the cluster does not spread out on the surface but forms an island, and the final state is reached after about 5 ps. At a cluster energy of 1 eV/atom (Fig. 2.26B), spreading of the cluster on the surface and mixing with the target surface layer are observed. The cluster forms an island about three layers high, as compared to 8 layers at 0.1 eV/atom. Initially, the kinetic energy of the cluster is sufficient to effect compression of the target through collective collisions (see (a) after 0.6 ps). At a bombarding energy of 10 eV/atom (Fig. 2.26C), compression of the target upon impact and mixing become pronounced. The surface is covered by an adlayer consisting of cluster and target atoms. In addition, sputtering of the target is observed although the energy is well below the threshold for single-ion sputtering. This finding is clearly in agreement with the experiments of Yamada et al. [2.318] described above (see Fig. 2.25). After 10 ps the target is still hot and the atoms have not settled at regular lattice positions. At the highest energies in these simulations (30 eV/atom, (c) and (d) of Fig. 2.26C), the regime in which crater formation occurs is reached; sputtering and reflection of cluster atoms are now substantial. Apparently, in this energy range, pronounced surface damage is created, reminiscent of features discussed in Sect. 2.2.1.

In view of experimental and simulation results such as those depicted in Figs. 2.23–2.26 and related observations (see e.g. [2.333, 355, 356]), it appears reasonable to expect that cluster-ion bombardment of surfaces may provide access to a rather wide realm of new and possibly surprising effects both in terms of fundamental ion–solid interaction processes and in terms of technologically important applications.

**(A)**

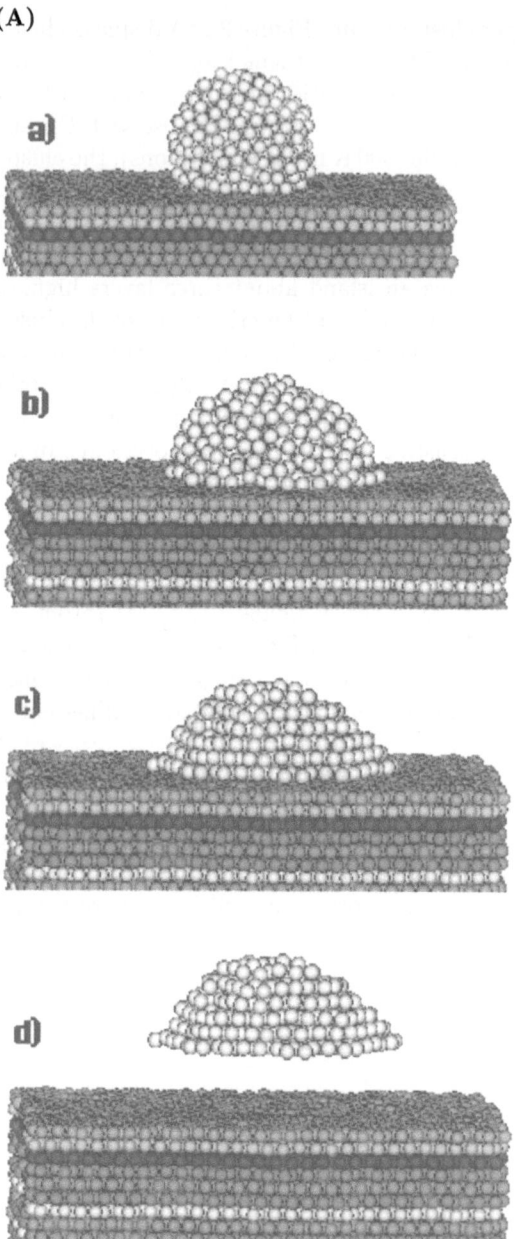

**Fig. 2.26A–C.** Snapshots from MD simulations for bombardment of a Cu(111) surface with an Al$_n$ cluster ($n = 504$) of various kinetic energies. (**A**) At a cluster energy of 0.1 eV/cluster atom after (a) 1 ps, (b) 2 ps and (c) 10 ps. (d) is also at 10 ps but with the cluster atoms lifted above the surface to demonstrate that no mixing with the target has occurred

**(B)**

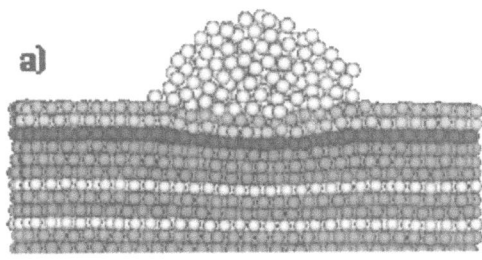

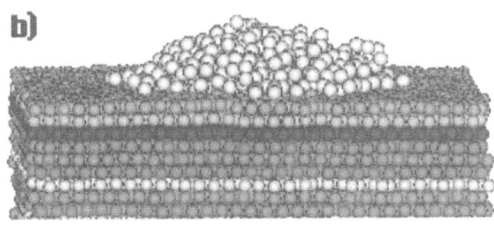

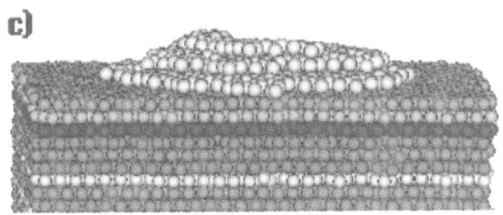

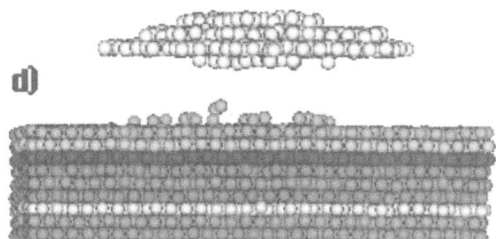

**Fig. 2.26A–C.** (Continued.) **(B)** At a cluster energy of 1 eV/cluster atom after (a) 0.6 ps, showing a cut through the crystal, (b) 1 ps, (c) 10 ps and (d) 10 ps but with the cluster atoms lifted above the surface to demonstrate that mixing with the surface layer has occurred

(C)

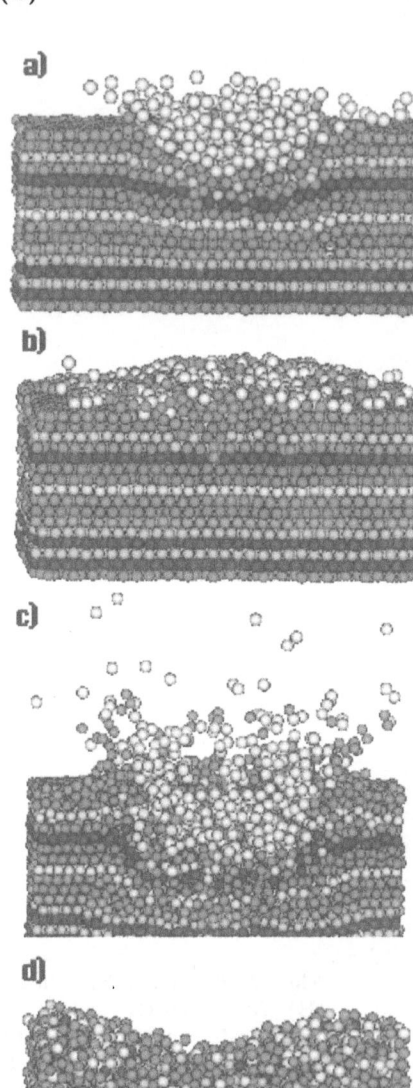

**Fig. 2.26A–C.** (Continued.) (**C**) At a cluster energy of 10 eV/cluster atom after (a) 2 ps and (b) 10 ps; at a cluster energy of 30 eV/cluster atom after (c) 5 ps and (d) 10 ps. From [2.354]

# 3. Composition Changes
# in Multicomponent Materials

While the basic principles of ion irradiation and sputtering are best illustrated for (idealized) monatomic targets, such materials do not exist and most samples are composed of two or more elements. Ion bombardment of such specimens produces generally a wide variety of different processes, which often are of considerable complexity and strongly related to each other [3.1–3]. As individual constituents may be transported within the solid at different rates [3.4], irradiation of multicomponent targets with energetic ions is usually accompanied by composition changes. In the context of defect production in solids due to collision cascades (cf. Sect. 2.2) it was noted that different *timescales* are relevant: a collisional phase was distinguished from a cooling (migration) phase. In the former, atomic relocation produces defects and effects sputter ejection, whereas in the latter, defect migration may lead to their partial annihilation. For monatomic systems only this final configuration is of importance. Conversely, in polyatomic materials collision-induced atomic transport in the first phase [3.4–6] and defect-assisted transport in the second phase [3.7–10] may give rise to a permanent compositional change of the solid [3.1–3, 9–11]. These modifications typically depend on the specimen temperature via the mobility of point defects.

These compositional variations in the target may also exhibit widely different *length scales*: (i) Effects solely due to (preferential) sputtering [3.2] should be confined to the outermost surface layers, i.e. to the depth of origin of sputtered species (see Sect. 2.3). (ii) Processes induced directly by the displacement cascade (collisional mixing) are usually associated with the ion's damage range in the target. (iii) An even more extended depth scale can be involved if defect migration at elevated temperatures is dominant [3.7, 8, 12, 13]. Experimental data indicate that situation (i) has rarely been observed in its pure form, although probably with the exception of systems with large mass differences of the individual components (see Sect. 3.3.1). More typical [3.1–3] are length scales defined by case (ii) and, at high temperatures, the length scales are defined most often by case (iii).

Sigmund [3.14, 15] has proposed the division of ion-bombardment effects in multicomponent targets into *primary* and *secondary* processes. The former are concerned with the sputtering behavior of a target with a given composition. Sputtering from polyatomic systems is preferential in the sense that the probability for an atom to be sputtered is species-dependent [3.3, 4]. Preferential sputtering is determined by several factors: The binding of an atom at the surface (or in the bulk) is element-dependent. The differences can be large for specific systems, but might still amount to a factor of two in metal alloys (cf. Sect. 3.1.2). Conversely, the differences are vanishingly small in isotopic mixtures (i.e. elemental samples composed of the various isotopes of that element), a fact used to investigate the *mass dependence* [3.3, 16] of preferential sputtering (Sect. 3.4). The preferential ejection of one species modifies the composition at the surface [3.1, 2] and results in a sputtered flux whose stoichiometry deviates, in the limit of low fluences, from that of the surface. This is one of the mechanisms resulting in compositional changes at the surface; it will be discussed thoroughly in the following (Sect. 3.2.1).

Secondary processes deal with further changes of the target composition due to effects initiated by the ion irradiation. They tend to alter the near-surface region over a depth range extending far beyond the topmost surface layers influenced by preferential sputtering [3.2, 3]. The major secondary processes are the following (a more detailed description is given later in this chapter): (i) *Collisional mixing* occurs during the displacement cascade in the solid, which effects relocation of atoms and the production of point defects. As the energy sharing in the collision cascade is species-dependent through the cross sections it may result in different ranges within the solid. (ii) At elevated temperatures these defects are mobile and give rise to *radiation-enhanced diffusion*. (iii) Preferential coupling of an alloy element to these defect fluxes results in radiation-induced segregation. (iv) In thermal equilibrium, *Gibbsian segregation* may occur, i.e. preferential segregation of one species to free surfaces (or interfaces), in order to reduce the free energy of the alloy. (v) Furthermore, *ion implantation* of the primary species changes the initial composition of the multicomponent target. Note that processes (iii) and (iv) are preferential but result in changes of the stoichiometry not necessarily related to preferential sputtering. This distinction between primary and secondary processes, which tends to emphasize the importance of sputtering for composition changes, will also be adopted in this work as the information on such compositional modifications presented here refers, to a large extent, to data obtained from sputtering. Obviously, other divisions of the processes according to the aforementioned time or length scales could also appear meaningful.

An important feature of the ion irradiation of multicomponent targets is the limit of steady state upon prolonged bombardment, i.e. at high fluences [3.3, 12]. Such a stationary state can only be reached if the initial composition is homogeneous in the bulk and if material can be transported only over finite distances. The steady state, then, results in a *sputtered-flux composition identical with the bulk stoichiometry*, a direct consequence of the conservation of matter.

Bombardment-induced changes in the surface and near-surface composition have been determined for numerous multicomponent materials and are compiled in various reviews [3.1–3, 17–19]. While the possible occurrence of the above-mentioned processes was noted, their complexity and synergistic nature often prevented a complete understanding of the experimental findings. Investigations covering a wider range of experimental parameters improved this situation considerably. Extensive theoretical modeling elucidated many of the pertinent processes. Lam and Wiedersich [3.7–9] have set up a comprehensive model of bombardment-induced surface composition changes which comprises both the primary and the secondary effects mentioned above. In a recent review emphasizing primarily the sputtering of multicomponent materials, Sigmund and Lam [3.3] formulated a theoretical scheme that accommodated all pertinent processes of sputtering and compositional changes. This approach is based on a nonlinear integro-differential equation generalized from an earlier version [3.15] that described composition changes due to preferential sputtering in the absence of migration. An explicit treatment of radiation-produced defects and defect-assisted processes has been added. Another important ingredient [3.12, 13] is that allowance is made for pressure relaxation. Irradiation-induced composition changes give rise to density and pressure gradients that cannot persist in a stable material. Therefore, rapid relaxation is taken into account in order to ensure stability of the target and the fulfillment of a packing condition at all fluences.

This chapter on compositional changes in multicomponent materials provides a description of the primary and (to a lesser extent) of the secondary processes in Sects. 3.1 and 3.2, respectively; it relies heavily (in both outline and nomenclature) on the recent work of Sigmund and Lam [3.3] as there appears to be little that can be added to that lucid exposition. With these conceptual means at hand, some results (of experiments and computer simulations) on ion irradiation of alloys (Sect. 3.3) and isotopic mixtures (Sect. 3.4) are presented.

## 3.1 Preferential Sputtering

The key quantity [3.14, 15] describing the sputtering of multicomponent targets is the partial sputtering yield $Y_i$ of component (element) $i$, giving the average number of sputtered $i$ atoms per incident particle. The partial yields $Y_i$ may, in principle, be determined in various ways (see Sect. A.1 for a description of experimental techniques): (i) Detection of sputtered atoms (or molecules) by mass-spectrometric techniques; while obviously mass-selective, this requires some kind of post-ionization (via interaction with electrons or photons) in the case of neutral species. Sputtered ions, on the other hand, make up a small fraction of the ejected flux and the relevant ionization processes are poorly understood, i.e. they may introduce some discrimination of unknown magnitude and direction (to some extent, this may hold also for some post-ionization schemes). (ii) Collection of the ejected flux on suitable collectors and subsequent analysis by surface-analytical techniques. Here distortions may occur both in the collecting step (species-dependent collection efficiency) and/or in the analysis (again some species dependence of detection efficiencies). In passing it is noted that changes in the surface composition can be monitored by these very same techniques.

As in the case of elemental targets, for polyatomic specimens sputtered atoms may also originate from a finite (albeit shallow) depth from the surface. The partial sputter yield $Y_i$ can then be defined [3.15] by considering the contributions from different depths:

$$Y_i = \int_0^\infty dx \sigma_i(x) N_i(x) \ , \tag{3.1}$$

where $N_i(x)$ is the atomic density (average number of $i$ atoms/volume) at depth $x$ and $\sigma_i(x)$ characterizes the sputtering of an $i$ atom at a depth $x$ (escape or ejection function). Since $\sigma_i(x)$ has the dimensions of an area it has been called the sputtering cross section [3.3, 15]. In general, it will decrease rapidly with increasing depth $x$.

A somewhat different description has been employed by Betz and Wehner [3.1] and by Lam and Wiedersich [3.7–10] for targets characterized by layers of unique thickness. The partial sputtering yield $Y_i$ in terms of different layers (index $l$) is given by

$$Y_i = \sum_{l=1}^\infty y_i^{(l)} c_i^{(l)} \ , \tag{3.2}$$

where $l = 1$ is the top surface, $c_i^{(l)}$ is the fraction of atoms of type $i$ in layer $l$ and $y_i^{(l)}$ is the so-called component yield of species $i$ from the $l$th layer. Taking into account the contributions to the sputtered flux from different layers, $y_i^{(l)}$ can be expressed in the form

$$y_i^{(l)} = \beta_i^{(l)} y_i \ , \tag{3.3}$$

where $\beta_i^{(l)}$ is the sputter fraction of the $i$ atom from the $l$th layer $\left( \sum_{l=1}^\infty \beta_i^{(l)} = 1 \right)$ and $y_i$ is the component sputtering yield, i.e. the yield per unit concentration of $i$ atoms. This approximation assumes that the values of $y_i$ are independent of the depth and the local composition at this depth. The partial sputtering yield is

$$Y_i = \sum_{l=1}^\infty \beta_i^{(l)} y_i c_i^{(l)} \ . \tag{3.4}$$

Furthermore, since sputtered atoms originate quite often from only the two outermost layers ($l = 1, 2$), (3.4) reduces to

$$Y_i = y_i \left[ \beta_i^{(1)} c_i^{(1)} + \beta_i^{(2)} c_i^{(2)} \right] = y_i \left[ \left( 1 - \beta_i^{(2)} \right) c_i^{(1)} + \beta_i^{(2)} c_i^{(2)} \right] \ . \tag{3.5}$$

In this description, "true" preferential sputtering would occur whenever $\beta_i^{(l)} \neq \beta_k^{(l)}$ and/or $y_i^{(l)} \neq y_k^{(l)}$; note, however, that essentially no information is available as to the relative values of the sputter contributions $\beta_i^{(l)}$ and $\beta_k^{(l)}$.

These component yields are then essentially equivalent to the sputter cross sections $\sigma_i$ defined above. Both quantities thus characterize the primary effect of preferential sputtering, i.e. the difference in the efficiency of removing an $i$ atom from the surface ($y_i \neq y_k$); the atom density $N_i$, (3.1) and the atomic fraction $c_i$, on the other hand, can be perturbed by any of the secondary effects mentioned above. The discretization expressed by (3.2) faces difficulties in describing density variations caused by composition changes.

### 3.1.1 Theoretical Concepts

In the theory of linear collision cascades in an infinite medium two quantities were found important [3.4] for the derivation of partial sputtering yields: the recoil density and the particle flux. The former represents the energy distribution of recoil atoms of type $i$ when set in motion, (2.25). The particle flux, on the other hand, reflects the energy distribution of the $i$ atoms under stationary irradiation conditions [3.3]; to a first approximation, it is inversely proportional to the stopping power of an $i$ atom. Hence, the smaller the stopping power, the greater the contribution of an atom to the particle flux, cf. (2.30) and its associated arguments. Conversely, the recoil density does not depend on the slowing down of recoiling atoms. From the particle flux the number of atoms crossing a surface plane (at $x = 0$) can be estimated, provided some assumptions [3.20] regarding the position of this plane relative to the collision cascade are made. These assumptions apply also in the case of monatomic targets. Furthermore, the analytical theory refers to energy distributions in the asymptotic limit where the particle energy is much smaller than the incidence energy. Similarly to the monatomic case (Sect. 2.1.1), the collision between two target atoms $i$ and $j$ can be described by a power cross section (see (2.9)):

$$d\sigma_{ij}(E, T) = C_{ij} E^{-m} T^{-1-m} dT , \tag{3.6}$$

where $E$ is the energy of an $i$ atom before the collision, $T$ is the energy transferred to a $j$ atom and both $C_{ij}$ and $m$ are species-dependent parameters determined by the interatomic potential that governs the collisions. The exponent $m$ ($0 < m < 1$) depends slightly on $E$. For an infinite, random medium this results in a ratio of particle fluxes $J$ at an energy $E_0$

$$\frac{J_i}{J_j} = \frac{N_i}{N_j} \frac{S_{ji}(E_0)}{S_{ij}(E_0)} , \tag{3.7}$$

where $S_{ij}$ is the stopping cross section of an $i$ atom interacting with a $j$ atom (see (2.12)),

$$S_{ij}(E) = \frac{1}{1-m} C_{ij} \gamma_{ij}^{1-m} E^{1-2m}, \tag{3.8}$$

and $\gamma_{ij} = 4M_i M_j / (M_i + M_j)^2$. It can be seen from (3.3) that the particle fluxes deviate from stoichiometry by a factor $S_{ji}(E_0) / S_{ij}(E_0)$. It was shown [3.4, 16] that this factor is independent of $E_0$ in the above approximation and is given by

$$\frac{S_{ji}(E_0)}{S_{ij}(E_0)} = \left(\frac{M_j}{M_i}\right)^{2m} \tag{3.9}$$

for a screened Coulomb interaction. For the low-energy collisions a Born–Mayer potential has been adopted that yields [3.14]

$$C_{ij} = \frac{\pi}{2} \lambda_m a^2 \left(\frac{M_i}{M_j}\right)^m (2A_{ij})^{2m} , \tag{3.10}$$

with $A_{ij} \cong 52\,\mathrm{eV}\,(Z_i Z_j)^{3/4}$ and $a = 0.0219\,\mathrm{nm}$. It was demonstrated [3.3, 4, 16, 21], furthermore, that this nonstoichiometry has its origin in two effects which contribute equally to the total deviations of the flux: first, lighter atoms have a larger cross section for being hit, i.e. more of them are set in motion; second, lighter species have a lower stopping power and thus a larger penetration depth, i.e. they travel further. For both reasons they contribute a larger share of the particle flux. The above equations can be extended to polyatomic systems, as shown by several studies [3.4, 16, 22].

For a monatomic target the energy- and angle-integrated yield of sputtered particles was shown (cf. (2.30), (2.33)) to scale $\propto U^{2m-1}$, where $U$ is the surface binding energy. This should translate into a dependence $\propto U_i^{2m-1}$ for the $i$th component of a multicomponent system, and therefore the ratio of partial yields in a binary sample (containing elements A and B) is given by [3.3, 14]

$$\frac{Y_A}{Y_B} = \frac{N_A}{N_B} \left(\frac{M_B}{M_A}\right)^{2m} \left(\frac{U_B}{U_A}\right)^{1-2m} . \tag{3.11}$$

This theoretical approach predicts, therefore, that preferential sputtering depends on the mass and binding-energy differences of the species involved and on the magnitude of $m$. Equation (3.11) predicts preferential sputtering of the lighter and/or more weakly bound atom. Because of the exponent, the effect of binding should dominate except in the case of large mass differences. There should be, on the other hand, no dependence on the emission angle and only a minor one (through $m$) on the emission energy of the sputtered atoms. Also, the type and energy of the projectile should have no influence on the preferential ejection. The magnitude of $m$ is not known exactly for the pertinent energy range of collisions (10 to 100 eV); this is due to the limited information available on the interaction potentials in this energy regime. For Born–Mayer scattering $m = 0.11$ has been derived [3.23]. For softer interaction potentials as are employed in some computer simulations [3.24], the value may come close to $m \sim 0.2$. (For widely different masses, e.g. $M_i \gg M_j$, $m_i \neq m_j$ with $m_i > m_j$.) Clearly, theoretical predictions of the magnitude of preferential sputtering hinge critically on the value chosen for $m$. Conversely, precise experiments may be applied to determine $m$ via (3.11).

The effects of preferential sputtering are usually thought to be most easily investigated for isotopic mixtures (elemental samples of two or more isotopes) since for these systems binding differences will be small and perhaps negligible [3.3]. Thus, the problem reduces basically to one of mass differences; the latter, however, are generally small and so will be any preferential sputtering effect for isotopes. Much larger effects can be expected for multielemental specimens, with the added complexity of the influence due to binding-energy differences; frequently, the latter will dominate preferential sputtering for these systems. For isotopic systems it has generally been assumed that in (3.11) the term relating to binding energies can be neglected and the yield ratio for two isotopes ("1" and "2") of an element is determined only by their different masses:

$$\frac{Y_1}{Y_2} = \frac{N_1}{N_2} \left( \frac{M_2}{M_1} \right)^{2m} . \tag{3.12}$$

Recently, it was argued [3.3] that even for isotopes minute differences in the effective binding energy may arise from the mass dependence of the zero-point energy; these authors give the approximation

$$\frac{\mathrm{d}U}{U} \cong \frac{3}{2} \frac{\hbar\omega_\mathrm{D}}{U} \frac{\mathrm{d}M}{M} , \tag{3.13}$$

where $\omega_\mathrm{D}$ is the Debye frequency and $3\hbar\omega_\mathrm{D}/2$ is the zero-point energy. The lighter isotope has therefore the smaller binding energy. Additional (minor) effects due to Gibbsian segregation and radiation-enhanced diffusion are conceivable which might modify the validity of (3.12).

Under prolonged particle bombardment the composition of the near-surface region is modified because of preferential sputtering and any of the secondary processes mentioned above that might occur. Since for steady-state conditions the emitted flux must be identical in composition to the bulk of the irradiated specimen, the surface composition adjusts to compensate for the preferential ejection of one species. From (3.11), the stationary surface concentration $N_i^s$ in a binary system is given by [3.2, 10]

$$\frac{N_\mathrm{A}^s}{N_\mathrm{B}^s} = \frac{N_\mathrm{A}^b}{N_\mathrm{B}^b} \left( \frac{M_\mathrm{A}}{M_\mathrm{B}} \right)^{2m} \left( \frac{U_\mathrm{A}}{U_\mathrm{B}} \right)^{1-2m} \tag{3.14}$$

and for *isotopic systems*, assuming $U_1 = U_2$,

$$\frac{N_1^s}{N_2^s} = \frac{N_1^b}{N_2^b} \left( \frac{M_1}{M_2} \right)^{2m} . \tag{3.15}$$

Here $N_i^b$ and $N_i^s$ are the bulk and surface densities, respectively, with the term "surface" referring to the depth from which sputtered atoms originate.

This approach is based on the assumption that there exists no pronounced concentration gradient over this depth.

Using the definition of the partial sputtering yield in terms of individual layers, steady-state conditions (considering only two layers) would correspond to [3.9, 10]

$$\frac{\sum_{l=1}^{2} \beta_A^{(l)} c_A^{(l)}}{\sum_{l=1}^{2} \beta_B^{(l)} c_B^{(l)}} = \frac{c_A^b y_B}{c_B^b y_A} = \frac{c_A^b}{c_B^b} \left(\frac{M_A}{M_B}\right)^{2m} \left(\frac{U_A}{U_B}\right)^{1-2m}, \tag{3.16}$$

where $c^b$ denotes the concentrations in the bulk, and $M$ and $U$ are, respectively, the masses and the surface binding energies of the two constituents. The generalization to multicomponent systems has been shown to be possible. The effect of preferential sputtering on the near-surface composition of a binary alloy AB is schematically shown [3.10] in Fig. 3.1 for the case in which $y_A > y_B$. Both the sputtered flux of component A as a function of bombardment time (fluence) and the concentration of A versus depth are depicted. Since A is preferentially sputtered, for low fluences the concentration of A in the flux is larger than in the bulk; the preferential removal leads to a gradual decrease in the surface concentration of A. With increasing fluence,

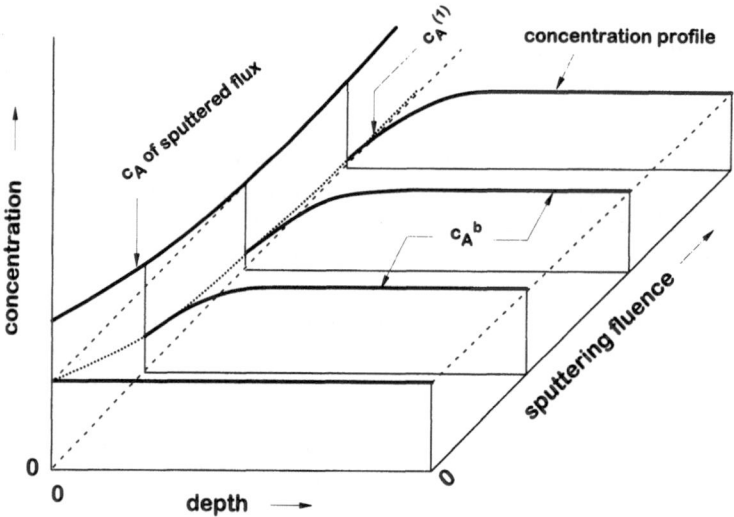

**Fig. 3.1.** Schematic representation of the effect of preferential sputtering on the time evolution of the sputtered-flux composition and the development of the internal concentration profile. Element A is sputtered preferentially, so it is enriched initially in the flux and, at steady state, it is depleted at the surface relative to the bulk stoichiometry. Depending on the actual depth scale and timescale considered, these evolution data may also include collisional mixing as a feeding mechanism that extends the compositional changes beyond the escape range of the sputtered species. From [3.10]

a steady state is eventually reached, and the composition of the sputtered flux is identical to that of the bulk.

Pure preferential sputtering and the associated composition changes as expressed by (3.11) and (3.14) are difficult to investigate experimentally, because of the frequent interferences from one or more of the aforementioned secondary processes. Experiments on isotopic mixtures are probably best suited for such investigations, but are often limited to systems featuring large relative mass differences. Such investigations are outlined in Sect. 3.4. Preferential sputtering has been addressed for random and crystalline targets by means of computer simulations. Standard Monte Carlo codes were employed [3.25, 26] to investigate both homogeneous and inhomogeneous target compositions. With suitable choices for bulk and surface binding energies, a wide variety of bombardment conditions has been simulated with respect to preferential-sputtering and collisional-mixing effects. Molecular-dynamics simulations are able to investigate crystalline systems, but quite frequently the statistics limit the irradiation conditions which can be studied. Results from such simulations for alloys and isotopic systems are discussed in Sects. 3.3 and 3.4, respectively.

### 3.1.2 Surface Binding Energies in Alloys

For elemental metals the surface binding energy $U$ has been identified with the cohesive energy $E_{coh}$. For alloys this approach might cause problems because the statistics of site occupancy can be important, as pointed out by Kelly [3.27]. Attempts were therefore directed to evaluating pairwise interaction of an atom with the remainder of the crystal. For pair potentials, and considering only nearest neighbors, the bond energy is related to $E_{coh}$ by [3.17, 27, 28]

$$E_{coh} = \frac{Z}{2} E_{bond} , \tag{3.17}$$

where $Z$ is the bulk coordination number. The surface binding energy of an atom is

$$U = Z_s E_{bond} = 2\frac{Z_s}{Z} E_{coh} ; \tag{3.18}$$

here $Z_s$ is the number of nearest neighbors in the surface. Thus, for the monatomic case, only one parameter is needed, namely the cohesive energy. For the (100) surface of an fcc crystal ($Z = 12$, $Z_s = 8$) the surface binding energy thus amounts to $U = 1.33 E_{coh}$. Values of $U$ for metals which exceed $E_{coh}$ have been advocated in several studies [3.27–32].

Gades and Urbassek [3.28], on the other hand, have stressed the importance of many-body interaction potentials for an adequate description of surface bonding in metals. This is due, largely, to the delocalized nature of metallic bonds. In studies of atomic emission processes and the influence of surface binding energies of alloys, they adopted a tight-binding potential. With this approach they derive, for the surface binding energy,

$$U = \frac{2Z_s}{Z} E_{coh} - f \left( E_{coh} - E_{vac} \right) , \tag{3.19}$$

where $E_{vac}$ is the energy to form a vacancy in the bulk and $f$ is a factor completely determined by the crystalline structure (e.g. $f \sim 0.28$ for the (100) surface of an fcc structure). Note that for pair potentials $E_{vac} = E_{coh}$ and (3.19) is identical with (3.18).

A somewhat similar treatment was proposed by Kelly and coworkers [3.27, 32]. These authors distinguish between the removal of half-space (kink-site) and in-plane atoms and argue that in the former case no surface vacancy is formed. Considering (hypothetically) the slow removal of an in-plane atom first to a half-space site before the ejection event proper, they derive for the binding energy of that atom

$$U = E_{coh} + E_{vac}^s \tag{3.20}$$

and assume that the surface-vacancy formation energy is similar to $E_{vac}^b$ for the bulk. With $E_{vac}^b \sim 9.7 k_B T_m$ ($T_m$ is the melting temperature), they find [3.32] $U \sim 1.33 E_{coh}$.

For an alloy system AB and a pair-potential interaction for the surface binding energies of A and B, the following equations are derived [3.28]:

$$\begin{aligned} U_A &= c_A U_{AA} + c_B U_{AB} , \\ U_B &= c_A U_{BA} + c_B U_{BB} , \end{aligned} \tag{3.21}$$

where $U_{AA}$ and $U_{BB}$ are the surface binding energies in elemental A and B, respectively, and

$$U_{BA} = \frac{Z_s}{Z} \left( E_{coh}^A + E_{coh}^B - \Delta E_{BA} \right) \tag{3.22}$$

is the binding energy of a single surface atom B in an otherwise pure A element. $\Delta E_{BA}$ is typically related to the heat of mixing $h_m$ or the heat of solution. (Exchanging the subscripts results in an analogous equation for the other species.) As $\Delta E_{BA}$ is usually considerably smaller than the cohesive energies, it can be neglected in (3.22), giving

$$U_{BA} = U_{AB} = \frac{Z_s}{Z} \left( E_{coh}^A + E_{coh}^B \right) \tag{3.23}$$

and

$$U_A = \frac{Z_s}{Z} \left[ E_{coh}^A + E_{coh}^B + c_A \left( E_{coh}^A - E_{coh}^B \right) \right] . \tag{3.24}$$

This indicates that the surface binding energy of a component varies linearly with its concentration between the value of the pure element and the mean of the two components, $(U_A + U_B)/2$. These results for the surface binding energies were obtained by Kelly and others [3.2, 17, 27, 28].

The corresponding expression for a many-body interaction then reads (cf. [3.28])

$$U_{BA} = \frac{Z_s}{Z} \left( E_{coh}^A + E_{coh}^B - \Delta E_{BA} \right) - f\left(R_{BA}\right)\left( E_{coh}^A - E_{vac}^A \right) , \qquad (3.25)$$

where $f\left(R_{BA}\right)$ again depends on the crystal structure (cf. [3.28]). $\Delta E_{BA}$ may be neglected, as above. The second term is a true many-body contribution. In contrast to the pair-potential case, $U_{AB}$ and $U_{BA}$ are not equal for many-body potentials. Thus, for a many-body interaction the surface binding energy of component A can be derived from (3.21), using (3.19) for $U_{AA}$ and (3.25) for $U_{BA}$. Gades and Urbassek [3.28] find that their many-body studies yield qualitatively similar results to previous pair-potential models. Specifically, they report that the surface binding energy of each species depends approximately linearly on the concentration and, secondly, is mainly determined by mono-elemental properties, in particular the cohesive energies; chemical energies of mixing exert only small effects. Furthermore, their results indicate that many-body interactions tend to lower the computed surface binding energy. Figure 3.2 depicts computed surface binding energies taken from this work [3.28]. Here, $U_{Ag}$ and $U_{Au}$ in a binary Ag–Au alloy are plotted as a function of the Ag concentration $c_{Ag}$ both for a pair interaction as expressed by (3.24) and for a many-body interaction, (3.25). The latter values are lower and exhibit, at least for this system, less variation with alloy composition. Gades and Urbassek also found [3.28] that preferential sputtering has only a small influence on the surface binding energies in their model. Conversely, Gibbsian segregation was found to enrich the weakly bound component at the surface and hence reduces the surface binding energy of all components.

In an MD simulation [3.33] of 1 keV Ar$^+$ sputtering of a Ni$_{0.5}$Cu$_{0.5}$ alloy, the computed energy spectra of the two components were fitted using (2.35); by these means, the surface binding energies $U_i$ were derived for Ni and Cu: $U_{Cu} = 3.28 \pm 0.13$ eV and $U_{Ni} = 3.97 \pm 0.17$ eV. As both values are noticeably smaller than the cohesive energies of pure Cu and Ni (3.51 and 4.45 eV, respectively), the significance of these data is not clear. Within the statistics, the computed values were found to be identical for the other alloy compositions Ni$_{0.1}$Cu$_{0.9}$ and Ni$_{0.9}$Cu$_{0.1}$.

Experimentally, data on surface binding energies have been determined from energy spectra of sputtered atoms [3.34–39], by fitting them to an expression $dY/dE_0 \propto E_0/\left(E_0 + U_i\right)^n$ using $n$ and $U_i$ as parameters. According to (2.35), the most probable energy should correspond to $U_i/2$. The available results suggest that $U_i$ may depend on alloy composition. Szymonski [3.36] found $U_{Zn}$ in Cu–Zn alloys to decrease with increasing Zn concentration toward the pure-Zn value, while $U_{Cu}$ decreases from the pure-Cu value to a considerably smaller value. Such composition dependences were also found in an Au–Ag alloy [3.34]; from the energy spectra obtained from this specimen under 6 keV Xe$^+$ ion irradiation the authors derived surface binding energies $U_{Ag} = 2.1$ eV and $U_{Au} = 3.3$ eV. The values from the corresponding

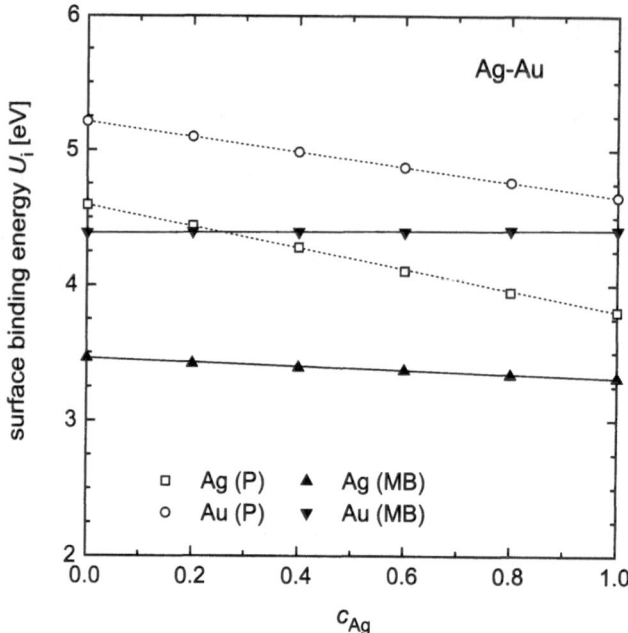

**Fig. 3.2.** Dependence of the surface binding energy on concentration in an Ag–Au alloy, calculated using a pair (P) and a many-body (MB) interaction. Data from [3.28]

elemental samples were $U_{Ag} = 3.1\,\text{eV}$ and $U_{Au} = 3.8\,\text{eV}$, which are very close to these elements' cohesive energies (2.95 and 3.81 eV, respectively). Conversely, the values for the alloy are considerably lower than the computed values; cf. Fig. 3.2. Energy spectra from Ni–W [3.37] and Cu–Li alloys [3.38] were evaluated in a similar fashion. Pronounced variations in surface binding energies were inferred from such measurements for metal atoms sputtered from the respective specimens upon exposure to various amounts of oxygen [3.39–45]. Typically, the energy spectra become broader, that is, $U_i$ is larger for an oxidized surface and the sputtering yield is reduced [3.46]. In a series of publications, Kelly [3.47–49] has evaluated these data from energy distributions in terms of surface-binding-energy differences in metals, metal oxides and related systems; he attempted to correlate this information with the available data on ion-bombardment-induced compositional changes in these materials. The energy spectra of atoms sputtered from alloys are discussed in Sect. 3.3.3.

Sigmund and Lam [3.3] have shown that the sputter cross section $\sigma_i(x)$ (and the component sputter yield $y_i$) is preferential because of the preferentiality of the displacement efficiency and of the surface binding energy $U_i$. The depth dependence is also preferential in the sense that atoms with lower stopping power or high binding energy may emerge from deeper layers.

## 3.2 Processes Effecting Composition Changes

### 3.2.1 Fluence Dependence and Stationary State

As outlined in Sect. 3.1.1, preferential sputtering causes a change in the surface composition after prolonged ion irradiation, and the stationary surface composition $N_i^s$ in a binary system (A–B) is given by [3.1, 2, 10]

$$\frac{N_A^s}{N_B^s} = \frac{N_A^b}{N_B^b}\frac{y_B}{y_A} \, , \tag{3.26}$$

where $y_i$ are the component yields defined previously. In the collision cascade theory their ratio is

$$\frac{y_B}{y_A} = \left(\frac{M_A}{M_B}\right)^{2m} \left(\frac{U_A}{U_B}\right)^{1-2m} , \tag{3.27}$$

as given by (3.16). For the pure primary effect of preferential sputtering, the compositional changes should be restricted to the depth interval from which atoms are sputtered. The above steady-state conditions for binary alloys can be generalized to multicomponent systems [3.50]. The transient behavior towards the stationary conditions of the surface composition and the sputtered flux are schematically depicted in Fig. 3.1 [3.10]. The flux is initially enriched in the component which is sputtered preferentially and, concurrently, the surface concentration of that element is reduced as compared to its original (and the bulk) value. Upon completion of this transient regime, the ejected flux and the surface composition remain unchanged; then, the stoichiometry of the flux is identical to that of the bulk. The time (i.e. the fluence) required to reach this condition should correspond, for pure preferential sputtering, to the fluence necessary to remove a surface layer with a thickness comparable to the depth of atom ejection (one to three monolayers).

Experimental investigations [3.1–3] of compositional changes of the surface or of the emitted flux generally indicate that most results cannot be explained solely by the occurrence of preferential sputtering. Typically, one or more of the following findings were noted and found incompatible with pure preferential sputtering:

(i)   The composition changes were much larger than predicted by the analytical theory or were in the wrong direction.

(ii)  Conflicting results were found when employing techniques with different information depths (e.g. ion scattering and field-ion microscopy, which probe the outermost surface layer, and Auger and photoelectron spectroscopy, which monitor a more extended depth range).

(iii) Changes in the surface stoichiometry were observed to persist in some systems even after termination of ion irradiation.

(iv)  Pronounced differences in the angular emission spectra of the two species were reported, pointing to strong composition gradients in the surface region.

(v)  The depth over which composition changes occurred was much larger than the outermost atomic layers, extending some tens to hundreds (or even thousands) of nanometers into the bulk. Similarly, the fluence required to establish steady-state conditions far exceeded the expected values necessary to sputter a few monolayers.

These observation are therefore indicative of the presence of (at least one of) the secondary processes; these tend to spread out the modifications in the solid to greater depths. Consequently, higher fluences are required to reach equilibrium conditions. Such transitions have been studied by experiments and by computer simulations, mostly for alloys but also for isotopic mixtures (see below).

In addition to preferential sputtering, various chemical and physical processes can change the near-surface composition of solids which are subjected to ion bombardment. Athermal effects are correlated with the collision cascade and include collisional mixing, ion implantation and the relaxation of the target to a stable state. Experimentally, athermal processes are related to the ion fluence (ions/area) of the irradiation, but are independent of the ion current density. Thermally activated processes such as Gibbsian segregation, radiation-enhanced diffusion and radiation-induced segregation may be rather important even at temperatures where they would be negligible in the absence of ion irradiation. Bombardment-induced defects may migrate over distances that are large compared to the size of an individual collision cascade. Thus, this kind of bombardment-induced composition change will depend on the ion current density and on the target temperature. Typically, more than one of these processes may be operative simultaneously and the compositional changes are then rather complex.

Many versions of rate equations have been set up to model composition changes in multicomponent systems under ion bombardment; depending on sophistication (and the application they were devised for), these approaches incorporate a varying number of the aforementioned primary and secondary effects. Sigmund and Lam [3.3, 8, 9, 12, 13] have succeeded in establishing a unified description which covers, in rather general terms, the pertinent processes. In the following, this description will be referred to repeatedly.

## 3.2.2 Collisional Mixing

In the course of a displacement cascade, relocation of target atoms occurs and point defects are created. In multicomponent materials this process leads to composition changes [3.2]; it has been termed "collisional mixing" (CM) and comprises recoil implantation and cascade mixing. While recoil implantation [3.51–56] is an ion–atom knockon event and transports atoms preferentially in the beam direction, cascade mixing is a random (spatially isotropic) process resulting from the movement of high-order recoil atoms in the target. As recoil implantation depends pronouncedly on the ion and target masses, it is inherently strongly preferential. Its relevance for composition changes in the near-surface region was proposed early [3.51–53, 57, 58], but many apparently conflicting results were reported [3.59–62]. A more coherent picture evolved only gradually. Theoretical work by Sigmund [3.53, 56] has shown that for recoil implantation in multicomponent targets, angular scattering of the slowing-down atoms is of decisive importance for the direction of implantation. At *low* energies this scattering causes the *heavier* species to be implanted preferentially. At higher energies the influence of this effect decreases and the lighter atoms are preferentially recoil-implanted. As direct ion–atom collisions are relatively rare events, in this energy range cascade mixing will dominate relocation processes in the solid. On the other hand, recoil implantation might be an important source of surface composition changes at low ion-beam energies [3.56], where cross sections are large.

At low energies, displacement cascades, as discussed in the context of damage production in Sect. 2.2, are often characterized by short replacement sequences along close-packed atomic rows, with little or no defect creation; still, the number of site-changing atoms and, hence, the amount of atomic mixing can be significant. Conversely, MD simulations of collision cascades at energies of a few keV and more (see Sect. 2.2) indicate the formation of a molten (highly disordered) region in the core of the cascade; most of the mixing occurs in this central region during the cooling phase ($> 10^{-12}$ s). At this stage, *thermodynamic forces* may strongly influence the extent of mixing even at low temperature. The significance of such quantities as the heat of mixing will be outlined below.

Cascade mixing was envisioned as a possible means of generating metastable phases of materials by ion irradiation at energies of some 100 keV [3.63, 64]. This prospect triggered (in the early 1980s) many experimental and theoretical studies. The discussion of this high-energy work is beyond the scope of this book, but comprehensive reviews exist [3.64–67]. At energies of a few keV, collisional mixing is of importance in sputter depth profiling; there, it tends to degrade the achievable depth resolution (see below). Various physical models of and experimental results on cascade mixing have been discussed in the literature. Theoretical treatments and computer simulations of cascade mixing have utilized various approaches [3.67–79]. Collisional mixing was frequently approximated within the framework of a diffusion (random walk) model [3.69, 70, 74, 75]. This is essentially an approximation to the isotropic displacement cascade where the successive knockon processes are described by random relocations over a small step length. Note, however, that in collision cascades some long-range relocation is also possible, which may cause tails at large depth. Also, steep concentration gradients in the near surface can evolve and appropriate boundary conditions have to be incorporated into the model. From the solution of the respective diffusion equation, an original $\delta$-type marker of atoms of type $i$ is broadened by cascade mixing into a Gaussian distribution with variance [3.69, 70, 74]

$$\Delta\sigma^2 = 2D_i^{\mathrm{CM}}t .$$
(3.28)

The diffusion coefficient $D_i$ is related to the site-exchange ("jump") frequency $\Gamma_i$ and the mean square jump distance $\langle r_i^2 \rangle$:

$$D_i^{\mathrm{CM}} = \frac{1}{6}\Gamma_i \langle r_i^2 \rangle .$$
(3.29)

As the value of $\Gamma_i t$ is related to the number of recoils created in the cascade with an energy in excess of the displacement energy $E_{\mathrm{d}}$, it can be written as

$$\Gamma_i t = \frac{F_{\mathrm{D}}(x)\,\Phi}{N\,E_{\mathrm{d}}} ,$$
(3.30)

where $F_{\mathrm{D}}$ is the energy deposited in nuclear collisions per unit depth, $N$ is the target density and $\Phi$ is the ion fluence. Then

$$D_i^{\mathrm{CM}}t = \frac{1}{6}\frac{F_{\mathrm{D}}(x)\,\Phi}{N\,E_{\mathrm{d}}}\langle r_i^2 \rangle ;$$
(3.31)

$\langle r_i^2 \rangle$ can be identified with the mean square range of a recoil atom at energy $E_i$. Using power cross sections, $\langle r_i^2 \rangle \propto E_i^{4m}$. As most recoils contributing to cascade mixing have energies slightly above $E_{\mathrm{d}}$, this energy dependence can be neglected and a constant mean square recoil range can be adopted [3.3]; its absolute magnitude is still difficult to determine at these low energies. A lower limit would be given by the separation of stable vacancy–interstitial pairs ($\sim 1\,\mathrm{nm}$). The preferentiality implicit in $r_i$ may be substantial.

Lam and Wiedersich [3.7–9] approximated the diffusion coefficient for collisional (or displacement in their terminology) mixing by $D_i^{\mathrm{CM}} \cong b^2\eta K/6$. In this relation $b$ is the nearest-neighbor distance, $K$ is the defect production rate (displacements per atom per second, dpa/s) and $\eta$ is proportional to the number of atoms changing sites (i.e. making jumps of length $b$) per Frenkel pair created. Values of $\eta$ in the range $10^2$ to $10^3$ have been found in experiments on ion-beam mixing. The coefficient $D_i^{\mathrm{CM}}$ is temperature-independent and therefore collisional mixing can be considered an athermal process. Employing appropriate numbers, these authors derive values of $D_i^{\mathrm{CM}}$ of the order of $10^{-16}\,\mathrm{cm}^2/\mathrm{s}$; this is several orders of magnitude larger than the thermal diffusion coefficient at low temperatures. It appears therefore that collisional mixing is a dominant atomic transport mechanism in multicomponent targets at low temperatures when vacancies are immobile, i.e. below $T \sim 0.2T_{\mathrm{m}}$. It may spread compositional changes produced in the outermost surface layers to greater depths. Sigmund and Lam [3.3] have emphasized that collisional mixing is an efficient feeding mechanism for preferential sputtering, despite the fact that the displacement energies are much larger than the surface binding energies ($E_{\mathrm{d}}/U \sim 5$–$10$), and stress the pronounced concentration gradients which may be involved.

The foregoing description of collisional mixing corresponds to a purely "ballistic" picture and the similarity to a thermal random-walk process was noted. Although this model was successfully applied to analyze data on ion–beam induced mixing, the measured mixing rates in many experiments [3.66, 80, 81] involving marker layers or bilayer samples were often greater than what is predicted by (3.28) and (3.31), even when accounting for the uncertainties associated with a proper choice of the recoil range. This finding was ascribed to the occurrence of "chemical" effects which produce an additional broadening in excess of the pure "ballistic" effects. Cheng et al. [3.66, 80, 81] invoked a chemical driving force due to a chemical-potential gradient, namely the heat (enthalpy) of mixing $\Delta H_m$. They proposed a model of mixing based on Darken's analysis of chemical interdiffusion [3.82]. Darken showed that diffusion equations must be modified when diffusion occurs in nonideal solutions, leading to a diffusion coefficient modified by a factor $(1 - 2\Delta H_m/k_B T)$. Experiments [3.66, 81] corroborate this argument by showing that the broadening (expressed by $2Dt/\Phi$; see (3.31)) scales linearly with $\Delta H_m$. While negative values of $\Delta H_m$ would enhance intermixing, positive values tend to produce the opposite effect, i.e. a demixing. In a series of papers, Kelly and Miotello [3.83–85] have investigated these possibilities and have stressed the fact that because of the typically small values of $\Delta H_m$ ($< 1\,\text{eV}$) as compared to the recoil energies ($\approx E_d$), "chemical guidance" (as they call it) would dominate in the late stages of the collision cascade. For ballistic cascade mixing (i.e. neglecting any chemical effects), the intermixed flux $J_i^{CM}$ across a fixed plane within the target is computed as

$$\Omega J_i^{CM} = -D_i^{CM}\left(\frac{\partial c_i}{\partial x}\right)\,,\tag{3.32}$$

where $\Omega$ is the average atomic volume in the alloy and $D_i^{CM}$ is the displacement-induced interdiffusion coefficient. Kelly and Miotello [3.83–85] have shown that for mixing effects induced by chemically guided ("g") defects, (3.32) must be expanded to become

$$\Omega\left(J_i^{CM} + J_i^g\right) = -D_i^{CM}\left(\frac{\partial c_i}{\partial x}\right) - c_i M_i^g\left(\frac{\partial \mu_i}{\partial x}\right)\tag{3.33}$$

for a nonideal solution. Here $M_i^g = c_i D_i^g/RT$ is the mobility and $\mu_i$ is the chemical potential. They further indicate that the total diffusive flux $J_i^t$ follows [3.84, 85]

$$\Omega J_i^t = -\left[D_i^{CM} + D_i^g c_i\left(1 - c_i\right)\frac{2h_m}{k_B T}\right]\left(\frac{\partial c_i}{\partial x}\right)\,,\tag{3.34}$$

where $h_m$ is the reduced heat of mixing, defined as $\Delta H_m = c_i\left(1 - c_i\right)h_m$. Their analysis [3.84, 85] shows that when $\Delta H_m$ is sufficiently positive, there is a strongly reduced solubility, i.e. a miscibility gap arises. The latter situation was also observed experimentally. From the evaluation of these data in the light of the above discussion, the authors conclude that for typical diffusivity ratios ($D^g/D^{CM} \sim 0.1$) and $h_m = -0.5\,\text{eV}$, mixing can be enhanced by a factor of four as compared to the purely ballistic case.

Ion-beam mixing at high energies (some hundreds of keV) has been explored as a means to create metastable mixtures of solids, with some possible applications emerging recently. At low energies (a few keV and less), collisional mixing is of great importance with regard to the depth resolution achievable in sputter depth profiling by surface-analytical techniques [3.86–89]. A very large number of such low-energy mixing experiments were carried out to elucidate the influence of ion-beam and target parameters on the degree of mixing. Typically, these were done at ion energies of a few keV or less [3.86–93]. In this energy regime information about collisional mixing must be extracted from the depth-profile data, usually by monitoring either the ejected particle flux (in mass-spectroscopic techniques) or the near-surface composition (in electron or ion-scattering spectroscopies) when sputtering through a sharp bilayer interface or a thin marker layer. These procedures have several inherent difficulties [3.86–89]. Apart from possible instrumental and specimen related artifacts (such as inhomogeneous sample erosion, adsorption of gas-phase species on the receding surface, development of surface roughness during the erosion, etc.), the measured distribution(s) of a marker species (or the elements of a bilayer sample) is (are) a convolution of the internal distribution caused by collisional mixing and the preferential ejection of this (these) species from the surface. Another problem relates to the magnitude of the mixing effect at low energies. As the broadening scales with the deposited energy [cf. (3.31)], it could be considerably smaller than at high energies. These complexities have been outlined by Wittmaack, both in the general context of ion-irradiation-induced broadening effects in sputter profiling for thin film analysis [3.87] and for the more specific situation encountered in SIMS experiments [3.86, 94] using reactive primary ions such as $O_2^+$. A recent review by Zalm [3.89] on ultrashallow depth profiling employing SIMS elucidates many of the pertinent questions associated with collisional mixing.

The above-mentioned interfering effects are well illustrated in simulations by King and Tsong [3.90]. These authors used a depth-dependent diffusion coefficient $D(x)$ essentially identical to that defined in (3.31). Assuming a low concentration for the intermixing species (in order to neglect nonlinear terms and to model a specific experimental situation), they establish the relevant diffusion equation, taking into account appropriate boundary conditions. Numerical computations based on this model produced, for thin marker layers in a matrix, the internal impurity distribution and the sputtered-flux profile of the impurity as a function of the erosion time. The sputter profile exhibits the shape commonly observed in experiments [3.86, 87]: as the erosion front approaches the marker the yield of the impurity rapidly rises, reaching a maximum at a depth lying before the original marker position. Beyond this depth the impurity yield decays with a slope which is typically exponential and much shallower than the uprise slope. This exponential decay,

$$I_i(x) \propto \exp(-x/\lambda_i) \;, \tag{3.35}$$

can be explained by the following observation: As sputtering proceeds beyond the interface between the marker and the substrate, intermixing of the marker (tracer) atoms results in a stationary internal distribution of the marker atoms $i$ which is independent of the original marker distribution; this feature is also evident from computations [3.72, 90]. The shape of this "mixing profile" is determined by the extent of atomic relocation and the surface recession rate. Wittmaack has proposed [3.95] the introduction of a concentration-independent mixing profile of the species $i$, $p_i(x')$, where $x'$ is the depth from the instantaneous surface. The depth-dependent concentration of intermixed marker atoms is related to $p_i(x')$ by $c_i(x, x') = N^{-1}n_i^A(x)p_i(x')$, where $n_i^A(x)$ is the areal density (atoms/cm$^2$) of marker atoms still retained in the specimen by the time sputter erosion has proceeded to the depth $x$ and $N$ is the atomic density of the substrate.

In the case of a dilute system [3.95], the characteristic decay length $\lambda_i$ depends on the depth-dependent ejection function of species $i$ (cf. (3.1)) and on the mixing profile $p_i(x')$. The value of $\lambda$ increases with increasing width of the mixing profile. Most often $p_i(x')$ will extend over a considerably larger depth than the escape function, the former being related to the ion range whereas the latter corresponds roughly to the depth of origin of the sputtered species. For $p_i(x')$ approximately constant in this near-surface region, $p_i(x' = 0) = p_i^s$, the decay length $\lambda_i$ can be approximated by $\lambda_i \cong 1/r_i p_i^s$, where $r_i$ is the ratio of the component sputtering yields of impurity $i$ and the matrix m, $r_i = y_i/y_m$, and $p_i^s$ is the mean value of $p_i$ at the surface. Hence, the reciprocal decay length describes the efficiency with which the residual content of marker atoms is removed from the specimen relative to the atoms of the host matrix. According to the aforementioned computations of King and Tsong [3.90], $p_i^s$ depends on $r_i$ in such a way that if $r_i$ increases (i.e. stronger preferential sputtering of the impurity), the value of $p_i^s$ decreases. As a consequence, even pronounced variations of $r_i$ may have only a moderate effect on $\lambda_i$. This indicates that detailed information about the internal distribution of the $i$ atoms is difficult (or even impossible) to extract from the sputter profile alone.

The predicted shape of the marker profile, namely the steep leading edge, the shift of the maximum to shallower depths and the exponential trailing slope, is corroborated by numerous experimental data (obtained mostly by SIMS [3.86, 87], as this technique provides a large dynamic range) and by the theoretical approaches based on transport equations [3.67, 72]. Figure 3.3 exemplifies such a depth profile of a $\delta$-marker layer. The sample consisted of several Si $\delta$-doping layers in GaAs grown by molecular-beam epitaxy. The two $\delta$-layers depicted in Fig. 3.3 have nominal areal densities of $4 \times 10^{13}$ and $4 \times 10^{12}$ Si atoms/cm$^2$, respectively. Depth profiling was performed with SIMS using a 14.5 keV Cs$^+$ ion beam at an incidence angle of $\sim 25°$ and detecting negative secondary ions. In agreement with theoretical studies of ballistic mixing of tracer elements [3.72], the profiles have an asymmetric

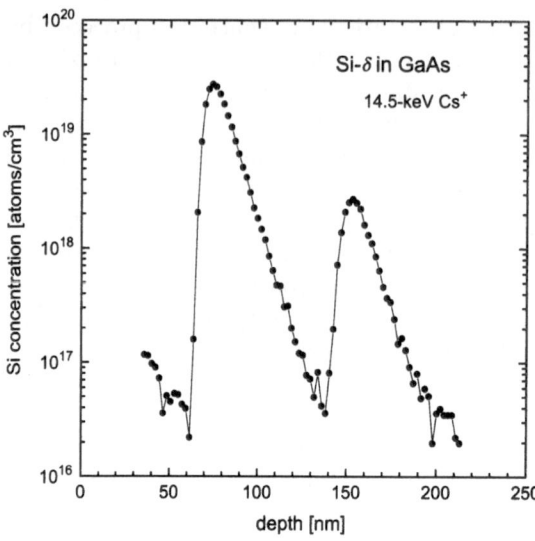

**Fig. 3.3.** Concentration depth profile of Si $\delta$-doping layers in MBE-grown GaAs; the nominal areal densities of the two dopant layers are $4 \times 10^{13}$ and $4 \times 10^{12}$ Si atoms/cm$^2$. The data were obtained using SIMS and 14.5 keV Cs$^+$ primary ions

shape, with the leading slope considerably steeper than the trailing slope. Defining exponential decay lengths $\lambda_L$ and $\lambda_T$ in analogy to (3.35) for the leading and trailing parts of a marker profile, values of $\lambda_L = 1.3$ nm and $\lambda_T = 8.5$ nm were drived for the Si tracer layer shown in Fig. 3.3. Employing 8 keV $O_2^+$ ions for depth profiling of the same specimen results in $\lambda_L = 1.1$ nm and $\lambda_T = 2.5$ nm, whereas a 2.5 keV $O_2^+$ ion beam effects a further reduction of the slopes, with $\lambda_L = 0.71$ nm and $\lambda_T = 1.4$ nm. The values obtained with oxygen-ion sputtering are in close agreement with previous reports on similar specimens [3.94]. Generally, very low values of $\lambda_L$ can be obtained in sputter profiling of sharp doping distributions by using low bombarding energies (values as low as $\lambda_L \sim 0.4$ nm have been reported). With respect to ion-energy dependence, $\lambda_T$ increases more rapidly with increasing energy than does $\lambda_L$. As will be exemplified below, for other bombarding species (e.g. inert gases or cesium) the decay length at the trailing edge $\lambda_T$ is essentially independent of the primary-ion mass whereas the value of $\lambda_L$ decreases with increasing mass. A rather detailed account of these experimental findings is given in [3.89, 94].

Data similar to those depicted in Fig. 3.3 have been reported in [3.86, 94, 96–100], to name but a few. Experiments at low energies have demonstrated that decay lengths and the extent of intermixing increase with increasing impact energy, but might be strongly dependent on the impurity–matrix system. As an important parameter influencing these processes, the heat of mixing was proposed. For impurity atoms in silicon bombarded with oxygen ions or in the presence of an enhanced oxygen pressure in the vicinity of the surface, another approach [3.101] proposed a scenario in which these dilute impurities segregate towards or away from the oxygenated surface; this may depend on whether they form stronger or weaker bonds with oxygen than do silicon atoms.

As mentioned above, collisional mixing due to low-energy ion irradiation is most often studied by monitoring the broadening of an interface between two different layers or of a $\delta$-like marker during *sputter* depth profiling, as shown in Fig. 3.3. Quite generally, such an approach faces problems that become apparent on considering the concurrence of sputtering and collisional relocation, as pointed out by Wittmaack [3.87]. For impurity atoms of element $i$ sputtered from a sample of element $j$, the partial sputtering yield $Y_i$ can be described by an expression like (3.1) or (3.2). Upon sputtering to a depth $x$, the concentration distribution of the $i$ atoms $c_i(x, x')$ (where $x'$ is the depth from the instantaneous surface) may differ from the original distribution $c_i(x)$ because of relocation processes and preferential sputtering. Since the component sputtering yield $y_i(x)$ (cf. (3.2)) is usually not known, $c_i(x)$ cannot be derived directly from the measured depth profile, that is, from monitoring $Y_i$ or the surface concentration $c_i^{(s)}$.

In order to eliminate the possible contribution of (preferential) sputtering to the measured interface broadening, Wittmaack and Poker [3.98] devised an experiment that investigated sputter depth profiling of thin layers composed of *different isotopes* of one single element. Specifically, they manufactured (by ion-beam deposition) specimens that contained alternating layers of the Si isotopes $^{30}$Si and $^{28}$Si. These specimens were irradiated with a variety of ion species ($Ne^+$, $Ar^+$, $Xe^+$, $Cs^+$, $N_2^+$ and $O_2^+$) at energies between 2 and 10 keV. During depth profiling through these layers, the sputtered secondary ions were monitored by SIMS. The authors derived from the profiles, for the different bombarding conditions, the broadening of the interfaces between the layers of different Si isotopes. They found a relatively sharp rise of the signals (characterized by the width $\delta$) and a moderately slow exponential fall-off at the trailing edge, with a decay length $\lambda$ as defined by (3.35). This asymmetric shape is rather typical for this kind of depth profiling at keV energies. Figure 3.4 compiles their data [3.98] for the front-end width $\delta$ and the decay length $\lambda_T$. These parameters were defined in such a way that $\delta$ corresponds to the depth interval in which the normalized intensity rises from the 1% to the 50% level, whereas $\lambda_T$ represents the decay of the signal by a factor of $e^{-1}$. It is obvious from Fig. 3.4 that $\delta$ for inert-gas and Cs ions at a fixed energy exhibits a decrease with increasing ion mass. Conversely, the decay length $\lambda_T$ does not show any variation with the projectile mass or, for that matter, with the sputtering yield.

Wittmaack and Poker [3.98] were able to describe the above finding of the mass dependence of $\delta$ in terms of ballistic mixing utilizing the random-walk approach of Andersen [3.70], which yields an expression for the interface broadening essentially identical to (3.30). This model of collisional mixing, on the other hand, predicts a dependence of the broadening on the sputtering yield ($\propto Y^{-1/2}$) which was not observed in the values of the decay

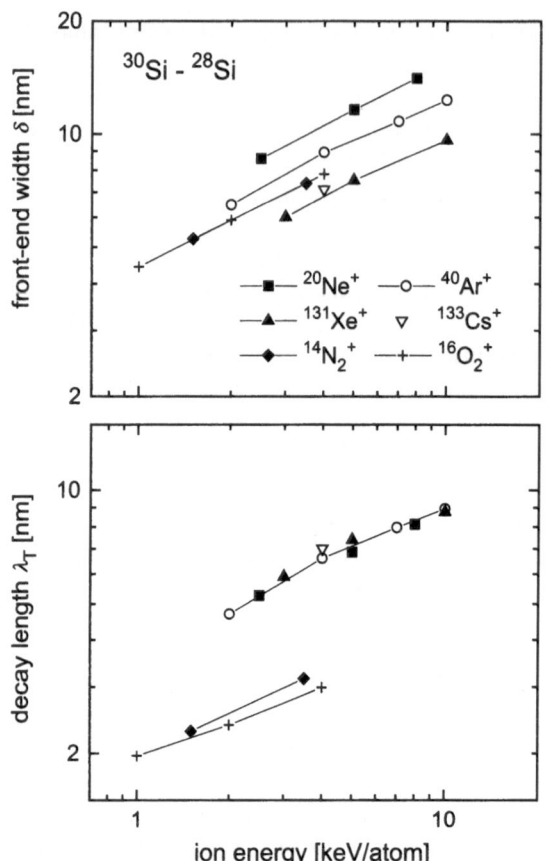

**Fig. 3.4a,b.** Results from collisional mixing of isotopically purified Si layers ($^{30}$Si–$^{28}$Si–$^{30}$Si) bombarded by the ion species indicated. Depicted is (**a**) the front-end width $\delta$ and (**b**) the decay length $\lambda$ derived from the interface broadening in sputter depth profiles through the layers as a function of ion energy. Data from [3.98]

lengths. The authors rationalize the projectile-independent values of $\lambda_T$ with an internal mixing profile $p_i(x')$ whose surface value $p_i^s$ does not vary for the rare-gas and Cs ions. The interface broadening observed under $N_2^+$ and $O_2^+$ bombardment is much smaller than expected in comparison to the other ion species (see Fig. 3.4). This observation was attributed to a beam-induced nitride and oxide formation, which results in a reduced depth of mixing in these altered specimens. Comparing their data with previous results, the authors conclude that the decay length for several impurities in Si is smaller than for Si in Si. For 4 keV Xe$^+$, for example, $\lambda_T = 7$ nm for $^{28,30}$Si in Si, whereas $\lambda_T = 4.5$ nm for $^{16}$O and $\lambda_T = 2.5$ nm for both $^1$H and $^{35}$Cl. On the other hand, $\lambda_T = 10$ nm for $^{12}$C in Si [3.86, 87, 98]. Furthermore, the impact-energy dependence of $\lambda_T$ is not identical for these species.

The observation of large element-dependent variations in the decay length (or in any other parameter characterizing mixing-induced broadening) triggered attempts to introduce chemical driving forces (such as the heat of mixing) for low-energy bombardment conditions also. Substituting several (unknown) parameters in the model originally developed by Johnson et al. [3.80, 81] to describe high-energy mixing, Zalm and Vriezema [3.102] derived a rather simple expression for collisional mixing at keV energies. In this scheme, broadening depends solely on the beam energy, the beam's projected range, the sputtering yield and an averaged cohesive energy. Although they reported an agreement of the model predictions with a (limited) set of experimental data, the validity of this approach was questioned. Wittmaack [3.94] has argued that the simplistic assumptions are not justified; in particular, the shape of the damage distribution may have a significant influence on the decay length (as shown experimentally) and can hardly be subsumed in the ion range.

Cheng et al. [3.93] attempted to correlate observed broadening effects in metallic bilayers with the respective values of the heat of mixing in these systems, thus extending the successful high-energy approach to ion energies of 1–2 keV. The data exhibit the expected relationship with the chemical properties of the elements, but the limited number of specimens investigated precludes more general conclusions. In a similar experiment, albeit using a wider range of bilayer systems, King et al. [3.91] studied interface broadening induced by 4 and 5 keV Ne$^+$ bombardment. They observed mixing efficiencies that are larger than those obtained in comparable high-energy experiments, but that scale in the same way with thermal diffusion constants. They concluded therefore that the same processes are responsible for collisional mixing.

MD simulations by Gades and Urbassek [3.78] studied the mixing induced by 0.5–2 keV Ar ions in metals and metallic bilayers and its dependence on material properties. This was made possible by the use of many-body interaction potentials of the tight-binding form, in which the cohesive energy and the heat of mixing can be independently assigned [3.78]. In accordance with the aforementioned model of Cheng et al. [3.66, 80], these simulations show the interface mixing to exhibit an inverse-square dependence on the cohesive energy; in contrast to the linear dependence on the heat of mixing predicted by the phenomenological model [3.66], the simulations [3.78] produce a non-linear variation of the interface broadening with $\Delta H_{\mathrm{m}}$; the broadening increases rapidly as $\Delta H_{\mathrm{m}}$ becomes more negative.

In summary, the understanding of collisional mixing at keV energies is far from being comprehensive, despite a rather extensive accumulation of data over the last two decades. Unfortunately, many of the results appear poorly related to each other, because of the widely varying instrumental and sample conditions. In addition, possible artifacts introduced by the latter have come under scrutiny only recently [3.94].

### 3.2.3 Gibbsian Segregation

When an alloy evolves towards equilibrium, a thermodynamic driving force causes composition changes at the surface in order for the system to minimize its surface free energy. This thermally activated segregation can be described by means of the so-called Gibbs adsorption equation [3.103]. At equilibrium this phenomenon effects compositional modifications at the very surface of the solid, i.e. in the outermost one (or possibly two) atomic layers. Thus, in addition to preferential sputtering, Gibbsian segregation (GS) also tends to change the composition of the surface region from which atoms are sputtered, either enriching or depleting the component $i$, depending on whether the heat of adsorption for this element is negative or positive, respectively [3.103]. If the atom exchange (thermal or radiation-enhanced) is sufficiently fast, the concentration ratio in the outermost layer, $c_i^{(1)}$, will approach the thermal equilibrium value. For a binary system [3.7–9, 104–106]

$$\frac{c_A^{(1)}}{c_B^{(1)}} = \frac{c_A^b}{c_B^b} \exp\left(-\frac{\Delta G_S}{k_B T}\right) , \tag{3.36}$$

where $T$ is the temperature of the specimen, $k_B$ is the Boltzmann constant and $\Delta G_S$ is the segregation free energy, $\Delta G_S = \Delta H_S - T \Delta S_S$, with $\Delta S_S$ and $\Delta H_S$ being the entropy and heat (enthalpy) of segregation, respectively; $\Delta G_S$ is associated with the exchange of an atom in the surface phase with an atom in the adjacent layer of the bulk phase. Lam and Wiedersich [3.7, 8] established a relation for the equilibrium flux of atoms into the surface layers which depends on the jump frequency of atoms from the bulk to the outermost atomic layer $\nu^{b(1)}$ and that for the reversed jump direction $\nu^{(1)b}$. The equilibrium corresponds to a net flux $J_A^{GS}$ of A atoms into the surface layer [3.8]

$$\Omega J_A^{GS} = \xi \left(\nu_A^{b(1)} c_A^b c_B^{(1)} - \nu_A^{(1)b} c_A^{(1)} c_B^b\right) , \tag{3.37}$$

where $\xi$ is the atomic layer thickness. The ratio of these frequencies is derived from the equilibrium condition and is found to be [3.8]

$$\frac{\nu^{(1)b}}{\nu^{(b)1}} = \exp\left(\frac{\Delta G_S}{k_B T}\right) . \tag{3.38}$$

Thus for a surface-segregating element ($\Delta G_S < 0$), the activation energy for a backward surface-to-bulk jump is effectively increased relative to the migration energy in the bulk. Since these atom jump frequencies are proportional to the concentrations of point defects, Gibbsian segregation can be strongly enhanced by irradiation (and the concurrent mixing-induced motion of atoms) even at temperatures where the concentration of thermal vacancies is smaller than that of radiation-induced defects [3.9, 10]. In fact, the importance of GS for ion-irradiation-induced composition changes was proposed [3.107–112] in many experiments carried out at low temperatures (at room temperature or even below).

Generally, the element segregating to the surface is depleted in the subsurface region. At low temperatures, this depletion is spread out mainly by collisional mixing and the altered layer should not exceed the damage range of the ions. At temperatures of significant vacancy mobility, point defects can escape from the damage region and these radiation-enhanced or radiation-induced processes produce altered layers extending much deeper into the target [3.8, 9, 113].

As indicated by (3.36), values of $\Delta G_S$ can be derived from measurements of the surface composition at different temperatures in the absence of ion bombardment. Such experiments have been carried out for a number of binary alloys using techniques that are sensitive to the outermost layer, such as ISS and field-ion microscopy [3.114–119]. These values are compiled in [3.48] and cover a range from about 0.06 eV to 0.5 eV. Usually, the segregating species either has the weaker bond (e.g. Cu in Cu–Ni) or is oversize (e.g. Au in Au–Cu). For thermally activated Gibbsian segregation and the (unusual) situation of high diffusional mobility, the composition would exhibit a spike at the surface exactly one atomic spacing wide. Then, the concentration gradient between the first and the underlying layers would be determined by (3.36), i.e. by an equilibrium segregation coefficient $K^{eq}$ with [3.48, 106]

$$K^{eq} = \exp\left(-\Delta G_S / k_B T\right) . \tag{3.39}$$

For the more common case of a limited mobility, the composition profile would exhibit a spike at the surface, but the layers below would be depleted in that element [3.48, 106] (see below). This shape is produced by the requirement that under steady-state bombardment, sputtering must be stoichiometric but segregation will still occur. From experiments, the ratio of the composition in the first and in the second (and deeper) layers might be described as

$$\frac{c_A^{(1)}}{c_B^{(1)}} = \frac{c_A^{(l \geq 2)}}{c_B^{(l \geq 2)}} K . \tag{3.40}$$

The composition in layers $l > 1$ may follow a diffusion-like profile as proposed in [3.120]. $K$ can be evaluated either from experimental composition profiles, by means of a comparison of annealed and subsequently irradiated samples, or, with some ambiguities, from angular distributions of sputtered atoms. Kelly [3.48] has analyzed such data and reports values in the range $1.2 \leq K \leq 5.4$. For ambient temperatures the inequality $K \ll K^{eq}$ holds, suggesting that the role of bombardment is not simply one of bringing the system to equilibrium at normally inaccessible temperatures. This is also an indication that GS under ion irradiation is not necessarily an equilibrium process, but depends on the interplay between ("chemically guided" [3.47, 48]) atomic jumps and the surface recession. Kelly argued that for most experimental situations both processes might be of similar magnitude.

The combined effect of preferential sputtering, collisional mixing and Gibbsian segregation is schematically illustrated [3.10] in Fig. 3.5. It can be seen that the surface stays enriched in the segregating species with increasing sputter time (fluence); however, the concentration is reduced. In the subsurface region a depletion of this species develops concurrently. Owing to the synergistic nature of Gibbsian segregation and preferential sputtering, it is

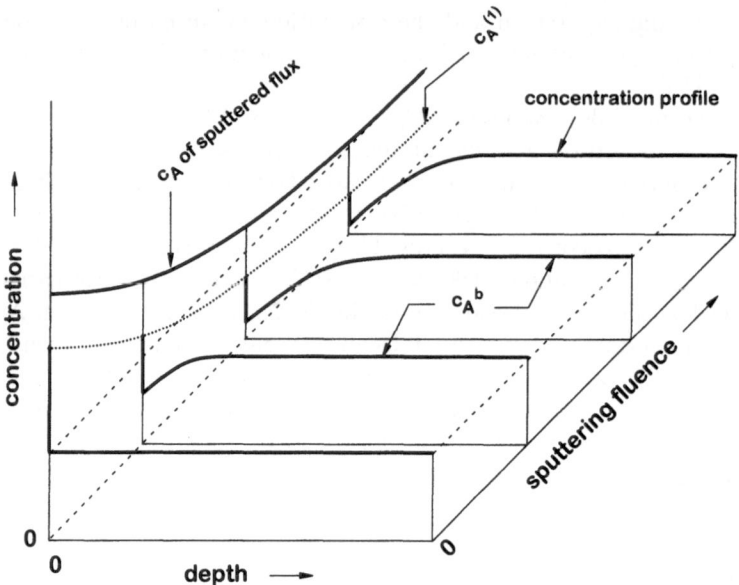

**Fig. 3.5.** Schematic representation of the time evolution of the sputtered flux and the near-surface altered layer as in Fig. 3.1, but including in addition to preferential sputtering (and collisional mixing) the effect of Gibbsian segregation. Component A both is sputtered preferentially and segregates to the surface. From [3.10]

difficult to isolate information on any single one of them from experiments measuring the sputtered flux or the surface composition as a function of the bombarding fluence. Nevertheless, correlations between the preferentially sputtered and the segregating component have been established by Kelly [3.48, 104, 105] for various alloys.

Andersen and coworkers have observed that the angular distributions of atoms sputtered from alloys are influenced by (strong) composition gradients at the surface (Sect. 3.3.3) and have ascribed the latter to the occurrence of Gibbsian segregation during irradiation. The excess Cu depletion in the subsurface region of a Cu–Ni alloy sputtered at high temperatures was explained in terms of preferential sputtering induced by Gibbsian segregation [3.121].

The identification of GS as a major cause of pronounced surface enrichment apparently resolved conflicting experimental observations [3.1, 17]: the description of compositional changes solely in terms of preferential sputtering failed, mostly because of the fact that the small differences in mass and/or binding energy produced, according to (3.14), enrichment effects too small to be compatible with the experimental results.

### 3.2.4 Radiation-Enhanced Diffusion

At elevated temperatures, between 0.2 and $0.6T_m$, thermally activated diffusion processes become dominant. In this regime, radiation-induced point defects are mobile and their concentrations can exceed the thermodynamic equilibrium values by many orders of magnitude [3.122–124]. As the average diffusion coefficients of the atoms in the alloy are functions of the defect concentrations $c_v$ and $c_i$, for vacancies and interstitials, respectively, atomic diffusion is strongly enhanced by irradiation. The diffusion coefficients then are given by [3.7–9, 125, 126]

$$D_A^{RED} = d_{Av}c_v + d_{Ai}c_i \,,$$
$$D_B^{RED} = d_{Bv}c_v + d_{Bi}c_i \,, \tag{3.41}$$

where $d_{pq}$ ($p = A, B$ and $q = v, i$) are the diffusivity coefficients, which contain the kinetic characteristics of the diffusion of the elements A and B via vacancies and interstitials. These coefficients depend on the specific jump distances, the coordination numbers and the effective exchange jump frequencies of the given element–defect pair. In the low-temperature range of radiation-enhanced diffusion (RED), the concentration of excess point defects is predominantly limited by mutual recombination [3.7–9]; then, the radiation-enhanced diffusion coefficient $D^{RED}$ is proportional to the square root of the product of the defect production (displacement) rate $K$ and the jump frequency of vacancies (the slower moving species) $\nu_v$: $D^{RED} \propto (K\nu_v)^{1/2}$. The jump frequency is related exponentially to the corresponding activation energy for motion of vacancies, $E_v$: $\nu_v \propto \exp(-E_v/kT)$. Therefore, the slope of an Arrhenius plot corresponds to one-half of the migration energy of vacancies. On the other hand, annihilation of defects at extended sinks (e.g. grain boundaries) dominates at higher temperatures and the radiation-enhanced diffusion coefficient will become temperature-independent [3.123]. Thus, the diffusion coefficient is proportional to the ion flux in the collisional-mixing regime and in the sink annihilation regime of RED, but scales with the square root of the flux in the recombination-limited regime of RED [3.8, 9, 123, 125–127]. At temperatures above $\sim 0.6T_m$, the concentration of thermal vacancies is larger than that induced by irradiation and, hence, normal thermal diffusion is dominant.

Figure 3.6 exemplifies the possible different contributions to the steady-state diffusion coefficient during ion irradiation. At very low temperature, the contribution to diffusion from thermal motion is below that for the athermal collisional mixing (horizontal line, labeled "mixing" in Fig. 3.6), primarily because the steady-state defect density is high and limits the number of jumps a defect can make between creation and annihilation. In the regime of radiation-enhanced diffusion, but still at moderate temperature and/or very low sink density, the defect concetration is limited by recombination of vacancies and interstitials and the slope of $D^{RED}$ exhibits the aforementioned slope of one-half of the activation energy (dotted line, "radiation-enhanced"). At higher temperatures and higher sink densities, the lifetime of excess defects is reduced, and therefore $D^{RED}$ is diminished relative to its value for the recombination-limited case. With sink annihilation dominating, $D^{RED}$ becomes temperature-independent. $D^{RED}$ is then proportional to the displacement rate as in the mixing-dominated temperature range. At very high temperatures, diffusion is dominated by the thermal vacancy concentration and the slope of $D$ in Fig. 3.6 is due to thermal diffusion (dashed line, "thermal"). Detailed mechanisms and rate equations of radiation-enhanced diffusion have been discussed in [3.123, 125–127].

Composition changes in multicomponent samples effected by ion bombardment in the presence of RED extend typically over a depth range that exceeds by far the penetration range of the incident ions. This has been shown in many experiments [3.116–119, 128]. While radiation-enhanced diffusion tends to smooth out concentration gradients, experimental results have shown [3.129] that radiation-enhanced diffusion rates can decrease at elevated temperatures. Thin Ag layers embedded in Ni depth profiled by sputtering at 600 K exhibited much sharper interface widths than in corresponding data obtained at 300 K. The authors suggested that at the higher temperature the

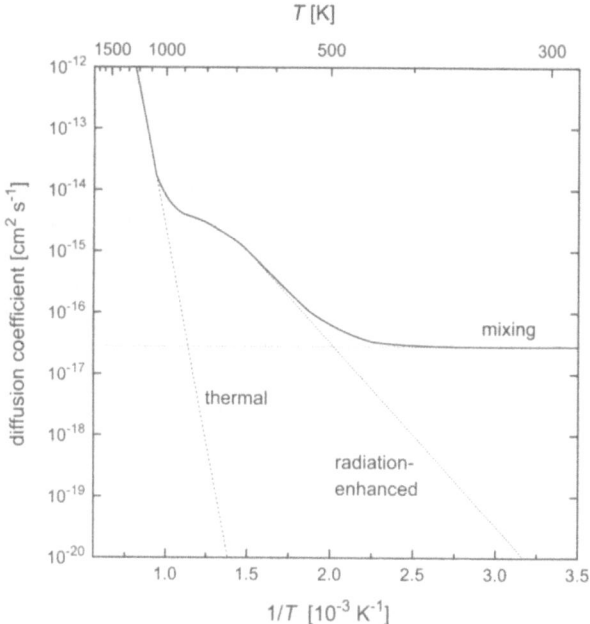

**Fig. 3.6.** The steady-state diffusion coefficient during ion irradiation as a function of the inverse specimen temperature. The contributions are due to collisional mixing (*horizontal line*), radiation-enhanced diffusion (*dotted line*) and pure thermal diffusion (*dashed line*). The specific values [3.125] are representative for nickel at a displacement rate of $10^{-3}$ (displacement per atom/s)

defects responsible for diffusion are annealed. Kelly and coworkers [3.48, 85] have proposed that because of the positive heat of mixing in this (and other [3.130]) systems, mixing will decrease when the temperature increases.

### 3.2.5 Radiation-Induced Segregation

Concurrently with RED, radiation-induced segregation occurs in a multicomponent sample during irradiation, because the same point defects that are responsible for radiation-enhanced diffusion can also induce segregation of the alloy components; this process is widely called radiation-induced segregation (RIS). At sufficiently high temperatures, bombardment-induced defects migrate over long distances in the solid before being eliminated by mutual recombination and/or annihilation at extended sinks. The spatial separation between defect production and elimination results in persistent defect fluxes: the defects can flow either toward the surface or from the peak damage region toward regions of lower defect density. Since these defect fluxes are associated with fluxes of atoms, any preferential association of defects with a particular alloy component results in a preferential defect–solute flux coupling and leads to a local redistribution of the elements within an initially homogeneous alloy. While radiation-enhanced diffusion drives the system toward a compositional equilibrium, radiation-induced segregation tends to move the alloy away from a spatially uniform composition state, producing local concentration gradients and, frequently, nonequilibrium precipitates. Radiation-induced segregation thus results from persistent radiation-induced defect fluxes and the coupling of an alloy component to these fluxes. Most often, the latter are the result of spatial nonuniformity in defect annihilation and/or defect production. Several authors [3.8–10, 125, 131] have pointed out that these fluxes are usually preferentially coupled to a particular solute element, giving rise to radiation-induced segregation.

The physical origin of a preferential coupling between defect and solute fluxes is as follows [3.9]: A gradient in vacancy or interstitial concentration causes a defect flux associated with a flow of atoms. A vacancy flux induces an atom flux equal in magnitude but opposite in direction, $J_v = -J_a^v$, whereas an interstitial flux corresponds to a flux of atoms of the same magnitude and direction, $J_i = J_a^i$. The total defect fluxes can be partitioned into partial fluxes occurring via different alloy components, for example A and B in a binary alloy:

$$J_i = J_A^i + J_B^i \,,$$
$$J_v = -\left(J_A^v + J_B^v\right) \,. \tag{3.42}$$

Quite often, the various alloy components do not participate in the defect flow exactly in proportion to their (local) concentrations, that is, $J_A^i/J_B^i \neq c_A/c_B$ and/or $J_A^v/J_B^v \neq c_A/c_B$. If, for example, interstitials migrate preferentially via A atoms and vacancies preferentially exchange with B atoms, the resulting preferential coupling between defect and solute gives rise to radiation-induced segregation.

It has been noted [3.7–9] that radiation-induced segregation only occurs within the temperature range where annihilation of mobile defects at extended sinks is dominant (depending on defect production rate, between 0.2 and 0.6 $T_m$). Outside this range radiation-induced segregation is unimportant, because defect recombination reduces long-range migration of defects at lower temperatures, and efficient back-diffusion prevents the buildup of gradients at higher temperatures. It has also been pointed out [3.10, 132] that radiation-induced segregation and Gibbsian segregation of an alloy component may occur in the same or in opposite directions. For example, in Ni–Si alloys the observed Si enrichment at the surface [3.117] can be effected by either of these processes, whereas Gibbsian segregation causes Cu enrichment in the first layer of Ni–Cu alloys [3.116], but radiation-induced segregation gives rise to a pronounced Cu depletion in the subsurface region of this alloy [3.124].

## 3.3 Data from Alloys

Bombardment-induced changes in surface and near-surface composition have been measured for numerous alloy systems. Compilations of these results can be found in reviews by Betz [3.18], Betz and Wehner [3.1], Andersen [3.2] and Shimizu [3.19]. Less comprehensive data sets are contained in some other publications [3.47, 48, 105]. Whereas preferential sputtering was identified early [3.133–136] as a potential source of composition changes at the surface, the relevance of the other processes outlined in Sect. 3.2 was recognized rather gradually [3.136–139]. Hence, the first models [3.140–144] of irradiation-induced composition changes considered only sputtering and diffusion. The possible contribution of Gibbsian segregation was noted in the context of experiments carried out at elevated temperatures and was included in more elaborate theoretical descriptions. A rather comprehensive model of beam-induced surface composition changes was developed by Lam and Wiedersich [3.7–9]. This phenomenological approach, covering all of the aforementioned processes, requires a large number of input parameters; although several of them are not known very well, the computed composition data agreed quite satisfactory with the corresponding experimental results. Sigmund and Lam [3.3, 12, 13] formulated a theoretical scheme that accommodated the processes described in Sect. 3.2. More recently, this scheme was successfully applied to investigate composition changes in binary alloys for a range of sample temperatures where all of these processes can attain their full importance.

A variation of the surface and near-surface composition of an alloy upon ion bombardment can be monitored by any of the surface-sensitive techniques presented in Sect. A.1. Depending on the specific information volume, the depth into the target that is accessible may range from the single outermost atomic layer at the surface up to some nanometers. Obviously, these different depth ranges must be accounted for in the evaluation of any alteration in surface composition. Information (sometimes more indirect) on compositional variations can be obtained by monitoring the flux of sputtered species from the surface; it will reflect changes in composition of the outermost monolayers from which the particles are ejected. Transient yield changes during irradiation and differential yield data (angular, energy and mass distributions of sputtered species) provide insight into processes occurring in this region. In this section, examples of data from both of these approaches are presented.

While experimental data have been obtained at steady-state conditions (i.e. for an altered surface composition), computer simulations are usually carried out in the zero-fluence limit, i.e. for the pristine state of the target. Compositional modifications of the latter can then be introduced separately. Eckstein, Möller and coworkers have performed a considerable number of simulations utilizing the binary-collision-approximation code TRIM [3.25, 26, 61]. Apart from static computations (the zero-fluence case), they also carried out dynamic simulations [3.145]; in these the target composition is modified by taking into account ion implantation of the bombarding ions, ballistic atomic relocation processes and (preferential) sputtering from the surface. Most of these data are compiled in a recent monograph by Eckstein [3.146] and in another review [3.25].

In binary systems with large mass ratios (e.g. TaC, WC), the simulations [3.25, 26] show that the lighter element is preferentially sputtered and that this effect is more pronounced for low energies and/or light bombarding species such as H, D and He. As expected, this causes a corresponding enrichment of the heavier species at the surface. Even for the pristine (unaltered) surface, the lighter atom is ejected in the forward direction and has a higher average emission energy. Furthermore, lighter atoms tend to originate from greater depths in the target. Results for systems with less pronounced mass differences indicate that the magnitude of this preferentiality reduces in this case.

### 3.3.1 Near-Surface Composition Changes and Profiles

The discussion of the various primary and secondary effects emphasized that all of them can contribute, to a variable extent, to compositional modifications in the near-surface region of a solid exposed to ion irradiation. These effects obviously may depend on a number of parameters associated both with the conditions of ion bombardment (ion species, energy, range) and with the specific characteristics of the target (formation and migration energies of defects, enthalpy of segregation, site-exchange frequencies). Since the latter, especially, are often unknown, the contribution of individual processes to composition changes is usually difficult to assess unambiguously.

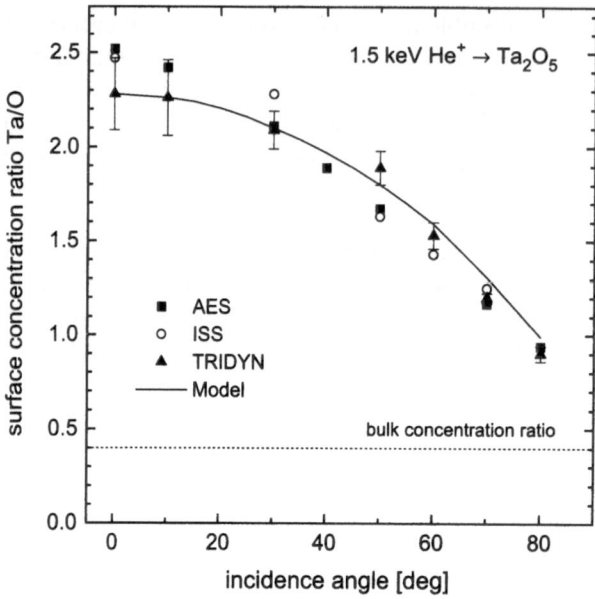

**Fig. 3.7.** The dependence of the steady-state surface concentration ratio Ta/O on the incidence angle (relative to the surface normal) for 1.5 keV He$^+$ bombardment of Ta$_2$O$_5$. The various values have been derived from AES and XPS measurements, from TRIDYN computer simulations and from a model (see text). Data from [3.151]

Pure preferential sputtering can best be studied by ion bombardment of isotopic mixtures; as discussed in Sect. 3.4, only mass differences should determine preferential ejection. Data on such systems will be presented there. Another class of targets where mass effects can often be explored unperturbed by other processes consists of targets with a large mass difference bombarded at low energies and/or with light ions [3.147–151]. Figure 3.7 exemplifies this situation, depicting results from 1.5 keV He$^+$ bombardment of Ta$_2$O$_5$; the steady-state surface concentration ratio $c_{Ta}^s/c_O^s$ as determined by low-energy AES and ISS is plotted as a function of the He$^+$ ion incidence angle [3.151]. The experimental data are compared with the outcome of computer simulations using the dynamic code TRIDYN [3.145]. The solid line in Fig. 3.7 was obtained by a model that assumes the component sputtering yields (cf. (3.2)) to be independent of the actual surface composition and uses the initial partial sputter yields derived from the simulations (for details see [3.151]). The various data exhibit good overall agreement and show a drastic Ta enrichment ($c_{Ta}^s/c_O^s$) at the surface for near-normal ion impact; with increasing incidence angle this enhanced Ta concentration is reduced but the bulk stoichiometry is never reached. The agreement between AES and ISS indicates that no concentration gradient exists in the outermost layers as this would produce diverging concentration values because of the different information depths of the two techniques.

Furthermore, the congruence with the computations is evidence for the presence of pure ballistic effects (preferential sputtering, collision mixing), as only these are modeled by the simulation. According to the authors [3.151], the surface binding energies are nearly identical for the two components, so the data in Fig. 3.7 essentially represent the mass effect on preferential sputtering (and on CM). The pronounced dependence on the incidence angle also provides some clue as to the details of the ejection processes. Since for 1.5 keV $He^+$ ions a full collision cascade would not evolve and sputtering would mostly proceed via low-generation recoils, the sputter ejection of heavy Ta atoms by the light He ions is very inefficient, in particular at near-normal incidence. At oblique incidence the kinematics for the emission of Ta atoms become more favorable and therefore the surface enrichment at steady state is less pronounced. These changes are reflected also in the total sputtering yield and the characteristic fluence required to reach equilibrium: low yield and large fluence at near-normal ion incidence, and vice versa for oblique ion impact.

The occurrence of (bombardment-enhanced) Gibbsian segregation at room temperature and below was investigated by Li et al. [3.152, 153] in a *nondestructive* way. The authors bombarded a $Cu_{0.81}Pt_{0.19}$ alloy with $Ar^+$

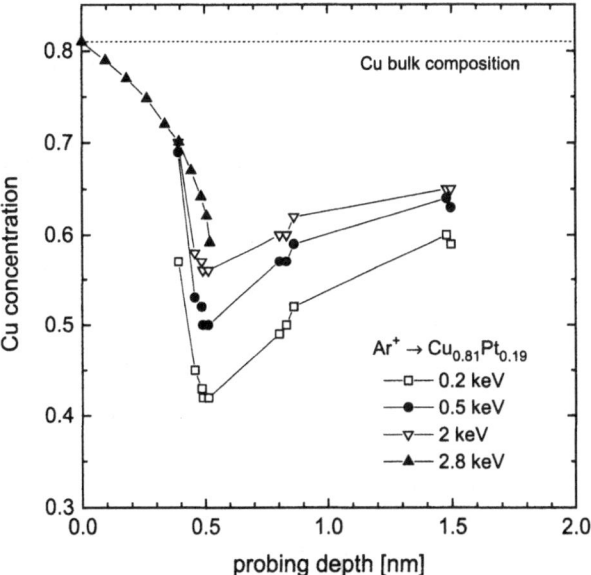

**Fig. 3.8.** Steady-state composition profiles of a $Cu_{0.81}Pt_{0.19}$ specimen under irradiation with $Ar^+$ ions of the given energies. The concentrations for the 2.8 keV data were obtained from angle-resolved Auger electron spectroscopy, whereas for the other bombarding conditions Auger electrons of ten different energies (i.e. probing depths) were employed to derive the composition. The experiments were performed at (or slightly below) room temperature. Data from [3.152, 153]

ions in the energy range from 0.2 to 2.8 keV; they studied the near-surface composition changes using Auger electrons of ten different energies (i.e. different probing depths) and emission-resolved AES. While the former method covers a depth range from $\sim 0.39$ nm to $\sim 1.5$ nm, the latter is sensitive to the topmost two surface layers (about 0–0.5 nm). Some of these data [3.153] are depicted in Fig. 3.8. The results show that over the accessible depth Cu is depleted relative to the bulk concentration, but in the outermost surface layer the composition is close to stoichiometric. This surface spike is indicative of concurrent preferential sputtering of Cu and segregation of Cu to the surface. Li and coworkers [3.153] also observed these features for a bombarding temperature as low as 190 K. These findings are in general agreement with similar observations made much earlier, but most of those utilized sputter profiling to derive the depth-dependent composition. Conversely, the investigations by Li and coworkers circumvented the possible further modifications that this approach (of sputtering) may induce.

Using very-low-energy ion bombardment by 50 eV $Ar^+$, Bartella and Oechsner [3.154, 155] determined the internal composition distribution in a $Ni_{0.79}Mo_{0.21}$ alloy induced by a preceding ion irradiation with $Ar^+$ ions of 40 to 2000 eV energy. For the lowest prebombarding energy the outermost surface is composed of pure Mo because of strong preferential sputtering of Ni in this near-threshold regime. For energies beyond 500 eV, this enrichment is reduced to about 40 at%, but the subsurface region is still enriched in Mo (up to 80%). Computer simulations [3.156] modeling ion irradiation of the same Ni–Mo binary system produced close agreement with the experimental results.

More recently, Cao [3.157] performed similar experiments on $Ni_{0.7}Mo_{0.3}$ and several FeSi alloys. He monitored the stationary near-surface composition changes induced by 1–5 keV $Ar^+$ bombardment using AES and low-energy sputtering ($Ar^+$ ions of 120 and 180 eV were employed for $FeSi_2$ and $Ni_{0.7}Mo_{0.3}$ samples, respectively). The results for $FeSi_2$ and $Ni_{0.7}Mo_{0.3}$ are depicted in Fig. 3.9. Pronounced deviations from the bulk composition are observed over a depth range that correlates roughly with the ion's damage range. Hence, collisional mixing is operative, giving rise to atomic relocations over this depth. As Si (from $FeSi_2$) [3.158] and Ni (from NiMo) [3.159] are sputtered preferentially, their concentration in the *outermost layer* is reduced relative to the bulk value. This reduction is slightly more pronounced at the lower $Ar^+$ energies. The distinct subsurface depletion of both Si and Ni indicates that, in addition, Gibbsian segregation plays a role. The segregation of these components of the alloy to the surface (and their preferential loss from there) results in a reduction in the concentration of the corresponding atoms in the *subsurface layers*, as the alloy attempts to approach equilibrium. This depletion is only weakly dependent on ion energy. The apparent similarity of the profiles shown in Fig. 3.9 to the schematic depiction in Fig. 3.5

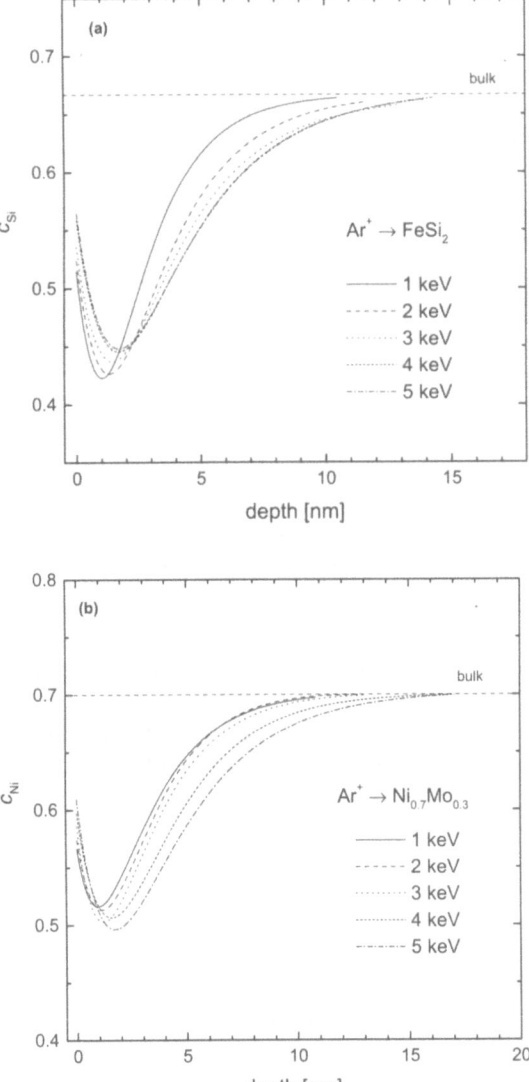

**Fig. 3.9a,b.** Depth profiles of the compositional changes induced in (**a**) FeSi$_2$ and in (**b**) Ni$_{0.7}$Mo$_{0.3}$ by Ar$^+$ ion bombardment at the energies indicated. The Si (a) and Ni (b) concentrations were derived by AES depth profiling using 120 eV (a) and 180 eV Ar$^+$ ions (b) for erosion. Data from [3.157]

is noted. The author [3.157] has reproduced the observed depth distributions by applying a model based on Ho's approach [3.142].

Similar composition profiles were reported for a variety of binary alloys under different irradiation conditions. Reference to these studies has been made repeatedly in several reviews of alloy sputtering [3.1, 2, 19, 47, 48]. At elevated temperatures, the steady-state subsurface depletion of the surface-segregating component can be become quite severe. This was nicely demonstrated by several pioneering investigations. Sigmund and Lam [3.3] provided an account of the more recent work and indicate, furthermore, some conflicting observations.

The influence of the surface orientation (and hence surface free energy) and the temperature on Gibbsian segregation has been studied in detail using $Pt_xNi_{1-x}$ single crystals of different orientations. Using ISS and AES, Varga and coworkers [3.160–162] have investigated $Pt_{0.1}Ni_{0.9}$ and $Pt_{0.25}Ni_{0.75}$ crystal surfaces of different orientations in order to elucidate surface properties and possible changes in surface composition. For a $Pt_{0.25}Ni_{0.75}$ crystal target exposed at room temperature to 500 eV $Ar^+$ bombardment up to steady state [3.162], the topmost atomic layer of the (100) surface exhibits a Pt enrichment $\left(c_{Pt}^{(1)} = 0.37\right)$, while the (110) surface is slightly enriched in Ni $\left(c_{Ni}^{(1)} = 0.82\right)$; see Fig. 3.10a. On this surface the second layer is strongly enriched in Pt, since *Ni atoms segregate into the topmost layer*, where they are sputtered preferentially. After annealing at 970 K, the (110) surface consists nearly exclusively of Ni atoms and the second-layer Pt enrichment is reduced somewhat because of diffusion into the bulk (Fig. 3.10b). Conversely, on the $Pt_{0.25}Ni_{0.75}(100)$ surface *Pt is the segregating component*; annealing at 970 K changes the Pt concentration in the outermost layer only minimally (to $c_{Pt}^{(1)} = 0.39$). The (111) surface appears to show similar features after sputtering and annealing. Hence, after $Ar^+$ ion bombardment at room temperature, the $Pt_{0.25}Ni_{0.75}$ (100) and (111) surface show Pt enrichment in the topmost layer; conversely, the (110) surface exhibits a slight Ni enrichment. While Ni atoms are sputtered preferentially, this effect is coupled to segregation and causes Pt enrichment on the (100) and (111) surfaces where Pt is the segregating species. By contrast, Ni atoms segregate at the (110) surface and, because of sputtering, the second layer becomes strongly enriched in Pt. The authors propose several possible causes for such an oscillating depth composition. Whereas differences in surface free energies are negligibly small in the case of Pt and Ni, a size effect favors an enrichment of the larger Pt atoms at the surface of Ni-rich alloys. Furthermore, an ordering tendency in PtNi alloys was found that can be explained by a preference for *unlike* bonds, allowing for alternations in the concentrations of the two components.

The same group [3.160, 163–165] observed, using scanning tunneling microscopy (STM), lattice mismatch dislocations on PtNi(111) surfaces induced by the differing lattice constants of the bulk and altered layers. The density of these dislocations was correlated with the Pt enrichment. From the analysis of atomically resolved STM micrographs, the dislocation depth and the corresponding thickness of the Pt-enriched layer were evaluated. The latter amounted to three and five monolayers for surfaces sputtered with 500 eV $Xe^+$ and $Ar^+$, respectively. The subsurface mismatch dislocations become manifest through a network of "ditches" at the surface upon ion bombardment and annealing. The density of these ditches in the STM images and, hence, the density of the dislocations is a direct measure of the lattice constant

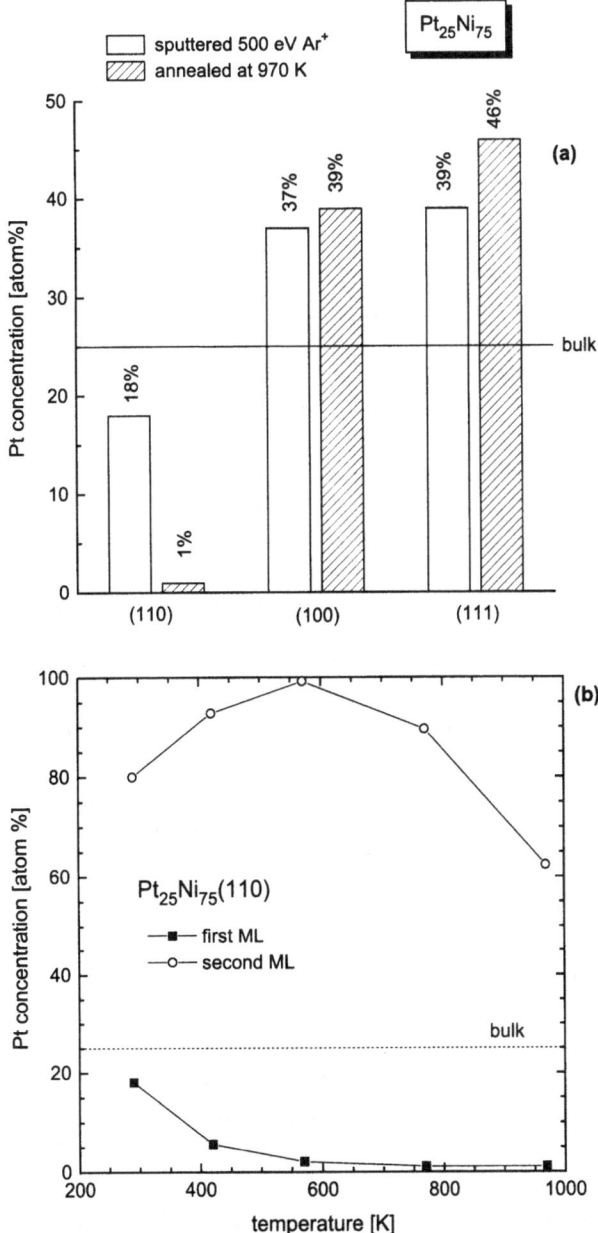

**Fig. 3.10a,b.** (a) First monolayer composition on different surfaces of $Pt_{25}Ni_{75}$ single crystals; the values after $500\,eV\;Ar^+$ bombardment at room temperature and after further annealing at $970\,K$ are compared. (b) First- and second-monolayer composition of $Pt_{25}Ni_{75}(110)$ determined by low-energy ISS versus annealing temperature (before each heating step, $500\,eV\;Ar^+$ bombardment was performed). Data from [3.162]

of the altered layer and provides, via Vegard's rule, a means to determine the average composition of the altered layer. While for the (111) surface the dislocation density does not exhibit any dependence on the thickness of the altered layer, on the (110) surface a strong influence is observed: for 500 eV $Xe^+$ and $Ar^+$ at an incidence angle of $\theta = 65°$ and for 500 eV $Ar^+$ at $\theta = 24°$, the authors found 3, 4.5 and 9 dislocations per 100 nm, respectively. They ascribe this to the fact that, unlike the case for (111), there exists no glide plane parallel to the (110) surface [3.163, 164].

The STM micrographs [3.160] shown in Fig. 3.11 exemplify the development of this dislocation network with irradiation fluence. A $Pt_{0.25}Ni_{0.75}$ surface bombarded with 500 eV $Ar^+$ ions at an incidence angle $\theta = 70°$ is shown for different ion fluences: at a fluence of $1.5 \times 10^{16}$ cm$^{-2}$ only a few individual ditches are visible at the surface, indicating that the surface composition is still close to that of the bulk. With increasing fluence, the surface becomes enriched in Pt (cf. Fig. 3.10a) and the density of dislocations increases, as is seen in Figs. 3.11b,c. The ditch pattern frequently exhibits a hexagonal shape that is aligned with the crystallographic directions. Annealing of the sample at 910 K (Fig. 3.11d) results in a reduction of the number of dislocations in accordance with the lowering of the Pt enrichment observed by ISS. The authors stress that the study of these mismatch phenomena provides a new method to investigate the near-surface composition changes in an alloy due to ion irradiation. Obviously, this approach constitutes an interesting alternative to the usual technique of sputter depth profiling, which, inherently, may cause further modification of the surface (and its composition).

While GS was observed to occur even at room temperature (and occasionally some surface composition changes were found to continue beyond the termination of ion bombardment [3.108, 131]), the modified region at the surface was usually restricted to about the range of ion penetration. At elevated temperatures, however, a drastic increase of this altered zone is typically encountered, indicative of the additional processes such as RIS and RED. Lam and coworkers [3.116–119] have investigated compositional changes in a number of Ni-based alloys. Both the compositional changes in the outermost atom layer during sputtering and the steady-state concentration profiles were measured. The concentrations in the outermost layer were monitored by means of ISS using the same ions. The results for a Ni–25at%Pd alloy irradiated with 3 keV $Ne^+$ ions at sample temperatures ranging from 200°C to 700°C will be discussed in the following. The evolution of the Pd surface concentration $c_{Pd}^{(1)}$ during ion bombardment at various temperatures was monitored as a function of fluence [3.119]. (The samples were held for 1 h at the respective temperature before irradiation commenced and the different initial concentrations reflect equilibrium GS. From such data the authors derive values of $\Delta H_S$ and $\Delta S_S$ for this alloy.) The data demonstrated that at elevated temperatures the steady state is reached only very gradually: after $\sim 10^3$ s at 200°C and $10^5$ s at 500°C (the erosion rate in that experiment was $\sim 8.5 \times 10^{-3}$ nm/s). Depth

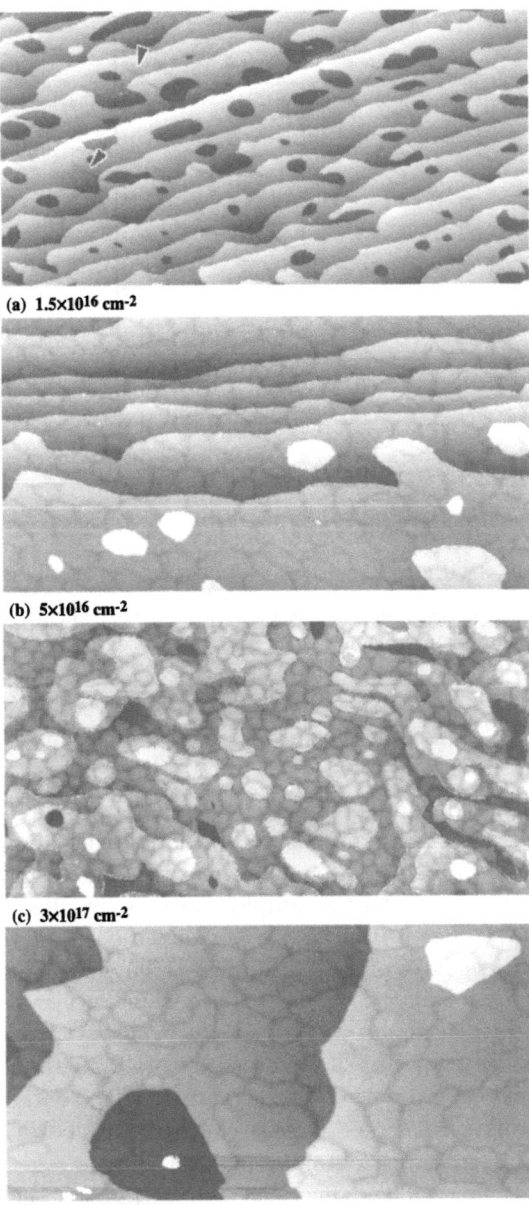

(a) $1.5\times10^{16}$ cm$^{-2}$

(b) $5\times10^{16}$ cm$^{-2}$

(c) $3\times10^{17}$ cm$^{-2}$

(d) annealed 910 K

**Fig. 3.11a–d.** (a)–(c) STM micrographs (size 400 nm × 200 nm) of a $Pt_{0.25}Ni_{0.75}$ surface after 500 eV $Ar^+$ ion bombardment at different fluences and annealing at about 770 K for 5 minutes. Note the increase of the dislocation density from very few isolated dislocations (*arrows* in (a)) towards a dense dislocation network at steady-state irradiation in (b) and (c). Micrograph (d) was taken after high-fluence bombardment and subsequent annealing to 910 K; in this case, the number of dislocations decreases again. From [3.160]

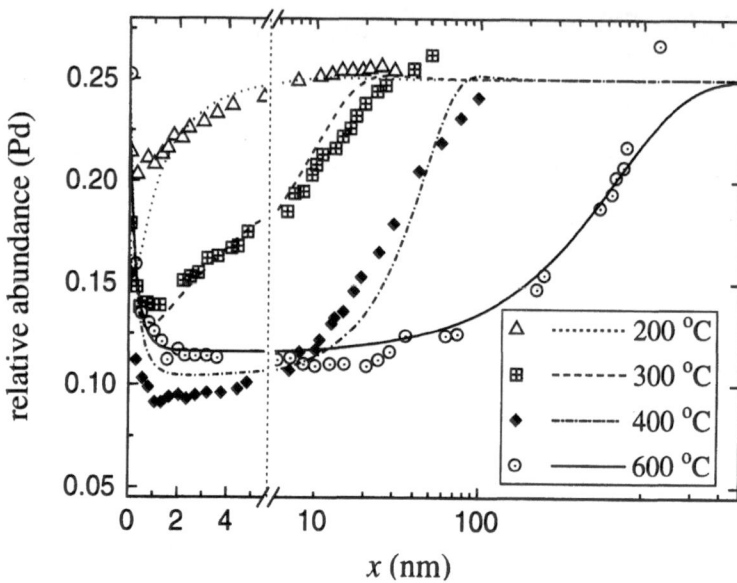

**Fig. 3.12.** Steady-state concentration profiles of Pd in a $Ni_{0.75}Pd_{0.25}$ specimen after bombardment with 3 keV $Ne^{+}$ ions at various temperatures. The experimental results (*symbols*) of Tang and Lam [3.119] were derived from ISS depth profiling using 3 keV $Ne^{+}$ ions with the specimen at room temperature; they are compared to calculated data of Sckerl et al. [3.12]

profiling of these altered surface regions was performed by Lam et al. [3.119] by rapidly cooling the sample to room temperature and employing again 3 keV $Ne^{+}$ for sputtering and ISS for composition determination. The results are depicted in Fig. 3.12. Not surprisingly, for the samples that were irradiated before at higher temperatures the composition changes extended deeper into the sample; for $T \geq 600°C$ a surface layer of about 1 μm has to be removed in order to reach the stationary (bulk) Pd concentration. The authors utilized a model developed by Lam and Wiedersich [3.7–9] that simulates surface modifications under ion irradiation incorporating all the processes briefly outlined in Sect. 3.2. The outcome of these computations (which, admittedly, contain a considerable number of fitting parameters) agrees reasonably well with the experimental results. Similar data were obtained by Lam and coworkers [3.116–118] for other Ni-based alloys. They generally demonstrate (i) that surface alterations may extend deep into the bulk, (ii) that defect production is an essential part of these modifications and (iii) that modeling is possible if the various processes are taken into account properly. From the latter, relevant parameters concerning defect formation and migration can be derived, as shown in those studies.

More recently, Sckerl et al. [3.12, 13] have set up a theoretical scheme to describe composition changes in alloys during ion bombardment at elevated temperatures. Based largely on previous work by Sigmund, Lam and coworkers [3.3, 15], this approach adds an explicit treatment of radiation-produced defects and defect-assisted processes. Specifically, it ensures mechanical stability of the target at all fluences by incorporation of pressure relaxation and the effects of preferential sputtering, collisional mixing, radiation-enhanced diffusion and Gibbsian and radiation-induced segregation are allowed for. The centerpiece is a nonlinear integro-differential equation generalized from earlier work [3.15]. Furthermore, the authors employ a random target structure as opposed to the layered medium used in previous treatments of composition changes. High-fluence composition profiles were determined and the dependence on various input parameters such as temperature and defect mobility was examined. A comparison with the data of Lam and coworkers [3.119] shown in Fig. 3.12 produced generally a good agreement. This is shown in Fig. 3.12 where the outcome of the computations of Sckerl et al. [3.12, 13] is depicted (solid and broken lines) for the corresponding temperatures.

### 3.3.2 Angular Distributions of Sputtered Atoms

An angular variation of the composition of the sputtered flux from multicomponent specimens was reported by Olson and Wehner [3.166, 167]. They observed that sputtering of Ag–Au, Cu–Ni and Fe–Ni alloys by $Hg^+$ or $Ar^+$ ions at energies below $300\,eV$ causes the lighter elements to be ejected preferentially in the direction of the surface normal. The authors suggested that, at these low energies, surface atoms bouncing back from underlying atoms may contribute significantly to sputter ejection. (Similar effects have been established in monolayer desorption experiments at low beam energies [3.147].) With increasing impact energy the enrichment decreases strongly. At $1\,keV$, preferential emission of Au (from Au–Ag) and of Ni (from both Ni-based alloys) along the surface normal was reported. This can be ascribed to steep concentration gradients at the surface, due to the enrichment of one component at the surface of these alloys by Gibbsian segregation.

Compositional gradients within the depth of origin of the sputtered atoms may influence the angular distribution of the sputtered species, as indicated by Sigmund et al. [3.15]. These authors proposed that the emission patterns should be narrower for those species for which a comparatively larger fraction of sputtered atoms originates from greater depths. This implies that in a binary system the species enriched in the surface layer will have a flatter angular distribution, while that of the depleted element is forward-peaked. This approach was taken by Andersen and coworkers [3.168–171] to derive the presence of segregation at the surface and to identify the segregating species. Specifically, they investigated different alloys (Cu–Pt, Ag–Au, Ni–Pt, $Cu_3Au$ and $Ni_5Pd$) at various temperatures and bombarding energies; they find [3.169, 170] preferential forward ejection of one element and ascribe this finding to the Gibbsian segregation of the other (the weaker-bound) component to the surface. The most detailed study was carried out for CuPt [3.170], with the $Ar^+$ ion energy covering the range from 1.25 keV to 320 keV. A distinct forward emission of Pt atoms was observed for $E \geq 40$ keV, whereas at 10 and 20 keV little angular variation of the yield ratio was found. At the lowest energies, a preferential ejection of Cu atoms at oblique angles dominates, while a slightly preferred Pt emission was found for intermediate and near-normal directions. The findings of Andersen and coworkers [3.169, 170] on CuPt are thus in accordance with the composition profile derived for this system by Li et al. [3.152, 153] (see Fig. 3.7). The contribution of the subsurface region (where Pt is enriched) to the sputtered flux results in a weakly preferential forward emission of Pt atoms.

Somewhat conflicting results were reported by Ichimura et al. [3.172] who investigated Cu–Ni, Co–Ni and Fe–Ni alloys under 3 keV $Ar^+$ bombardment at room temperature and 570 K. While preferential ejection of Ni in near-normal directions was observed for Cu–Ni at both temperatures, for Fe–Ni no difference in the angular distributions was found, in contrast to the aforementioned data of Olson [3.167].

Kang et al [3.173] measured angular distributions of Au and Cu sputtered from an Au–Cu alloy under 3 keV $Ar^+$ impact at room temperature. They did not observe any differences in the angular spectra, although ion scattering spectroscopy indicated a strong Au enrichment in the outermost layer and a depletion in the subsurface region induced by ion bombardment. Au is known to segregate at the surface of Au–Cu alloys.

Tombrello and coworkers [3.174, 175] determined the angular distributions of sputtered atoms from a liquid Ga–In eutectic alloy. In this system, Gibbsian segregation gives rise to an outermost layer that is virtually pure In. As expected, their data for $Ar^+$ bombardment showed that the In atoms had a $\cos^\beta \theta_0$ distribution, with $\beta = 1.80 \pm 0.1$, largely independent of ion energy. By contrast, the angular distribution of Ga was significantly narrower, with $\beta = 3.2 \pm 0.2$ in the ion energy range of 15–250 keV, and $\beta = 4.9 \pm 0.3$ at 3 keV. This increase at low energy was accompanied by an increase in the contribution of the topmost layer to the sputtered flux of atoms (see Sect. 2.3.1). MD simulations of the sputtering of this liquid Ga–In alloy were performed by Shapiro et al. [3.176] for 1.5 and 3 keV Ar bombardment. They determined yields and energy and angular distributions of the sputtered atoms and derived information on the depth of origin of the ejected atoms. The results generally corroborate the corresponding experimental findings [3.175].

Aoyama et al. [3.177] determined the emission distribution of the sputtered flux from GaAs under 1–3 keV $Ar^+$ irradiation, using a collector technique and analyzing the deposit by electron microprobe. As can be seen in Fig. 3.13, both the Ga and the As distributions could be fitted with $\cos^\beta \theta$ distributions, but $\beta$ was found to be 2.0 for Ga and 1.0 for As at 1 keV; both values slightly increased (to 2.5 and 1.5, respectively) at high energies. From these findings the authors [3.177] conclude that the As concentration is higher in the outermost layer but (strongly) depleted of As in the subsurface

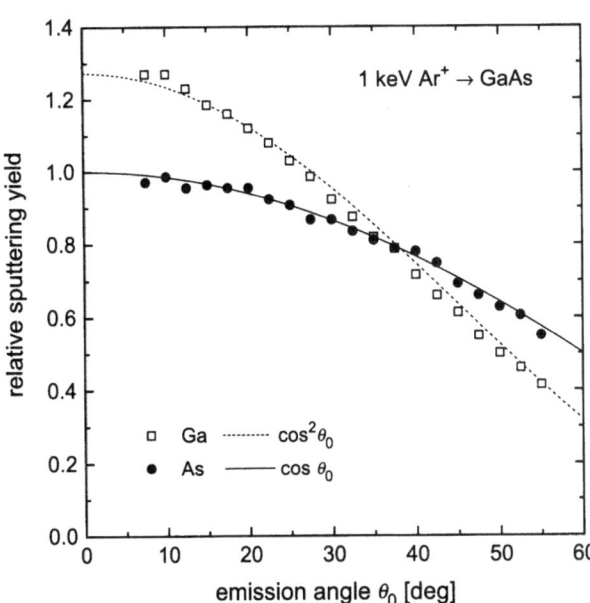

**Fig. 3.13.** Angular sputter-yield distributions of Ga and As atoms sputtered by 1 keV $Ar^+$ ions from GaAs. The *dotted* and *solid lines* are fits to the respective relative yield data; they are proportional to $\cos^2\theta_0$ for Ga and to $\cos\theta_0$ for As. The data refer to the stationary state. Data from [3.177]

region; this would fit the previous observations that the GaAs surface averaged over several monolayers (as seen, for example, by AES) is depleted of As (cf. Sect. 4.3). Such an oscillatory form of the composition profile would indicate that Gibbsian segregation plays an important role for this system. The same group also investigated [3.178] InP surfaces using $Ar^+$ and $Xe^+$ bombardment at two different energies (1 and 3 keV) and two sample temperatures (153 and 293 K). For this target the angular distributions of the two constituents were essentially identical to each other for all experimental conditions, but the magnitude of $\beta$ was found to vary with these parameters; at the lower temperature, $\beta$ was higher for both ion species at 3 keV impact (about 3.3) compared to 1 keV (2.3 to 2.6). At 293 K the development of some surface texture was observed and the fitted functions were considered [3.178] less reliable ($\beta \sim 2.1$ to 2.4). The similar angular distributions of In and P suggest that the specimen composition is homogeneous within the depth explored and that, contrary to GaAs, no segregation is occurring. (An evaluation of irradiation-induced composition changes in compound semiconductors is given in Sect. 4.4.)

Angular spectra of atoms sputtered from Co–Au, Cu–Be, Cu–Zn and W–Si alloys were recorded under 250 eV and 2 keV $Ar^+$ bombardment using SNMS by Wucher and Reuter [3.179]. While at the higher impact energy the angular distributions of the ejected atoms were rather similar (albeit strongly forward peaked, with a $\cos^3 \theta_0$ dependence), significant differences were seen at 250 eV: the lighter species exhibited a preferential ejection along the surface normal and this effect was most pronounced for Cu–Be. The latter finding is consistent in view of the fact that the light Be atoms are expected to be sputtered preferentially and hence Cu atoms may segregate to the surface of Cu–Be alloys during irradiation; these conditions result in an enhanced Cu concentration at the surface and a preferential ejection of Be in near-normal directions.

Employing a similar experimental setup, Bock [3.180] analyzed a $WSi_2$ alloy by means of SNMS and determined the angular ejection distribution of Si and W atoms and of some selected clusters (e.g. $Si_2$ and WSi). For steady-state sputtering, Si atoms are emitted strongly in near-normal directions. Figure 3.14 exemplifies this finding, depicting for 500 eV $Ar^+$ ion bombardment the ratio of partial yields $Y_{Si}/Y_W$ as a function of the polar emission angle. It is seen that the $Y_{Si}/Y_W$ ratio along the surface normal is enhanced by a factor about four as compared to oblique directions. This preferential forward emission of Si atoms is accompanied by an even stronger emission of $Si_2$ dimers along the surface normal (cf. Sect. 3.3.4).

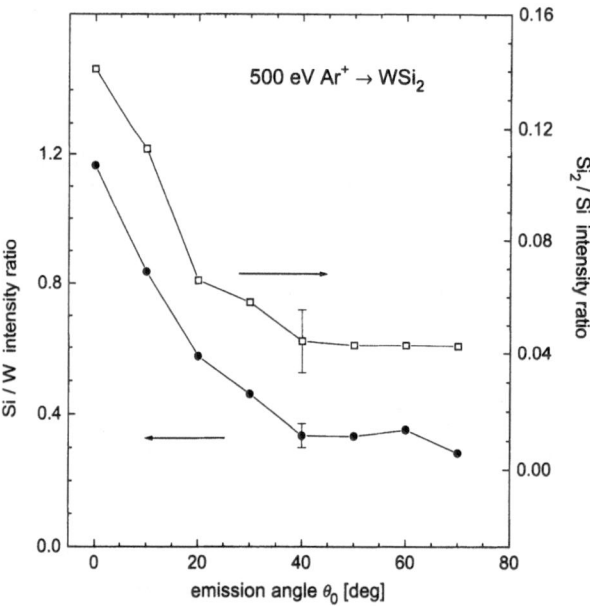

**Fig. 3.14.** The steady-state yield ratio $Y_{Si}/Y_W$ as a function of emission angle for a $WSi_2$ sample irradiated by 500 eV $Ar^+$ ions at normal incidence (*filled circles*) and the Si dimer-to-atom yield ratio $Si_2/Si$ (*open squares*). Data from [3.180]

### 3.3.3 Energy Spectra of Sputtered Species

According to (2.35), the energy spectra of sputtered atoms should peak at $U_i/2$ in the regime of collision cascade sputtering. As discussed in Sect. 3.1.2, experiments and computer simulations have been utilized to derive values of $U_i$ for the components in binary alloys by fitting measured and computed distributions with (2.34).

Szymonski and coworkers [3.34–36, 181] have used this approach intensively, recording energy spectra of several binary alloys. They used a mass-spectrometric technique to obtain mass-selective data of the ejected fluxes under stationary conditions. Apart from the Cu–Zn specimen [3.36] already discussed in the context of surface binding energies, Ag–Au [3.34], GaAs [3.35] and HfC alloys [3.181] were also analyzed. One of the objectives in these investigations was the identification of nonlinear (spike) effects in sputtering by heavy-ion (Xe in this case) impact. The authors proposed that a contribution from such a process was observed in the low-energy portions ($< 1\,\mathrm{eV}$) of the energy spectra of sputtered atoms. While the difficulties of accurate energy measurements below about $1\,\mathrm{eV}$ emission energy may be formidable, the authors observed distinct shifts in the peak positions of energy spectra depending on the alloy composition. As mentioned above, these were employed to establish concentration-dependent surface binding energies. For example, in an $Ag_{0.6}Au_{0.4}$ alloy [3.34] they find $U_{Ag} = 2.1\,\mathrm{eV}$ and $U_{Au} = 3.3\,\mathrm{eV}$, as compared with the values derived from the respective pure-element specimens of $3.1\,\mathrm{eV}$ and $3.8\,\mathrm{eV}$. For the case with the largest mass ratio (HfC), Szymonski [3.181] derived, by fitting the energy spectra to (2.34), values of the parameter $2m$; he established that it is identical for both components and close to zero, while the surface binding energies are $4.8\,\mathrm{eV}$ and $6.7\,\mathrm{eV}$ for C and Hf, respectively. The latter finding would indicate preferential sputtering of C, which is also the much lighter atom.

Vicanek et al. [3.182] studied energy partitioning and particle spectra in multicomponent collision cascades analytically. As a case study, they investigated a binary $Hf_xC_{1-x}$ compound and observed that the particle flux of the lighter species is overstoichiometric, while the flux of the heavy component shows only small deviations from stoichiometry. The energy spectra of sputtered particles, on the other hand, exhibited similar slopes for Hf and C atoms.

Oechsner and Bartella [3.37] determined $m$ from the slope of the recoil spectrum and used the expression $E_{max} = U/(2 - 2m)$ to derive values for $U_i$ in a Ni–W alloy and in the respective elemental specimens. In the alloy, the high-energy fall-off is steeper than in the pure metals. They correlated these data with surface composition investigations by AES.

Rather drastic changes in the peak positions were observed in the sputtering of oxidized metals as compared with the respective clean surfaces. In terms of $U_i$, an increase by up to a factor of 10 (e.g. in the cases of $Ba+O_2$ and $Ca+O_2$) was reported (see [3.27] for a compilation of pertinent data).

Frequently, the increase of $U_i$ for oxidized surfaces is accompanied by a reduction of the total sputtering yield.

Energy spectra from NiW and CuW alloys for bombarding energies in the range from 80 to 800 eV were recorded [3.183] by means of SNMS, utilizing a low-pressure plasma for post-ionization. (The energy distributions of elemental Cu depicted in Fig. 2.16 were obtained with the same instrument.) Owing to the high abundance sensitivity of the technique, energy spectra could be monitored over an intensity range of five orders of magnitude, despite the low impact energies and the correspondingly low sputtering yields. Figure 3.15 shows such distributions for neutral Cu and W atoms sputtered from $Cu_{0.53}W_{0.47}$. Two observations are of note: First, the most probable energy of the Cu atoms is lower than that of the W atoms, although their precise values cannot be derived from the spectra because of an insufficient energy resolution. These different peak positions are in agreement with the predictions of (3.25), on the basis of the cohesive energies of pure Cu ($E_{coh} = 3.54$ eV) and pure W ($E_{coh} = 8.68$ eV). Second, at a given bombarding energy the spectra of the Cu atoms extend to distinctly higher emission energies than those of the W atoms. This finding indicates that for these high emission energies the dominant sputter ejection is due to a primary knockon process from a reflected incident ion on its way out of the sample or another of the specific ejection events discused in Sect. 2.3.2 (see Fig. 2.14). Under such conditions, an $Ar^+$ ion can transfer a higher amount of energy to a Cu surface atom than to a W atom.

A comparison of the Cu spectra from the CuW alloy and from the pure Cu specimen indicates, furthermore, that the energy distributions of the Cu atoms sputtered from the CuW specimen exhibit a somewhat more gradual

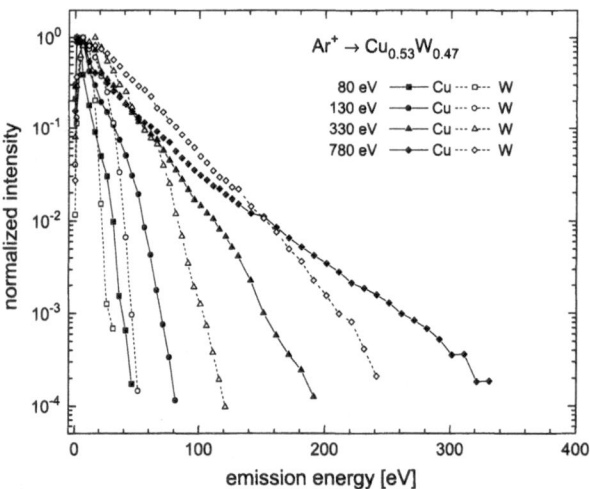

**Fig. 3.15.** Normalized energy spectra of Cu and W atoms sputtered from a hemispherical $Cu_{0.53}W_{0.47}$ sample by $Ar^+$ ions of the indicated energies

decay with increasing emission energy. Assuming the specific emission scenarios mentioned above to be operative, the presence of the heavy W atoms provides a higher energy for the reflected $Ar^+$ projectiles and, hence, a slightly higher amount of final energy transfer to the sputtered Cu atoms.

Energy distributions from multicomponent targets have frequently been recorded by detecting secondary ions. The interpretation of these data in terms of preferential ejection and changes in surface composition (and binding energies) appears rather difficult, however; this is largely due to the complexities of the ionization probabilities of sputtered ions: these are drastically different for different species and commonly exhibit some variation with emission energy (velocity), see Chap. 5.

### 3.3.4 Mass Distributions of Sputtered Atoms and Clusters

The composition changes induced at the surface by ion bombardment, in particular strong concentration gradients within the outermost atomic layers, may have a pronounced influence on the flux of sputtered species. Apart from the angular and energy distributions of the emission (see Sects. 3.3.2 and 3.3.3), the partial yields of atoms and clusters can also be affected. Such effects are expected to be very distinct at low irradiation energies. To validate this notion, Gnaser and Oechsner [3.184, 185] sputtered various binary alloys (CuZn, NiW and CuW) with $Ar^+$ and $Xe^+$ projectiles in the energy range from 30 eV to 1000 eV. Emission-angle-integrated yields of emitted neutral atoms and clusters were determined as a function of bombarding energy by means of secondary-neutral mass spectrometry. For selected impact energies, the transients in the partial yields were monitored. Composition changes in the near-surface region and the fluence dependent evolution of the atomic yields were also investigated for $Ni_{0.92}W_{0.08}$ by means of a binary-collision computer code [3.184], using Ar and Xe projectiles and an energy range comparable to that of the experiments.

The experiments were carried out in the secondary-neutral mass spectrometer described in Sect. A.1. In this arrangement ions are extracted from the post-ionization plasma and accelerated onto the target, which is biased negatively (with respect to the plasma) to effect sputtering. The electron component is utilized to post-ionize the sputtered species on their way through the plasma. The plasma potential ($+30 \pm 2$ V relative to ground) was determined from Langmuir probe measurements, thus establishing rather accurately the ions' impact energies. Hemispherical samples were manufactured from high-purity, polycrystalline alloys of $Cu_{0.63}Zn_{0.37}$, $Cu_{0.53}W_{0.47}$ and $Ni_{0.92}W_{0.08}$. They were mounted on the sample holder of the instrument in such a way that their convex surface faced the entrance aperture of the mass spectrometer and were completely immersed in the plasma, and homogeneous, normal-incidence ion bombardment was achieved for all impact energies. The hemispherical specimen geometry ensures an integration over all emission angles, despite the limited geometrical acceptance angle of the spectrometer. Hence,

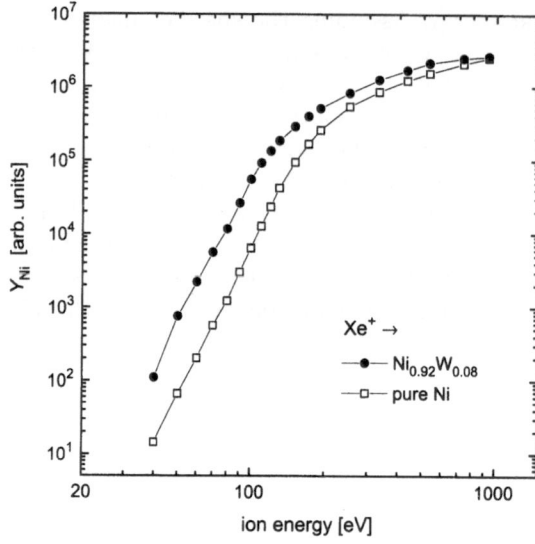

**Fig. 3.16.** The partial Ni sputtering yield $Y_{Ni}$ versus $Xe^+$ ion energy for Ni atoms ejected from pure Ni and from a binary $Ni_{0.92}W_{0.08}$ sample

the oblique emission of sputtered species at low ion energies [3.186] does not influence the measured yield. Such an arrangement was first employed [3.187] for the homogeneous deposition of high-temperature superconducting thin films from a spherical sputter target.

Figure 3.16 shows the yield of Ni atoms sputtered from a $Ni_{0.92}W_{0.08}$ alloy as a function of $Xe^+$ ion energy. The data [3.185] refer to steady-state conditions. As Ni and W atoms constitute the dominant species emitted, their yields are directly proportional to the partial sputtering yields $Y_{Ni}$ and $Y_W$. Because of the extreme sensitivity of the mass-spectrometric technique, $Y_{Ni}$ can be monitored to energies as low as 40 eV. In this low-energy range the partial yield of Ni is considerably higher for the NiW target than for the pure specimen, while the yields merge towards an energy of about 1 keV. Apparently, in this single-knockon regime the presence of heavy W atoms facilitates the reversal of the momentum of incoming $Xe^+$ ions and thus enhances the ejection of Ni atoms. This is corroborated by results obtained for sputtering of Cu atoms from the CuW alloy and from pure Cu.

The authors [3.184, 185] studied also the partial sputtering yields $Y_{Ni}$ and $Y_W$ for steady-state sputtering of the NiW alloy versus the incidence energy for Ar and Xe projectiles, comparing experimental results and data from the computer simulations. Both data sets demonstrate that the W yields decrease more pronouncedly than $Y_{Ni}$ towards lower impact energies. In fact, the simulations indicate a threshold for the sputtering of W atoms by Ar on the order of 50 eV, while for Ni the corresponding value lies beyond the energy range investigated (< 40 eV). For Xe impact the corresponding threshold values are found to be somewhat higher; a similar observation is derived from computations on elemental samples using TRIM [3.24]. Comparison of the experimental results and the simulation data for Ar bombardment reveals that the partial yields of Ni agree (within a factor of about 3) in the energy range from 70 eV to 1 keV; at lower energies, however, the experimental values of $Y_{Ni}$ decrease more strongly. A similar result is found for $Y_W$: here the agreement for energies between 100 eV and 1 keV is even better (a factor of $< 2$), but again a strong divergence between experimental and simulation data is observed for lower energies. Qualitatively the same conclusions can be drawn from a comparison of the data due to Xe impact. A possible reason for the prominent discrepancies at near-threshold energies might be the breakdown of the validity of the binary-collision approximation inherent in the simulation code. If multiple interactions are included in the atom collisions, as will occur in this energy regime, an increase of the sputtering yield is anticipated since specific ejection mechanisms might become dominant [3.188–190]. Another, possibly more serious, uncertainty concerns the choice of the binding energies entering the simulations.

In addition to the atomic species, Gnaser and Oechsner [3.184, 185] determined the intensities of various neutral species sputtered from the aforementioned alloy systems as a function of ion impact energy. All data were recorded at steady-state conditions, i.e. for a near-surface composition which may deviate pronouncedly from the bulk value. In addition to those of the atomic species, the signals of molecules could also be measured to very low energies ($< 100$ eV). As the ionization and dissociation cross sections of the various species are unknown, the data do not necessarily represent the actual partial yields; on the other hand, these are not expected to depend on the impact energy; therefore, the given yield-versus-energy curve for any single species reflects the dependence of the partial sputtering yield on the bombarding energy.

In studies of low-energy sputtering of pure metals [3.191, 192], a distinct feature was recognized in the emission of small molecules (dimers and trimers): down to very low impact energies ($< 100$ eV) the yields of small $n$-atom clusters were found to scale with the $n$th power of the average sputtering yield $Y$ (see (2.42) and (2.44)). Such a dependence was predicted by the double-collision model of cluster formation (Sect. 2.3.4). For a binary system at low ion energies (i.e. when mostly atoms are sputtered), the yields of small $n$-atom clusters $A_n$ would scale with the $n$th power of the average number of A atoms sputtered, $Y_A$:

$$Y_{A_n} \propto (Y_A)^n .\tag{3.43}$$

Owing to the statistical nature of the formation mechanisms governing the emission of small molecules, heteronuclear species are also expected to obey a correlation similar to (3.43). Therefore, the yield of a molecule AB should be proportional to the product of the partial yields $Y_A$ and $Y_B$:

$$Y_{AB} \propto Y_A Y_B .\tag{3.44}$$

Figure 3.17 shows such a dependence for $Ar^+$ bombardment of the $Ni_{0.92}W_{0.08}$ alloy [3.184], depicting the yield ratio $Y_{NiW}/Y_{Ni}$ as a function of $Y_W$. An excellent linear correlation in accordance with (3.44) is found, covering about two orders of magnitude in yield variation.

In addition to those of the heteronuclear molecules, the yields of $Ni_2$ and $Ni_3$ sputtered from NiW also exhibit a dependence in agreement with (3.43), thus confirming the validity of the corresponding data [3.192] on pure nickel, and extending it to the Ni–W binary system. These observations may constitute an indication that the atoms forming, for example, $Ni_2$ do not necessarily originate from contiguous positions on the sample surface: since at bombarding energies of less than 100 eV, where (3.43) is still valid for $Ni_2$, the stationary surface concentration of nickel is 10% or less according to the simulation data; therefore, the average distance between any two Ni atoms on the surface should be fairly large. An alternative possibility would be that even at a low average concentration of one of the components, these atoms may agglomerate into islands of pure material.

In multicomponent systems that exhibit strong Gibbsian segregation, a pronounced concentration gradient will exist in the topmost one or two atomic layers, that is, over the depth from which sputtered species originate. Such a strong variation in concentration may influence the composition of (large) clusters. To investigate this possibility, Lill et al. [3.193, 194] carried out sputtering experiments on various Ga-based eutectic alloys (Ga–In, Ga–Al, Ga–Sn). To detect homo- and heteronuclear clusters sputtered from these specimens they utilized single-photon post-ionization of the neutral species in conjunction with time-of-flight mass spectrometry. On the basis of the finding of Tombrello and coworkers [3.174, 175] that in Ga–In alloys the first layer consists of 94% In (see Sect. 2.3.1), these experiments were devised to make

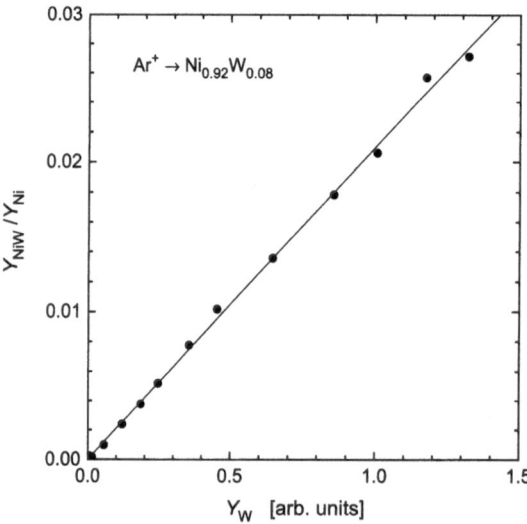

**Fig. 3.17.** The yield ratio of NiW molecules to Ni atoms as a function of the partial yield of W atoms, $Y_W$. The straight line is a linear least-squares fit to the data and agrees with the predictions of (3.44). Data from [3.184]

use of this natural labeling of the first atomic layer to investigate the depth of origin of sputtered clusters and its dependence on the nuclearity $n$ (i.e. the cluster size). Both for the Ga–In and for the Ga–Al eutectic, an increasing gallium content in the flux of sputtered neutral (and ionic) clusters was found for increasing nuclearity. This has been interpreted [3.193, 194] as the consequence of an increasing depth of origin for larger clusters and, furthermore, provides strong arguments against preferential unimolecular fragmentation controlling the composition of large clusters in these systems. The experiments also showed that the abundance distributions of the sputtered GaIn and GaAl clusters are purely statistical (i.e. binomial), reflecting the fact that differences in the photoionization efficiency and in the cluster stability are small. This is probably due to the similar electronic structure of these elements. Evaluating their results for the Ga–In and Ga–Al alloys, Lill et al. [3.193, 194] established the fraction $F_1(n)$ of atoms originating from the first atomic layer and comprising a cluster containing $n$ atoms. While $F_1(n) \sim 0.9$ for $n \leq 3$, $F_1(n)$ is decreasing with increasing $n$, amounting to about 0.5 for $n \sim 10$. The authors argue that the ejection volume grows *laterally* for $n = 1$–3 and *vertically* (with a constant number of atoms per layer) for $n \geq 7$. Very probably, the larger clusters are emitted in special ion-impact events, characterized by a large amount of energy deposited near the surface [3.195].

## 3.4 Sputtering of Isotopic Mixtures

As mentioned in Sect. 3.1.1, preferential sputtering can probably be studied most directly for isotopic mixtures, that is, for elemental samples of two or more isotopes. In fact, the aptness of this approach was realized some three decades ago [3.196] and this approach has been resumed time and again since then. It was common practice to neglect possible differences in binding energies for two isotopes of an element and to ascribe any effects solely to their different masses. As noted by Sigmund and Lam [3.3], this might not be justified a priori; they argued that differences in the effective binding energies may result from the mass-dependent zero-point energies; see (3.13). Such an effect could influence both the ejection from the surface and the relocation in the solid, e.g. due to Gibbsian segregation. Isotopic mass effects in diffusion are well known. Sigmund and Lam suggested also that recoil implantation might play a role, since the lighter isotope would be implanted preferentially from the surface into deeper layers. Apparently, the possible significance of these effects has, as yet, not been investigated.

Considering only mass differences, preferential sputtering effects are expected to be rather small since the relative mass differences of the elements accessible are small ($\sim 10\%$ with the exception of H and He). According to the theory of linear collision cascades (see (3.12)), the ratio of the partial sputtering yields of two isotopes ("1" and "2") of an element with masses $M_1$ and $M_2$ is given by

$$\frac{Y_1}{Y_2} = \frac{N_1}{N_2} \left(\frac{M_2}{M_1}\right)^{2m}. \tag{3.45}$$

$N_1$ and $N_2$ are the natural abundances of these isotopes. (The latter are very uniform among terrestrial specimens and can be taken from tabulations; see e.g. [3.197].) The nonstoichiometric emission of sputtered isotopes is often expressed in terms of the relative fractionation $\delta$, i.e. the relative deviation from these natural abundances,

$$\delta = \frac{Y_1}{Y_2} \frac{N_2}{N_1} - 1. \tag{3.46}$$

Because of the typical smallness of $\delta$, its value is often given in % or ‰ (permil ≡ parts per thousand). To facilitate the comparison of different experimental and simulation data, quite frequently (3.45) has been utilized as a suitable means of parametrization, considering the exponent $m$ as an empirical parameter [3.198]. To avoid confusion with the parameter characterizing the interaction cross section, this effective $m$ value is designated $m_{eff}$. From (3.45) and (3.46),

$$m_{eff} = \frac{1}{2} \frac{\log(\delta + 1)}{\log(M_2/M_1)}. \tag{3.47}$$

Clearly, $m_{eff}$ will depend on the irradiation conditions and may have values beyond the general validity of (3.45) and of $m$ as outlined in Sect. 2.1.1.

In complete equivalence to the situation found for alloys, prolonged ion irradiation will give rise to a change in the surface composition for isotopic mixtures. The isotope that is sputtered preferentially will be depleted at the surface and a stationary state is eventually established; then, the stoichiometry of the sputtered flux corresponds to the abundances in the target. Monitoring the flux during this transient to steady-state conditions provides another means of determining the fractionation $\delta$. Assuming an initially homogeneous composition of the target,

$$\delta = \frac{(Y_1/Y_2)_0}{(Y_1/Y_2)_\infty} - 1, \tag{3.48}$$

where $(Y_1/Y_2)_0$ and $(Y_1/Y_2)_\infty$ are the yield ratios in the limit of zero fluence ($\Phi \to 0$) and for equilibrium ($\Phi \to \infty$), respectively. This approach was utilized recently in several experiments to determine $\delta$ and $m_{eff}$ for various elements and bombardment conditions (Sect. 3.4.1). Experiments performed on the sputtering of isotopic mixtures [3.196, 199–209] were originally carried out at high bombarding fluences [3.196, 199–202]. These are generally limited by the fact that under equilibrium conditions the composition of the emitted particle flux will be identical to that of the bulk and any remnants of preferentiality will be found only in angular or emission-energy variations and/or in deviations of the surface (layers) from the bulk composition. The latter effect appears difficult [3.208] to analyze, however. By contrast, low-fluence experiments [3.203–207] sample (also) the ejected flux in the limit of zero fluence and specifically monitor the aforementioned gradual transient from the undisturbed state towards equilibrium to derive the amount of the preferential sputtering $\delta$; its value should be comparable directly with the-oretical predictions of preferential sputtering following from (3.45), namely $(M_2/M_1)^{2m}$. The derivation of preferential emission according to (3.48) is useful in that it requires only a relative isotope measurement and avoids any problems associated with an absolute determination of isotopic ratios. Experiments of this kind were performed by detecting the ejected species either directly by means of secondary-ion [3.203, 204] or secondary-neutral [3.205, 206] mass spectrometry or by analyzing the sputtered material de-posited on a suitable collector [3.207]. The results obtained in these ways produce a distinct preferential ejection of the lighter isotopes for a variety of elements.

The first experiments [3.196] addressing isotope effects in sputtering, per-formed in the early 1960s, investigated the sputtering of Li isotopes from lithium metal. The sputtered material was collected on quartz and then transferred to a surface ionization mass spectrometer for isotopic abundance determinations. For an incident fluence of $5.3 \times 10^{16}$ ions/cm$^2$, enrichments of the lighter isotope by 19‰ and 16‰ were reported for 5 and 20 keV Ar$^+$ bombardment, respectively.

Wehner and his coworkers [3.167, 199] carried out a collector experiment on isotope sputtering using very low bombarding energies (60–100 eV Hg$^+$). For high-fluence conditions these authors observed a strong angular varia-tion of up to 60‰ for the irradiation of Cu, Mo, W and U targets. These pronounced variations were considerably reduced (to $\leq 6$‰) when the im-pact energy was raised to 300 eV. Owing to the low energies employed in this experiment, recoil implantation and/or surface scattering of projectiles may dominate the observed isotope effects.

A different approach [3.200] was used to study changes in the isotopic composition of Mg produced by 2 keV He$^+$ impact. Contrary to most other experiments, the ejected flux was not investigated, but, rather, the remaining surface after bombardment. Thin layers (10–100 nm) of Mg were deposited onto a glass substrate and, after irradiation, were (partially) dissolved and analyzed in a mass spectrometer. The Mg remaining in the substrate was en-riched in the *heavier* isotope, indicating that again the *lighter* isotopes were preferentially emitted. The $^{25}$Mg/$^{24}$Mg ratio was enriched by up to 10‰ rel-ative to the normal value for fluences ranging from $10^{17}$ to $10^{18}$ cm$^{-2}$. Since the dissolved Mg layer was probably thicker than the layer in which isotopic abundance variations due to ion impact occur, a pronounced dilution of sput-tering effects can be expected. Quantitative conclusions on the magnitude of the initial enrichment produced in this experiment [3.200] are therefore hard to establish.

Russell et al. [3.201] measured the isotopic composition of Ca sputtered from a variety of Ca-bearing minerals by 130 keV N$^+$ and 100 keV N$_2^+$ beams.

Sputtered material collected on foils was analyzed by thermal-ionization mass spectrometry, providing a precision for the $^{40}Ca/^{44}Ca$ ratio of $\sim 0.3\%_{oo}$. At fluences of, typically, $1.5 \times 10^{17}$ cm$^{-2}$, the $^{40}Ca/^{44}Ca$ ratio of the collected material was greater than the normal value by $11\%_{oo}$–$21\%_{oo}$. At higher fluences, the enrichment of $^{40}Ca$ rapidly dropped to zero. Since the lowest fluences exceeded $\sim 10^{17}$ cm$^{-2}$, these measurements [3.201] did not actually determine the magnitude of the light-isotope enrichment for very low fluences but, rather, integrated the isotopic shift over a certain fluence range. The magnitude of the $^{40}Ca$ enrichment would, therefore, be anticipated to be smaller than the light-isotope enhancement measured in the present-day low-fluence experiments. It is difficult to extract an accurate fluence dependence of the enrichment from the data of Russell et al. [3.201]. For the sputtering of a plagioclase sample with $5.8 \times 10^{17}$ ions/cm$^2$ the collected material was enriched by $21\%_{oo}$ in the lighter isotope. This fluence is roughly a factor of 10 *higher* than typical fluences necessary to establish steady-state conditions in these experiments. This observation points to a strong *dilution* of the preferential sputtering effects reported by these authors [3.201].

More recently, Baumel et al. [3.204] analyzed preferential sputtering of B isotopes at low fluences by monitoring the flux of secondary ions. Employing comparatively high-energy projectiles (100 keV Ar$^+$ and Ne$^+$), these authors found an initial enrichment $\delta$ of the lighter isotope of $51.8 \pm 1.8\%_{oo}$ for Ar$^+$ and of $46.1 \pm 2.4\%_{oo}$ for Ne$^+$ for normal emission. The corresponding fluences to reach equilibrium were determined as $3 \times 10^{17}$ Ar$^+$/cm$^2$ and $2 \times 10^{18}$ Ne$^+$/cm$^2$, respectively. These rather high values apparently reflect the much larger penetration depth of the high-energy projectiles and a reduced sputtering yield as compared to the ion energies considered in this book. On the other hand, the values observed for $\delta$ fall in the range derived from several other low-fluence experiments (Sect. 3.4.1).

Ackermans et al. [3.208] determined the steady-state surface composition of a natural boron specimen by means of low-energy ion-scattering spectroscopy. Using 3.5 keV $^4He^+$ ions for sputtering, they observed a *depletion* of the *lighter* isotope ($^{10}B$) in the surface relative to the bulk, in accordance with the preferential sputtering of the lighter species. For the magnitude of the observed enrichment a value $N\left(^{10}B\right)/N\left(^{11}B\right) = 0.19 \pm 0.03$ was reported, while the bulk ratio was $N^b\left(^{10}B\right)/N^b\left(^{11}B\right) = 0.25$. In terms of (3.46), this indicates an enrichment factor $\delta$ for the lighter isotope of about 32%. Despite the large uncertainties, this huge value clearly has to be ascribed to the very light projectile, for which single-knockon events will govern sputtering.

Using secondary-neutral mass spectrometry, Bieck et al. [3.209] analyzed, for Ar$^+$ ions of 40 to 400 eV, the dependence of the isotopic flux on the emission energy. This was accomplished by energy-selecting the sputtered flux before detection in the mass spectrometer. For Cu and Mo isotopes the lighter isotope(s) are ejected with a somewhat higher average energy. Figure 3.18 exemplifies these data, showing the ratio of sputtered Cu isotopes

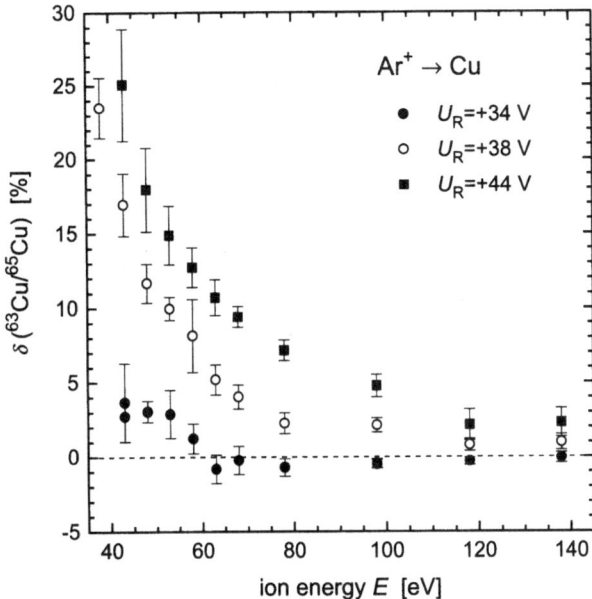

**Fig. 3.18.** The relative fractionation $\delta$ of sputtered Cu isotopes versus the $Ar^+$ bombarding energy. The parameter is the setting of the retarding potential in a retarding-field analyzer With increasing retardation, atoms sputtered with low energy are progressively suppressed; the deviation of the isotope ratio indicates a higher emission energy for the lighter isotope. Data from [3.209]

as a function of the bombarding energy for different retardation potentials. The effect is most pronounced at very low bombarding energies, while the energy fractionation becomes negligible for energies above $\sim 150\,\mathrm{eV}$. The data [3.209] show that for a low retardation voltage ($U_\mathrm{R} = +34\,\mathrm{V}$) which allows the transmission of the complete emission-energy range of the sputtered atoms, the isotopic ratio is very close to the natural abundance ratio (dotted line in Fig. 3.18) even for the lowest irradiation energies. With increasing retardation potential, the low-energy portion of the sputtered flux is blocked and a preferential detection of the lighter isotope is found. This finding indicates that the lighter isotope is emitted with a larger average energy and, therefore, is less efficiently suppressed by the retarding-field energy analysis. This effect is most drastic at the lowest bombarding energies (cf. Sect. 3.4.3).

Employing the same technique but a slightly different instrument, Bock [3.180] has recently measured the isotopic ratio of Cu isotopes as a function of emission angle for very low $Ar^+$ bombarding energies (below $100\,\mathrm{eV}$) under steady-state sputtering. Under this condition, the total emitted flux must be stoichiometric but angular variations may persist. These data show a preferential ejection of the lighter Cu isotope in the forward direction while at oblique angles the preferentiality is reversed to ensure complete stoichiometry. This isotopic enhancement becomes the more pronounced the lower the impact energy is (a relative enrichment of up to 10% at 90 eV). Although at higher energies the differences are reduced, even for 500 eV ion bombardment some angular-dependent isotopic fractionation exists.

### 3.4.1 Transient Variations of the Isotopic Flux Composition

The first experiments of this type monitoring the isotopic variations at very low fluences (about $1 \times 10^{15}$ ions/cm$^2$) were performed employing secondary ions [3.203]; specifically, Li$^+$ from LiAlSi$_2$O$_6$, Ti$^+$ from TiO$_2$ (rutile), Ga$^+$ from GaAs and Mo$^+$ from Mo metal (using specimens with natural isotopic composition and one enriched in $^{92}$Mo and $^{100}$Mo) were detected. The experiments were carried out on a secondary-ion mass spectrometer using 14.5 keV O$^-$ primary ions at an incidence angle of 26° off normal. The result of these first studies is the observation of a pronounced enrichment of the lighter isotope of each element in the initially sputtered flux relative to the steady-state isotope ratio. The magnitude $\delta$ (see (3.46)) of the enrichment is not constant but varies from element to element as discussed below. The initial $^6$Li$^+$/$^7$Li$^+$ ratio is substantially enriched in $^6$Li$^+$; the enrichment decreases monotonically to zero with increasing sputtering. In the limit of zero fluence the $^6$Li$^+$/$^7$Li$^+$ ratio is enriched by about 54 permil in the lighter isotope relative to the steady-state value ($^6$Li/$^7$Li $\approx 8.1 \times 10^{-2}$). The results for all samples and isotope ratios investigated in these experiments are compiled in [3.203].

To substantiate further the finding of a preferential light-isotope sputtering when monitoring *secondary ions*, experiments were performed [3.205, 206] which recorded the isotopic flux of *sputtered neutrals*. Again the transient behavior between a virgin state and equilibrium conditions was of interest; in addition, emission-angle-selective data of isotopic fractionation were recorded. The measurements were carried with a secondary-neutral mass spectrometer. In the separate-bombardment mode of operation, 5 keV Ar$^+$ ions were used as projectiles. In analogy with the data mentioned above, the isotopic composition in the sputtered flux of neutrals was monitored as a function of bombarding fluence. Thus, by comparing yield ratios of two isotopes for zero fluence and steady-state conditions, values of $\delta$ can be derived. For a Mo target, the data exhibit an enrichment of the lighter isotope in the flux which gradually diminishes until the steady state, established for a bombarding fluence of $6 \times 10^{16}$ ions/cm$^2$, is reached. The initial enrichment of the lighter isotope derived from several runs amounts to $\delta = 1.051 \pm 0.004$ and is thus in excellent agreement with the value determined from the secondary-ion experiment.

Similar results have been obtained for Ge: The preferential sputtering of the lighter isotope was derived again from the initial enhancement of this species and yields a value of $\delta = 1.052 \pm 0.005$ from a set of different runs. Not surprisingly, the magnitude of $\delta$ is essentially identical to that obtained for Mo since the mass differences of the respective isotopes are identical and the masses are also not too different. Steady-state sputtering was reached for a fluence of about $2 \times 10^{16}$ ions/cm$^2$. With a sputtering yield $Y = 2.5$ atoms/ion [3.206], this corresponds to an eroded depth of $\sim 11$ nm. Since in the present case the projected range is $\sim 5.5$ nm, a depth two times larger has to be eroded in order to establish sputter equilibrium.

Isotopic transients were recorded also in a collector experiment [3.207]. The sputtering targets were molybdenum foils nominally containing 50 at% each of $^{92}$Mo and $^{100}$Mo. Ar$^+$ and Xe$^+$ ion beams of 5 and 10 keV were used for sputtering. Ultrapure graphite sheets (0.13 mm thick) served as the collector material. Secondary-ion mass spectrometry was utilized for analyzing the Mo deposits on the collector foils. Figure 3.19 shows the enrichment $\delta$ as a function of the bombarding fluence for Xe$^+$ and Ar$^+$, for material sputtered into a direction $\sim 17°$ from the target normal. Large enrichments were observed at the lowest fluences in all cases [3.207]; this enrichment was as high as 53‰ for the 5 keV Xe$^+$ bombardment at a fluence of $3 \times 10^{14}$ ions/cm$^2$. The enrichment fell off rapidly with increasing fluence, approaching essentially zero at fluences of $\sim 2 \times 10^{16}$ ions/cm$^2$ for all but the 10 keV Xe$^+$ ion impact,

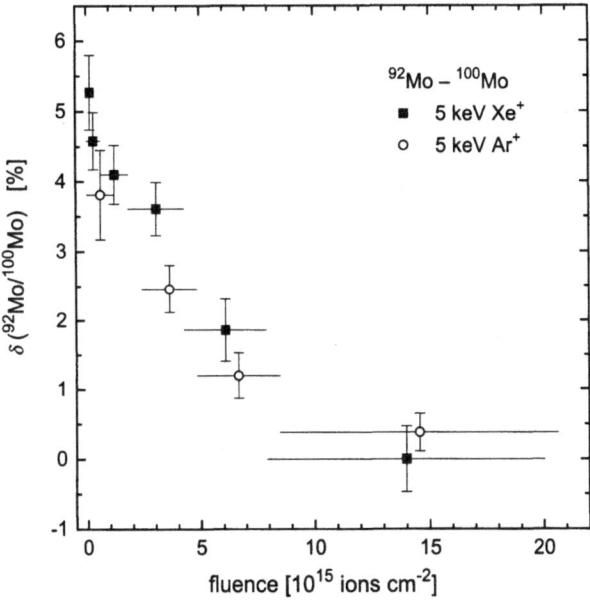

**Fig. 3.19.** Isotopic fractionation $\delta$ of the flux of Mo isotopes ($^{92}$Mo/$^{100}$Mo) sputtered into near-normal directions ($\theta_0 = 17°$) from a Mo specimen composed of roughly equal amounts of the two isotopes $^{92}$Mo and $^{100}$Mo versus the bombarding fluence for 5 keV Xe$^+$ and Ar$^+$ projectiles. Each symbol is plotted at the midpoint of the fluence range used for the corresponding collection; the horizontal bar through each symbol indicates this range. The uncertainties indicated by the vertical error bars are $\pm 2\sigma$. Data from [3.207]

which was not conducted to as high a fluence. The data are all qualitatively similar, exhibiting low-fluence enrichments ranging from 30‰ to 50‰, and having "decay constants" corresponding to fluences of about $5 \times 10^{15} \, \text{cm}^{-2}$. Within the experimental uncertainty, the two $\text{Ar}^+$ bombardments produced essentially identical results.

### 3.4.2 Emission-Angle-Dependent Isotopic Flux

The dependence of the isotope ratio on the emission angle was investigated using both sputtered-neutral mass spectrometry [3.206] and the collector experiment [3.207]. In the former case, values of $\delta$ were obtained for $\theta_0$ ranging from 0° to 80° (with a concurrent change of the incidence angle) for the irradiation of Ge with 5 keV $\text{Ar}^+$ ions. The initial enrichment relative to steady state, $\delta$, as a function of the emission angle $\theta_0$ for the $^{70}\text{Ge}/^{76}\text{Ge}$ isotope pair exhibited only a weak variation (less than 1%) for $\theta_0 \le 60°$, but decreased more pronouncedly for larger emission angles. It can be concluded that the initial isotopic enrichment (relative to equilibrium) in the sputtered flux is largely independent of the emission angle, with the possible exception of very oblique ejection directions from the surface normal.

In the collector experiment [3.207] the angle-selective measurements were carried out as described in the previous section. The angular dependence of the enrichment $\delta$ $(^{92}\text{Mo}/^{100}\text{Mo})$ was investigated for $\text{Ar}^+$ and $\text{Xe}^+$ ions with energies of 5 and 10 keV for several different bombarding fluences. The 5 keV $\text{Xe}^+$ bombardment exhibited strong preferential sputtering of the light isotope into directions close to the surface normal compared to oblique directions, even at the lowest fluences: $\delta$ was roughly 30‰ greater at normal emission angles than at oblique angles in this case. The 5 keV $\text{Ar}^+$ and 10 keV $\text{Xe}^+$ bombardments exhibited less pronounced, but still resolvable (at the 95% confidence level) preferential emission of the light isotope into normal directions over oblique directions at the lowest fluences: the near-normal enrichment was $\sim 18$‰ for 5 keV $\text{Ar}^+$ and $\sim 12$‰ for 10 keV $\text{Xe}^+$. The 10 keV $\text{Ar}^+$ bombardment, on the other hand, exhibited no statistically significant angular dependence of the preferential emission of the light isotope. Some caution should be exercised in making direct quantitative comparisons between the low-fluence data for the various projectiles because of the different fluence ranges and sputtering rates involved. The data for 5 keV $\text{Ar}^+$ impact at an ion fluence of $1.2 \times 10^{15} \, \text{cm}^{-2}$ are shown in Fig. 3.20; they are depicted together with results from computer simulations which are discussed in the following section.

For both the 5 keV $\text{Ar}^+$ and the 5 keV $\text{Xe}^+$ sputtering, the shape of the angular dependence of the light-isotope enrichment did not change drastically as the bombarding fluence increased: rather, the effect was essentially to reduce the magnitude of the enrichment by a nearly constant additive factor at all angles. The 10 keV bombardments appeared to produce a more pronounced change in the shape of the angular dependence as the fluence

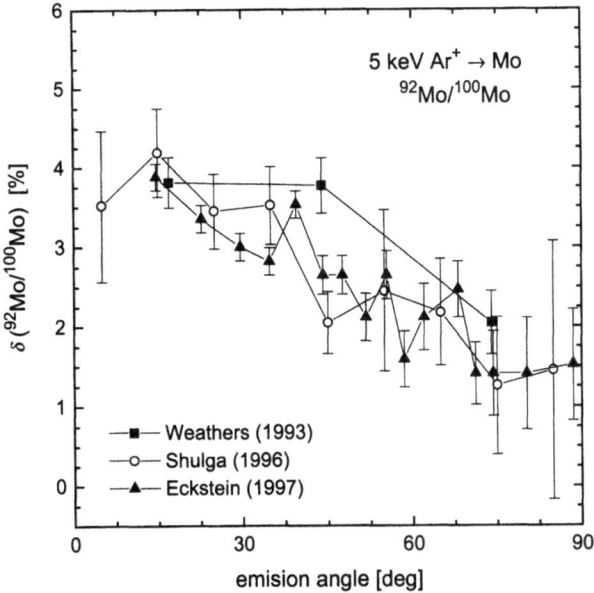

**Fig. 3.20.** Isotopic fractionation $\delta$ for the two isotopes $^{92}$Mo and $^{100}$Mo versus emission angle for 5 keV Ar bombardment of Mo. The results from experiments by Weathers et al. [3.207] are compared with those from computer simulations obtained with the TRIM code by Eckstein [3.221] and from MD simulations by Shulga and Sigmund [3.219]. The data refer to the low-fluence limit. The errors refer to $\pm 1\sigma$

increased, with the difference between the normal and oblique enrichments increasing with fluence. A detailed compilation of the angle-selective values of $\delta$ for the various projectile ions and energies is given in [3.207].

### 3.4.3 Computer Simulations of Isotope Sputtering

Computer simulations of isotope sputtering were carried out using both molecular dynamics (MD) algorithms and binary-collision approximation (BCA) codes. The MD simulations performed by Shapiro et al. [3.210–212] usually employed systems with artificially enhanced mass differences ("pseudo copper") in order to improve the statistical significance of the results: the latter was recognized as the major limitation of this approach [3.16]. The computations of this group were carried out for a variety of target configurations (two- and three-isotope crystals and liquids) and compositions (mass differences from 5% to 30%). The results, however, appear somewhat inconsistent: for example, for a two-isotope, *crystalline* sample [3.211] with a 10% mass difference of the isotopes, a value $\delta = 55.5°/_{oo}$ is found, which splits into values of 133.7°/$_{oo}$ and $-22.2°/_{oo}$ for emission angles $\theta_0 \leq 35°$ and $\theta_0 \geq 35°$, respectively. The same numbers for a three-isotope *crystalline* target [3.212] are 27°/$_{oo}$ for the total fractionation, 52°/$_{oo}$ for $\theta_0 \leq 40°$ and $-13°/_{oo}$

for $\theta_0 \geq 40°$. Surprisingly, a three-isotope *liquid* sample (with a 10% mass difference) [3.212] exhibits the same total fractionation ($\delta = 25°/_{oo}$), but the angular effect is reversed, i.e. $\delta = -14°/_{oo}$ for $\theta_0 \leq 40°$ and $\delta = 130°/_{oo}$ for $\theta_0 \geq 40°$. Obviously, crystal lattice effects in sputtering (see Sect. 2.3.3) do influence the angular dependence of the isotope fractionation in these simulations. The angle-integrated fractionation, on the other hand, exhibits a consistent preferential ejection of the lighter isotopes, in agreement with the experimental observations. Also, its magnitude ($\delta$ ranging from $27°/_{oo}$ to $62°/_{oo}$ for a mass difference of 10% in the various simulation runs) is compatible with the experimental findings. The data [3.211, 212] also indicate the light-isotope enhancement to decrease with increasing projectile mass and with decreasing impact energy. By contrast, no consistent trends in the overall magnitude of the fractionation are apparent in the experimental data [3.205–207] as a function of projectile mass and energy. Shapiro et al. [3.210] have pointed out that their simulations show a large momentum asymmetry in the collision cascades leading to sputtering, with the light isotopes having a significantly larger momentum component directed back toward the surface than the heavy isotope. The intuitive explanation for this behavior is that the light isotopes can undergo large-angle deflections in single collisions with heavy atoms in the targets, whereas the same mechanism is not available for heavy atoms, which must undergo several collisions to change direction by more than 90°. Hence, a larger fraction of the light atoms with initial momenta directed away from the surface will scatter back toward the surface. In pointing to the momentum asymmetry as the dominant source of the isotopic fractionation, these authors have indicated that the assumption of an isotropic velocity distribution in the linear cascade theory may be responsible for the discrepancy apparently existing between the theory and the simulations. The notion of a momentum asymmetry in the collision cascade has, however, been questioned recently by Sigmund and Sckerl [3.21].

Eckstein and Biersack [3.25] utilized the BCA code TRIM to simulate sputtering of boron isotopes. For 2 keV Xe projectiles they found $\delta = 31.4°/_{oo}$, in close agreement with (3.45), provided $m \approx 1/6$ is chosen: this value of the power-potential exponent represents the KrC interaction potential (see Sect. 2.1.1) employed in that code. On the other hand, these simulations yield $\delta = 63°/_{oo}$ for 2 keV $^4$He bombardment, thus demonstrating a distinct projectile dependence. In addition, the computations reveal a pronounced variation of $\delta$ for the Xe species with impact energy: for energies below $\sim 700$ eV the isotopic fractionation reverses sign, that is to say, the *heavier* B isotope is sputtered preferentially. These results are apparently due to the large mismatch in mass (between B and Xe) and/or the very oblique incidence angle (60°) used in these simulations.

Monte Carlo simulations were carried out for isotopic mixtures of both Ge and Mo by Urbassek and coworkers [3.20, 213, 214]. The authors reported [3.20], for the first time, a strong enrichment of the lighter component in the upper parts of the emission-energy spectrum (above $\sim 100\,\text{eV}$). By contrast, only a very minor dependence of $\delta$ on the emission angle is observed. For $5\,\text{keV}$ Ar bombardment of Mo the simulation yields $\delta = 29.5 \pm 1.4\%_{\text{oo}}$, and for $5\,\text{keV}$ Mo impact $\delta = 26.9 \pm 1.1\%_{\text{oo}}$ is found, thus establishing a distinct, albeit small projectile-mass dependence. For the Ge target very detailed simulations were performed [3.213, 214] which were aimed at following the evolution with fluence of the isotopic composition of the near-surface region; furthermore, the respective contributions of cascade mixing and recoil implantation to the atomic relocation process were investigated. The results showed that, with increasing bombarding fluence, light isotopes are depleted close to the surface, and are enriched deeper inside. Under steady-state conditions the lighter isotope is depleted at the surface by about $16\%_{\text{oo}}$ relative to the bulk value. While the fluence to reach equilibrium of $(2\text{–}3) \times 10^{16}\,\text{cm}^{-2}$ compares quite satisfactorily with the experimental findings, the zero-fluence enrichment factor $\delta = 3\%$ is distinctly smaller. Interestingly enough, the authors [3.213] demonstrate that by introducing an initial deviation of the surface stoichiometry from the bulk composition (i.e. increasing the light-isotope abundance at the surface from 20.7% to 21% and depleting the heavy-isotope one correspondingly) the simulation can reproduce the experiments both in the initial enrichment and in the fluence dependence. The possible mechanism which can cause such a stoichiometric distortion is not easily identified, however. A different computation [3.214], which evaluates analytically the relocation cross sections and solves the evolution equations numerically produces, again for the sputtering of Ge isotopes, a distinctly larger light-isotope enrichment ($\delta \sim 43\%_{\text{oo}}$). The reason for these discrepancies is not completely clear.

Using the TRIM code, Eckstein [3.215] carried out simulations for a Mo target and Ar and Xe projectiles of $5\,\text{keV}$ and $10\,\text{keV}$. These computations were specifically performed in response to the experimental data [3.207] presented in Sect. 3.4.2. For both of these projectiles and energies, the results of the simulation runs showed a convincing overall agreement with the corresponding experimental data: in particular, the simulations demonstrated the occurrence of angle-dependent isotopic fractionation in the limit of zero fluence and the persistence of these effects towards equilibrium sputtering conditions. As mentioned above, the total angle-integrated values of $\delta$ are in close agreement with (3.45) with the choice of $m = 1/6$. Within the very small statistical uncertainties of the computations ($2\sigma \sim 2\%_{\text{oo}}$), the values of $\delta$ are independent of the projectile species (Ar or Xe) and of the impact energy (5 or $10\,\text{keV}$).

Apart from the variations in fractionation related to the emission angle, the computations by Eckstein [3.215] also revealed a pronounced effect of emission energy: In accordance with Urbassek's data [3.20], preferential light-isotope sputtering increases drastically for ejection energies above $\sim 100\,\text{eV}$ (e.g. $\delta \sim 100\%_{\text{oo}}$ for $E_0 = 500\,\text{eV}$). Furthermore, the simulations demonstrate that collisions due to primary-knockon atoms, although they constitute only a small fraction of the total, produce a much larger isotope fractionation than those due to secondary-knockon atoms; for the energies and projectiles under consideration, the latter are the dominant type of interaction responsible for sputtering. By employing the dynamic version of the simulation code, which accounts for target composition changes, the author observed that steady-state conditions are reached for fluences of about $1 \times 10^{16} \, \text{cm}^{-2}$; this number is again in good agreement with the experimental data.

A dependence of the isotopic composition on the emission angle was also found in the Monte Carlo simulations of Zheng et al. [3.216, 217]. In a detailed study, these authors were able to reproduce quantitatively most of the experimental results of Weathers et al. [3.207], in particular the angular- and fluence-dependent isotopic fractionation. These data are also in close agreement with those of Eckstein [3.215] mentioned above.

The preferential sputtering of a series of model alloys CuX by 1 and 3 keV Ar and Cu atoms was studied via MD simulations by Gades and Urbassek [3.218]. They used pair potentials and many-body potentials of the tight-binding form. As component X they employed several model species: A *heavy* Cu with twice the natural mass and *weakly* and *strongly* bound Cu, with cohesive energies which varied between 50 and 200% of that of natural Cu. The preferential sputtering of Cu is well described by the analytical theory (see (3.45)), if the power exponent $m$ is derived from the appropriate potential. The distribution of sputtered heavy Cu atoms is found to exhibit a higher energy loss and more oblique emission angles; the authors [3.218] attribute this finding to the last collision that the particles experience in sputtering. The preferential sputtering of the more weakly bound species is more pronounced than in the analytical estimate (3.11). This feature, the authors conclude, may be connected to the anisotropy of the collision cascade in the vicinity of the surface; these effects are somewhat reduced at 3 keV bombarding energy. Furthermore, Gades and Urbassek report a shifting of the energy distribution of the strongly bound sputtered atoms to higher emission energies, but no binding effect was observed on the angular distribution or the depth of origin of the sputtered particles.

Shulga and Sigmund [3.198, 219, 220] carried out a very intensive study of isotope sputtering using binary-collision and molecular-dynamics simulation, emphasizing primary effects. They investigated $^{100}\text{Mo}$–$^{92}\text{Mo}$ and, to enhance statistics, artificial $^{100}\text{Mo}$–$^{50}\text{Mo}$ targets under Ar bombardment in the range from 100 eV to 100 keV. Apart from a careful evaluation of the influence of various modifications (a total of 18) applied to their standard setup on the computed results, they determined partial sputtering yields of different Mo isotopes as a function of bombarding energy, of the type of projectile (apart from Ar, B and U were also used in some runs) and of the emission angle and energy. Further investigations concerned details of the emission process (escape depth of sputtered atoms), recoil generation and the relation between knockon energy and recoil angle. On the basis of their results, Shulga and Sigmund [3.219] stress the possible limitations of binary-collision simulations, in particular for low energies where simultaneous collisions are often important. In addition, the target–target interaction potential and the surface barrier may produce noticeable effects on the yield ratios. At high bombardment energies (upper keV range) the yield ratio approaches an asymptotic limit, in agreement with the value of $m$ characteristic of the interaction potential adopted (see (3.45)). This is shown in Fig. 3.21, which depicts the isotopic fractionation $\delta$ and the derived value of $m_{\text{eff}}$ as a function of Ar impact energy [3.219]. At energies $\leq 2\,\text{keV}$, $m_{\text{eff}}$ strongly increases and reaches $m_{\text{eff}} \sim 0.7$ at $100\,\text{eV}$; at this energy the fractionation is $\delta \sim 12\%$. A dependence on the emission angle prevails even for the highest energies. These data are presented in Fig. 3.22 [3.219]: the yield ratio and $m_{\text{eff}}$ are plotted versus the polar emission angle for three different bombarding energies (0.1, 1 and 10 keV). Even for the higher energies, a distinct variation of the yield ratio is observed for varying ejection angle. In agreement with the experimental and previous simulation results, the lighter isotope is sputtered preferentially in the forward direction. (The values of $\delta$ obtained in these simulations for 5 keV Ar impact are included in Fig. 3.20.) The simulations indicate the importance of sputter emission from a greater depth: lighter isotopes have a higher probability to escape from deeper within the target as compared to the heavier species.

At low and near-threshold energies (around $100\,\text{eV}$ and below), sputtering is distinctly different from the keV regime and this is reflected also in the respective isotope data. The sputter preferentiality is strongly dependent on the type and energy of the bombarding species and varies dramatically with the emission angle. The physical processes leading to these observation are rather well known and have been studied by simulations; generally, they can be ascribed to the increasing dominance of primary recoil atoms in the sputtered flux and, furthermore, impact events resulting in sputtering may not be typical in that energy range. According to Shulga and Sigmund [3.219], the most obvious feature leading to nonstoichiometric sputtering lies in the collision kinematics of binary scattering (in the laboratory frame): the scattering angles for a light isotope scattered by a heavy one range from 0 to 180°, while a heavy isotope can be scattered by a light one only by less than 90°. This apparently holds even if many-body interactions are considered. These authors additionally note the possible importance of preferential recoil implantation at low irradiation energies and experiments performed under steady-state conditions. Shulga and Sigmund observe that the range of intermediate energies constitutes a smooth transition between the large isotope effects at low energies and the moderately small effects at high energies. They were able to fit their yield ratios with a power law of the form

$$Y_1/Y_2 \propto (M_2/M_1)^{2m} \left(1 + \text{const}/E\right) . \tag{3.49}$$

Such a fit is included in Fig. 3.21 (dashed line) for the depicted values of the fractionation. They conclude that the behavior at intermediate energies is determined by primary-knockon sputtering, which shows strong preferentiality

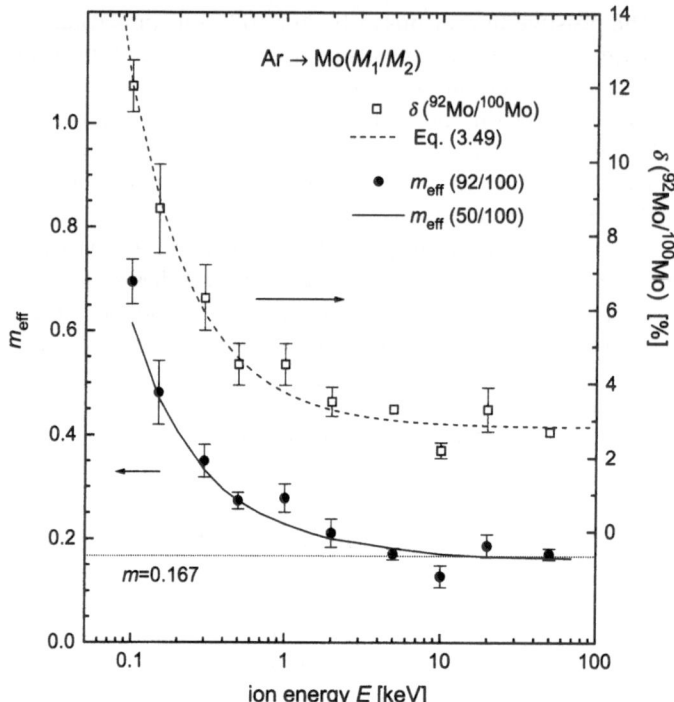

**Fig. 3.21.** Computer simulation data for the isotopic fractionation $\delta$ (*open squares*, right-hand scale) and the corresponding power exponent $m_{\text{eff}}$ (*closed circles*, left-hand scale) as a function of Ar impact energy for a 1:1 isotopic mixture of $^{92}$Mo and $^{100}$Mo. Also included are the data for $m_{\text{eff}}$ for a hypothetical 1:1 mixture of isotopic masses 50 and 100 (*solid line*) and a fit according to (3.49) to the values of $\delta$ (*dashed line*). Data from [3.219]

but has low probability, while cascade processes dominate but exhibit little preferential sputtering.

Employing vastly parallel computing, Eckstein and Dohmen [3.221] have recently performed the first simulations of isotope sputtering on a Mo target composed of the seven naturally occurring isotopes. They studied a wide range in energy (50 eV to 20 keV) and ion mass (all rare gases from He through Rn) in both static and dynamic (i.e. fluence-dependent) computations. Owing to their enormously enhanced precision (typically, $10^7$ to $5 \times 10^8$ incident projectiles were used for a calculation), very detailed investigations into the influence of the various bombardment and emission parameters became feasible. Furthermore, the relative contributions of the different types of recoils (primary or secondary, ion moving into or out of the target) were evaluated. The low-fluence data confirm the previous observations of a (strongly) preferential sputtering of the lighter isotope which is close to linear with the differences in isotopic masses. The magnitude of the fractionation is much larger for He

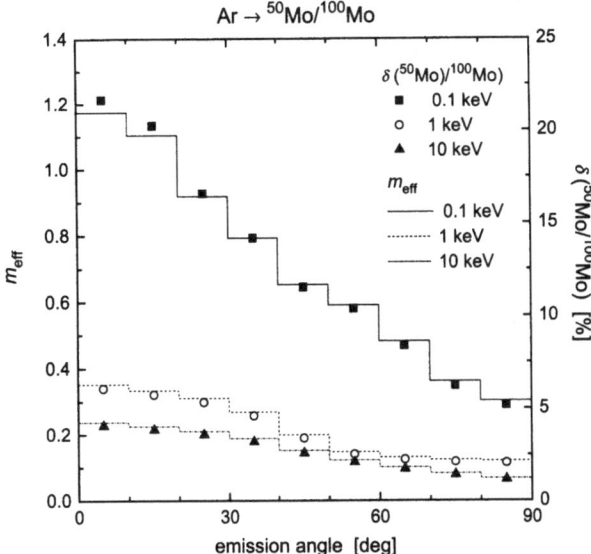

**Fig. 3.22.** Simulation data for the isotopic fractionation $\delta$ (*symbols*, right-hand scale) and the extracted power exponent $m_{\text{eff}}$ (*lines*, left-hand scale) versus emission angle for an isotopic 1:1 mixture of hypothetical $^{50}$Mo and $^{100}$Mo. Ar bombarding energies of 0.1, 1 and 10 keV were used in the computations. Data from Shulga and Sigmund [3.219]

ions than for the other species; the reason is that for He, primary-knockon processes are essentially the only ones which contribute to sputtering and, generally, these contribute the most to the preferential ejection. The results demonstrate a pronounced fractionation at low impact energies (e.g. 24% enrichment in the isotopic pair $^{92}$Mo/$^{100}$Mo at 50 eV) and for atoms ejected with high emission energies (again a consequence of primary-knockon contributions). The fractionation $\delta$ as a function of the emission energy is depicted in Fig. 3.23 for the $^{92}$Mo/$^{100}$Mo isotopic pair and several Ar bombardment energies [3.221]. As the emission energy approaches the maximum value for the sputtered atoms, the preferential ejection of the lighter isotope increases drastically. This is in qualitative agreement with the experimental findings for Cu shown in Fig. 3.18. The dependence of preferential sputtering on the emission angle is seen for all energies and projectiles, with the lighter isotope ejected preferentially in near-normal directions. The corresponding data for 5 keV Ar bombardment are included in Fig. 3.20.

The dynamic studies show transient changes in the partial sputter yields and in the surface composition with fluence. The fluence required to reach steady-state conditions is ion- and energy-dependent and can be reasonably well described by $\Phi_{\text{eq}} = NR/Y$, where $N$ is the target density, $R$ is the mean range of the projectiles and $Y$ is the total sputtering yield. For Ar in the energy range 100 eV to 20 keV, $\Phi_{\text{eq}} \sim (1-3) \times 10^{16}$ cm$^{-2}$. The altered

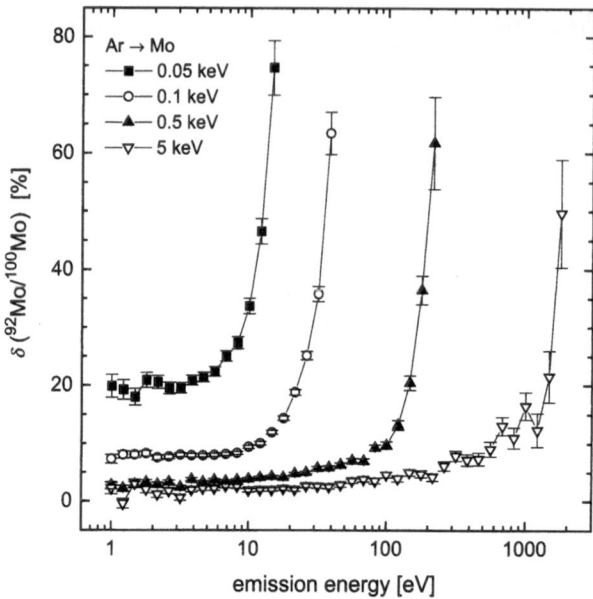

**Fig. 3.23.** Computer simulation data obtained with the TRIM code for sputtering of a Mo target of natural composition with Ar projectiles of the given energies. The isotopic fractionation (in percent) for the $^{92}$Mo/$^{100}$Mo isotope ratio is depicted as a function of the emission energy of the sputtered Mo atoms. The increasing positive fractionation with increasing ejection energy indicates that lighter isotopes are sputtered with a higher average energy. Data from [3.221]

composition distributions within the target at steady state are nonmonotonic for energies below 200 eV, with a light-isotope *depletion* at the surface and a slight *enrichment* in the subsurface region. The angular dependence of equilibrium sputtering persistently shows preferential light-isotope emission along the surface normal and heavy isotopes dominating at oblique emission directions. These data thus agree with the corresponding experimental observations.

### 3.4.4 Summary of Isotope Sputtering

From the low-fluence experiments on the sputtering of isotopic mixtures, the following features are established.

(i) The low-fluence isotopic fractionation derived from transient yield ratios is essentially identical whether secondary neutrals or secondary ions are monitored.

(ii) The magnitude of the fractionation exhibits no dependence on the target isotopic composition: the light-to-heavy isotope abundance ratio ranges from 2.7 for the Ge isotopes to about 1.5 and 1 for the natural and enriched

Mo specimens, respectively. In these cases the mass ratio is the same, as is the fractionation of the isotopes considered.

(iii) For the isotopic system studied most often, namely $^{92}$Mo/$^{100}$Mo, the near-normal light-isotope enrichment for the lowest fluences ranges from $\sim 30°/_{oo}$ to $\sim 50°/_{oo}$. While the various collector data cover this range of enrichment values, the mass-spectrometric data appear to fall at its upper end. However, no clear trends of a projectile or bombarding-energy dependence are discernible. Some of these differences may be associated with the specific fluence ranges used to derive the initial fractionation value. Furthermore, the angular range sampled by the various experiments is different. Conceivably, these differences might contribute to the variations of the low-fluence values of $\delta$ for the Mo isotopes.

(iv) An angular dependence of the isotopic fractionation is observed in the limit of low fluence: lighter isotopes exhibit a stronger preferential sputtering at near-normal emission directions than at oblique angles. This finding is seen most clearly in the collector experiments [3.207]: the magnitude of this variation of $\delta$ from $\theta_0 = 17°$ to $\theta_0 = 74°$ amounts to about half of the value of $\delta$ at $\theta_0 = 17°$ for most of the bombarding conditions. With increasing fluence this $\theta_0$ dependence of $\delta$ persists, and at steady-state conditions *light* isotopes are preferentially sputtered normal to the surface, while *heavy* isotopes exhibit preferential ejection at oblique emission directions. Again, the magnitude of this angular shift is moderate, ranging from $10°/_{oo}$ to $25°/_{oo}$ for the various projectiles and impact energies.

(v) The fluences required for the sputtered flux to reach the steady-state composition are $(1-2) \times 10^{16}$ ions/cm$^2$ for the heavier projectiles (Ar, Xe). In terms of eroded depth, this fluence would correspond to a layer thickness of at least one to two times the penetration depth of the projectiles in the respective targets.

Early theoretical approaches [3.222–224] to isotope sputtering produced some apparently controversial results [3.224]: first, the magnitude of the predicted isotope effects is composition dependent; second, depending on the element, preferential sputtering of either the lighter or the heavier isotope is observed; and third, isotope effects are found to exhibit a small nonlinearity in the mass dependence. The first two of these predictions are clearly in conflict with current experiments, while the third one is probably too small to be verified experimentally.

The conclusions of the linear collision cascade theory regarding the preferential sputtering of isotopic mixtures are summarized briefly in Sect. 3.1. The predicted scaling [3.4, 16] of the partial yields with $(M_2/M_1)^{2m}$ has been emphasized in essentially all comparisons with experimental and simulation data (neglecting possible differences in surface binding energies for the isotopes of an element; see (3.8)). The validity of the power-law cross section underlying the theoretical treatment was confirmed recently by Monte Carlo simulations of the particle fluxes in a homogeneous target [3.20]. Evidently, the magnitude of the parameter $m$, which characterizes the interaction potential, is crucial for a theoretical prediction of preferential isotope sputtering. A choice of $m \cong 0.11$ for the Born–Mayer interaction has been adopted [3.23]. Larger values would be applicable for softer interaction potentials. In fact, a value $m \approx 0.15$–$0.20$ appears suitable for the interaction potentials employed in several computer simulations [3.24, 146, 218, 219, 221]. Theoretical values of $\delta$ can be derived using these values of $m$. For some of the isotopic pairs investigated experimentally [3.203–207], these values agree rather well with the corresponding experimental values of $\delta$ (e.g. for Li and Ga isotopes), but are distinctly smaller for Ti and Ge. For the Mo isotopes which have been studied most extensively in the experiments, the theoretical value of $\delta$ is comparable in magnitude with all but one of the collector experiments [3.207]; it is smaller, however, than what is derived from the relative yield measurements.

Another long-standing disparity concerned the question of a possible dependence of isotope fractionation on emission angle. There exists a rather general consensus [3.2, 3] that such effects are expected at high (steady-state) fluence conditions, largely because of the development of an altered layer at the surface [3.15]. In the limit of zero fluence (i.e. for a homogeneous target), on the other hand, the analytical theory did not predict a variation of $\delta$ with the ejection angle. By contrast, an angular dependence was observed for low bombarding fluences in the experiments [3.207], with a magnitude comparable to the respective enrichment factor at near-normal emission. Although it has perhaps not been completely settled, some convergence has been achieved in this question of angular-emission effects. As Fig. 3.20 shows, experimental data [3.207] on the angle-dependent fractionation in the flux of Mo isotopes are, within the mutual statistical uncertainties, compatible with the results of computer simulations, of both the MD [3.219] and the binary-collision type [3.221]. Earlier experimental data [3.205] on isotope fractionation in Mo sputtering produced somewhat larger values of $\delta$, but the question as to a possible influence of the surface texture on the results was raised [3.219]. In fact, recent simulations [3.220] carried out for Mo single-crystal surfaces produced emission-angle-dependent values of $m_{\text{eff}}$ (and $\delta$) comparable in magnitude to the results of these experiments. The processes leading to these strong angle-dependent variations of isotopic fractionation in these simulations are not established [3.220]; hence, the question of the potential influence of the surface crystallographic orientation on the measured fractionation (in both [3.207] and [3.205], and possibly in other experiments) persists.

In the context of angle-dependent preferential sputtering, a momentum asymmetry in the collision cascade was observed in MD simulations and was held responsible [3.210, 211] for the isotopic fractionation effects reported in these studies. Sigmund and Sckerl [3.21] have recently reassessed this possibility theoretically. Two consequences of their investigations are directly relevant to the sputtering of isotopic mixtures and the aforementioned claims

[3.21]: (i) the light-isotope enrichment in the integrated flux is *diminished* by momentum asymmetry; (ii) the dependence on emission angle is such that the enrichment factor will *increase* with increasing emission angle. Both effects are expected to be weak, except possibly for low impact energies, high ion/target mass ratios and large surface binding energies. From these findings the authors [3.21] conclude that a momentum asymmetry in the collision cascade will be only of minor importance for isotope effects in sputtering, at least for the projectile species and energies investigated up to now.

Since most samples investigated experimentally are *not* binary systems, i.e. contain more than two components, the question was raised [3.225], as to what extent the presence of additional components (other isotopes or elements) might influence preferential sputtering. It was established theoretically that the preferential emission of the lighter isotope as expressed by (3.7) is also valid in a polyisotopic target [3.221]. Furthermore, the presence of a second element (say B) has only a minor impact on the sputter preferentiality of the isotopes of element A. Specifically, the admixture of O was found to induce a relative change of the enhancement factor for Mo isotopes by about 5% [3.22].

# 4. Ion Bombardment
# of Crystalline Semiconductors

Low-energy ions are used extensively in the manufacture of semiconductor devices [4.1]. Implantation of dopant atoms, cleaning of surfaces and metallization are all processes that may involve collisions of energetic ions with crystalline materials [4.2, 3]. The production of semiconductor devices with ever-smaller feature sizes [4.4] necessitates precise control of processing steps such as ion implantation, thermal annealing and ion-beam-assisted deposition. The defects and the strain created by the interaction of energetic species with the solid can have pronounced effects on the diffusion of dopant atoms; many of those mechanisms are not completely understood at present. Hence, detailed investigation of the processes related to the interaction of low-energy ions with crystalline semiconductors has become an issue of great importance. Naturally, the material studied most frequently is silicon.

Unlike most metals, crystalline semiconductors are known to amorphize readily under energetic-ion irradiation at room temperature [4.5, 6]. This crystalline-to-amorphous (c–a) transformation of the irradiated semiconductor lattice is critically dependent on the irradiation parameters and is generally controlled by a competition between damage accumulation and dynamic annealing. The mass of the bombarding ion species, its energy, the temperature of the specimen, the fluence and the flux rate all play largely interrelated roles in this transition [4.7–14].

Although much research has been devoted to these subjects, questions remain as to the fundamental mechanisms of the c–a transformation in semiconductors. While some models advocate a mechanism based on point defect accumulation [4.15], others proposed a mechanism based on the overlap of small, inter-cascade-produced amorphous zones [4.16]. Other models [4.17], somewhat intermediate between these extremes, consider damage events converting a portion of the crystal into a damaged state which can be transformed into the amorphous state by subsequent overlapping events. No single model, however, appears to account properly for all the experimental observations [4.6, 14, 18, 19]. Owing to their atomistic nature, molecular-dynamics simulations can in principle provide such information and have, in fact, been utilized extensively [4.20–28] to investigate ion irradiation effects in semiconductor materials.

The remarkable differences between defect production in metals on one hand and in semiconductors on the other was ascribed to the occurrence of specific phenomena: (i) Both the number of isolated Frenkel pairs and the number of defects leaving the displacement cascade are small in semiconductors. As shown by MD simulations [4.23, 24], this is apparently due to the fact that replacement collision sequences do not readily propagate in the diamond structure; the replacement sequences are few in number and short in length (2 to 3 atomic replacements) for both Si and GaAs. (ii) On the other hand, these simulations [4.23, 24] show that large "pockets" of unrelaxed amorphous material are produced by the displacement cascades: for a 5 keV cascade, these disordered regions contain an average of $\sim 800$ atoms, about seven times the number of displaced atoms predicted by Kinchin–Pease-type models (see Sect. 2.2). (iii) The kinetics of resolidification at the solid–liquid interface are slower in covalently bound materials than in metals; therefore, the liquid zone might quench to an amorphous region rather than crystallize. This might explain why the number of defects is generally larger than predicted by the modified Kinchin–Pease expression. Experiments and MD simulations show that by raising the sample temperature, complete recrystallization can be obtained since the vacancies and interstitials are never far separated.

The modification and, ultimately, destruction of the crystalline order usually exerts a tremendous influence on the electronic properties of the semiconductor material. These processes have been studied extensively for high ion energies (upper keV range), mostly as a means of controlled dopant introduction for electronic device applications [4.3]; on the other hand, data for low-energy ions ($\sim 100$ eV to a few keV) are more limited (despite the fact that, historically, low-energy ion bombardment followed by annealing was very early recognized as an effective means [4.29, 30] for preparing atomically clean surfaces of semiconductors, and these experiments predate most others in this field). At these low impact energies the surface is expected to play a significant role in the interactions that occur [4.31–35]. Furthermore, in this regime the ions are near the threshold for creating atomic displacements in crystals. Owing to the lower coordination of the surface atoms, an energy region might exist where surface displacements can occur without concurrently producing bulk defects [4.34, 36]. While bulk displacement energies are rather poorly known [4.37], there is very little information available at all on surface displacement energies [4.35, 36]. Closely related, of course, to these threshold values is the absolute number of displacements at the surface and in the bulk for low-energy ion irradiation. Recent work employing scanning tunneling microscopy has provided considerable information on defect formation on surfaces, and examples will be given in this chapter.

The possible importance of low-energy ion irradiation for epitaxial growth processes has been emphasized repeatedly [4.31–35, 38–42]. Silicon molecular-beam epitaxy at temperatures below 600 K in the presence of ion bombardment at energies of less than about 50 eV was found to produce an enhanced epitaxial thickness while still yielding high-quality films [4.41, 42]. Another recent example of a successful application of very-low-energy ion irradiation during MBE resulted in the suppression of three-dimensional island nucleation during GaAs growth on Si(100) [4.40]. A smoothing of growing Ge surfaces under concurrent Ar$^+$ ion bombardment was reported [4.34, 39]. Generally, these and other investigations stressed the influence that the presence of *surface defects* may have on the growth process. Conversely, the type of defects generated by ion impact (isolated adatoms and vacancies or island structures) will be determined, at least in part, by specimen-related parameters such as defect mobility and sample temperature.

This chapter first elucidates, essentially by means of data from scanning tunneling microscopy (STM) investigations, the formation of (isolated) surface defects for very low ion fluences (single-ion-impact regime) and the transition to the development of atomic-scale roughness associated with multilayer removal (Sect. 4.1). Extended structural modifications and the formation of an amorphized surface layer under low-energy ion bombardment are discussed in Sect. 4.2. Some aspects of the modification of electronic properties by ion irradiation are described in the context of surface-sensitive detection methods (Sect. 4.3). Finally, Sect. 4.4 provides some examples of ion-induced composition changes in the near-surface region of compound semiconductors, complementing thereby the more general description of such processes given in Chap. 3.

# 4.1 Surface Morphology: Defect and Adatom Formation

As noted above, surface and near-surface defects may decisively influence growth processes in ion-assisted thin film deposition [4.33]. At ion-bombardment fluences much lower than a monolayer equivalent ($\sim 10^{15}$ atoms/cm$^2$), the created defects will be separated spatially; this might open the possibility to study them as individual events. (In a deposition process this regime would relate to an ion-beam flux rate much smaller than the depositing-atom arrival flux.) With increasing fluence, single defects may agglomerate into larger defect structures (depending on their mobility), and a pronounced surface roughening may occur because of sample erosion. These features have been discussed for metals in Sect. 2.2. Examples illustrating similar effects for semiconductors are given in this section.

## 4.1.1 Single-Ion Impacts and Surface Defects

Low-fluence experiments on the reconstructed Si(111) (7×7) and Si(100) (2×1) surfaces of silicon were performed by Zandvliet et al. [4.43–46], by monitoring the surfaces by means of scanning tunneling microscopy. For 3 keV Ar$^+$ bombardment and fluences of $\leq 6 \times 10^{12}$ ions/cm$^2$, the defects created

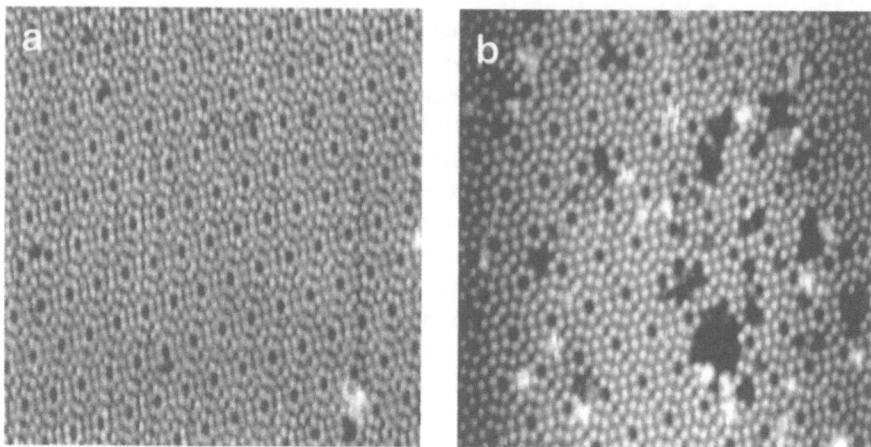

**Fig. 4.1a,b.** Empty-state STM images of a Si(111) (7×7) surface (**a**) prior to ion bombardment and (**b**) after bombardment by 3 keV Ar$^+$ ions at a fluence of $3 \times 10^{12}$ ions cm$^{-2}$. The scan area was $\sim 40$ nm×30 nm in (a) and $\sim 25$ nm×25 nm in (b). From [4.43]

by ion impact are mainly in the form of missing atoms. These vacancies tend to congregate to give the appearance of craters. In the vicinity of these craters, some bright features were found. Most probably these are adatoms (or adclusters) which originate from the emission process, possibly due to atoms displaced from nearby craters. The STM image of a clean Si(111) (7×7) surface prior to ion bombardment (Fig. 4.1a) reveals a low density of defects, which consist mainly of single missing adatoms. Figure 4.1b depicts the empty-state image of a 25 nm × 25 nm area after bombardment with 3 keV Ar$^+$ ions at an ion fluence of $3 \times 10^{12}$ cm$^{-2}$; this fluence corresponds to approximately 12 ion impacts on the imaged area. Some large "craters" are visible. From a detailed investigation of many images using different scan areas and ion fluences, Zandvliet et al. [4.43] concluded that the observed craters are associated with single-ion impacts on the surface. They identified three kinds of defect: (i) dark holes representing missing adatoms (the 7 × 7 unit cell has 12 Si adatoms); (ii) holes that are not totally dark but rather grayish; and (iii) holes having atom-size bright spots in the vicinity. According to the authors, the latter two kinds of defect are due to displaced atoms, i.e. Si adatoms not completely removed. The authors noted that the observed craters are much larger than the values estimated from a simple model of damage creation. Using current-imaging tunneling spectroscopy, they furthermore concluded that only the Si adatoms were removed in the cratered areas. This is in agreement with the average crater depth of $\sim 0.21$ nm measured in the empty-state images, which also suggests that only adatoms were removed. Although this observation would indicate layer-by-layer removal, the authors [4.43] emphasize that the STM data cannot preclude the possibility of sputter ejection from the second (or a deeper) layer. In particular, the rather open structure of the Si(111) (7×7) surface could favor such an emission process. Upon annealing (at 1020 K for 2 min) the ion-bombardment-induced defects on the Si(111) surface disappeared completely and the (7×7) reconstruction was totally restored.

Generally, the craters appear larger on the Si(111) (7×7) surface than on Si(100) (2×1), an observation ascribed to the larger size of the (7×7) unit cell. From the number of craters created, these authors conclude that at the initial stage of sputtering the yield may be higher than that determined at steady state. Alternatively, they suggest that some craters may be the result of surface relaxation to relieve the strain produced by subsurface damage effected by the incoming ion.

Bombardment experiments by Zandvliet, Tsong and coworkers [4.44–46] on Si(100) (2×1) surfaces, again using STM imaging, are summarized in Fig. 4.2. The pristine surface, Fig. 4.2a, is relative defect-free, with single missing dimers being the predominant defects. Figure 4.2b shows a filled-state STM image of the surface after irradiation by 3 keV $Ar^+$ at normal incidence with a fluence of $1.5 \times 10^{12}$ ions/cm$^2$, which translates into about 10 ion impacts on an area of the size shown. A higher concentration of random defects is now observed, some of which contain multiple missing dimers. Depth measurements of the defects by line scans yield an average depth of ~0.14 nm which is close to the single-step height on the (2×1) surface. The authors [4.44] evaluated the number of (additional) defects created by ion bombardment and found it to be six times larger than the number of ion impacts. They noted that this high yield is in distinct contrast to their observations on the Si(111) (7×7) surface, on which the ratio of defects (craters) to ions was about unity.

**Fig. 4.2a–c.** Filled-state STM images of Si(100) (2×1) surfaces taken at a sample bias of −2 V and a tunneling current of 0.5 nA. (a) Pristine surface before ion bombardment; the scan area is 28 nm×28 nm. (b) After bombardment by 3 keV $Ar^+$ ions at a fluence of $1.5 \times 10^{12}$ ions/cm$^2$; the scan area is 26 nm×26 nm. (c) Bombarded surface after annealing at 1025 K for 2 min; the scan area is 40 nm×40 nm. From [4.46]

Annealing the bombarded surface at 1020 K for 2 min produced an ordering of the defects on the Si(100) surface into line defects perpendicular to the dimer rows, as shown in Fig. 4.2c. The authors [4.44, 46] observed that the threshold temperature for this ordering lies between 770 and 870 K. The ordering of the defects into defect lines oriented perpendicular to the substrate dimer rows is driven by a short-range attractive interaction between defects located in adjacent dimer rows and a long-range repulsive interaction, which is related to the strain relaxation. Only upon annealing at very high temperatures (1220 K for 2 min) do these line defects disappear and the surface reverts to the original (2×1) state. Feil et al. [4.44] have carried out MD simulations to explain the shapes and sizes of the experimentally found random defects and to investigate the stability of ordered defects. These computations showed good agreement with the STM observations.

Further studies of bombardment-induced defects on Si(100) (2×1) surfaces showed different forms of vacancy ordering upon annealing. Because of their relatively high mobility (the diffusion of single vacancies along dimer rows has an activation energy of $1.7 \pm 0.4$ eV [4.47]), vacancies coalesce into vacancy islands; the equilibrium structure of these vacancy islands is, for low vacancy concentrations ($< 0.2$–$0.3$ ML), elongated in a direction perpendicular to the dimer rows of the upper terrace [4.48], whereas for a high vacancy concentration the equilibrium shape is rotated by 90°. Zandvliet has given a detailed account of vacancy ordering on Si(100) surfaces in [4.45].

The depths of the sputtered craters on the Si(111) and Si(100) surfaces determined [4.43] from the STM images are 0.21 nm and 0.14 nm, respectively; apparently, only the top-layer surface atoms are removed or missing. This finding does not necessarily indicate that sputtered atoms originate exclusively from the topmost layer. To some extent this might be due to the surface relaxation mentioned above. High-fluence experiments, however, provide strong evidence for layer-by-layer sputtering on Si surfaces (see below).

Surface defects created on Ge(100) by exposure to very-low-energy ($20 \leq E \leq 240$ eV) Xe$^+$ ions were examined by Cahill and coworkers [4.49, 50], employing STM imaging. These experiments were carried out at a sample temperature of 440 K, since Ge remains crystalline under ion bombardment at these low energies. Ion impacts generate defects (vacancies and adatoms) which nucleate and form vacancy and adatom islands. This is exemplified in Fig. 4.3, which depicts STM micrographs of Ge(100) surfaces bombarded with 20, 40, 130 and 240 eV Xe$^+$ ions at an incidence angle of 50° from the surface normal [4.50]. The ion fluences were chosen such that the removal of Ge atoms corresponded to $\sim 0.015$ ML for the 20 eV bombardment and to $\sim 0.2$ ML for the three higher energies. All surface images indicate adatoms are produced during ion irradiation. A second layer of an adatom island is formed on one of the islands in Fig. 4.3b. It is seen from Fig. 4.3a that for 20 eV bombardment the numbers of vacancies and of adatoms are comparable. Generally, for a fixed total number of created vacancies, the vacancy

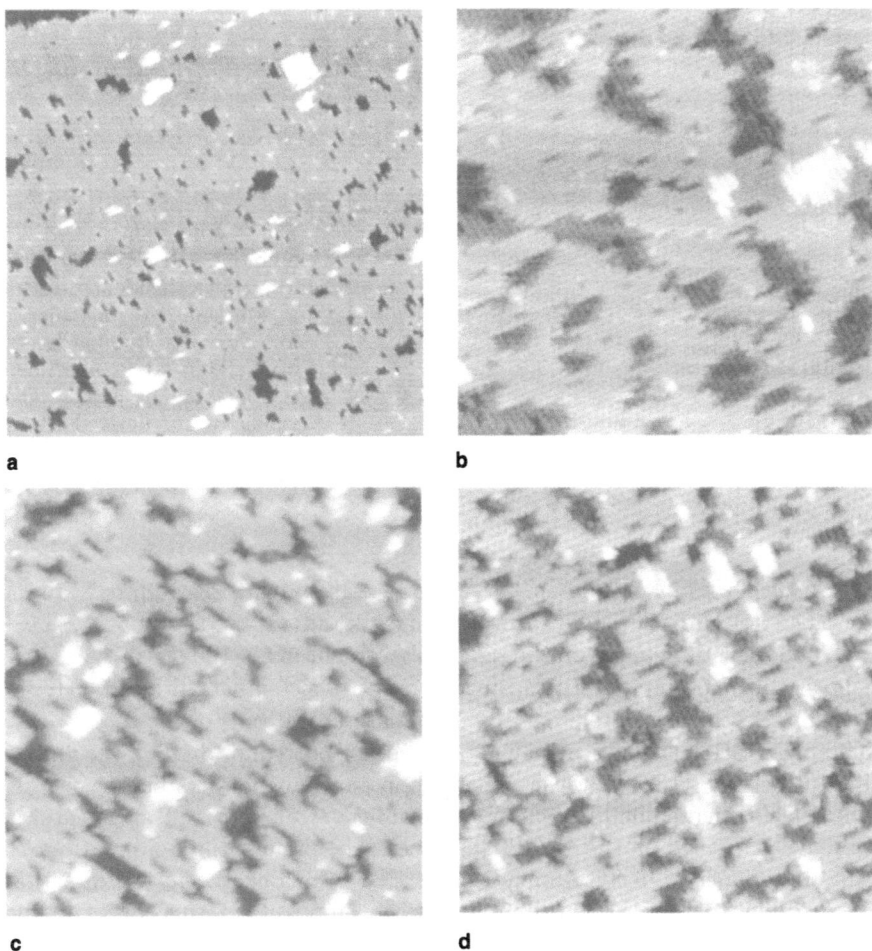

**Fig. 4.3a–d.** STM images of Ge(100) bombarded at 440 K with $Xe^+$ ions of energies
(a) 20 eV, (b) 40 eV, (c) 130 eV and (d) 240 eV. The scan area is 90 nm×90 nm in
(a) and 45 nm×45 nm in (b), (c) and (d). In the latter images the material removed
is roughly 0.2 ML, whereas it is about a factor of 30 smaller in (a). From [4.50]

island number density increases with increasing ion energy, whereas the vacancy island size decreases. The authors determined [4.50] from the STM
micrographs the yield of vacancies and adatoms as a function of ion energy.
These data are shown in Fig. 4.4. For an ion energy from 40 to 240 eV, the
ratio of adatoms to vacancies is roughly constant, $\sim 0.14 \pm 0.03$, but increases
to $0.85 \pm 0.17$ for 20 eV ion impact. This defect production includes the annihilation of adatom–vacancy pairs due to the finite defect mobility at 440 K.
Similarly, the authors derived the total sputtering yield in the investigated
$Xe^+$ ion energy range: $Y$ increases from $\sim 10^{-3}$ Ge atoms/$Xe^+$ at 20 eV
to about $Y = 0.7$ at 240 eV. At much higher bombarding energies (20 keV

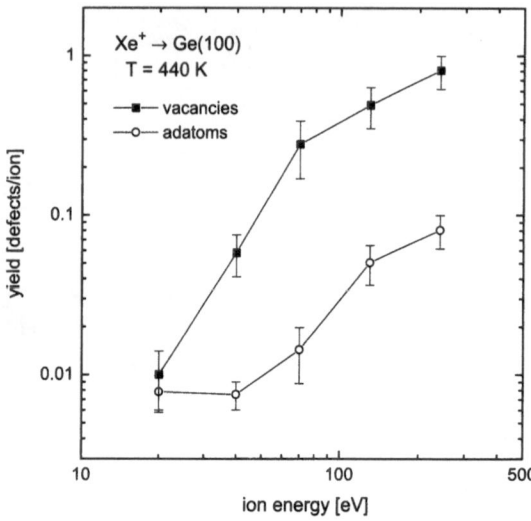

**Fig. 4.4.** The yield of adatom and vacancy creation per incident ion as a function of ion energy for Xe$^+$ ion bombardment of Ge(100) at a specimen temperature of 440 K. The yields were derived from STM images such as those shown in Fig. 4.3. Data from [4.50]

Ga$^+$), this group [4.51] observed very large craters (with sizes up to 85 nm$^2$) for single-ion impacts, albeit at a frequency of only $\sim 10^{-3}$.

A related study [4.52] on the ion bombardment of Ge(100) by 0.8 keV Ar$^+$ ions at specimen temperatures ranging from room temperature to 600 K revealed that vacancies preferentially annihilate at the ends of the dimer rows. This observation is identical to what has been reported for Si(100); see above. In contrast to Si(100), however, where upon annealing ion-irradiated surfaces line defects oriented perpendicular to the dimer rows are formed, such features were never formed on the Ge(100) surfaces. The preferential annihilation results in monolayer deep vacancy islands that are elongated in the direction of the dimer rows of the upper terrace, with the possible formation of an antiphase boundary. The authors suggest that these features can be a driving force for the development of surface roughness on Ge(100). The STM results show that, generally, the number density of adatom and vacancy islands decreases, whereas the average vacancy-island size and spacing increase with increasing temperature. This implies the occurrence of intralayer transport of vacancies and adatoms at temperatures as low as 380 K. From a detailed counting of the adatom cluster density upon annealing, the authors [4.52] concluded that ion bombardment at elevated temperature is *not* equivalent to ion irradiation at room temperature followed by annealing at the elevated temperature.

STM investigations of ion-bombarded GaAs(110) surfaces were carried out by Weaver and coworkers [4.53–56], using Ar$^+$ and Xe$^+$ ions with energies from 300 eV to 5 keV. Figure 4.5a shows an STM image from this work [4.54] taken after bombarding the surface at 300 K with 3 keV Ar$^+$ at normal incidence with a fluence of $5.4 \times 10^{12}$ ions/cm$^2$. Pits (craters) of one or a few missing atoms, adatoms ejected onto the surface and disordered regions are produced by ion irradiation. Most of the observed surface layer defects span 1–5 unit cells (the unit cell has a size of $0.4 \times 0.56 \, \text{nm}^2$) at low fluences ($10^{12}$ to $10^{13} \, \text{ions/cm}^2$) and room-temperature bombardment. The average pit size is 2.1 unit cells for 300 eV $\text{Ar}^+$ ions. This value increases slightly to about 3.1 unit cells for 3 keV $\text{Ar}^+$ or $\text{Xe}^+$ ion bombardment. The higher impact energies produce individual pits involving more than 20 unit cells in the lateral dimensions, with an unknown depth. Roughly equal numbers of Ga and As atoms were removed, with some of the ejected atoms displaced onto the surface as adatoms. The average crater size increased moderately with ion energy, while the sputtering yield was essentially constant for $\text{Ar}^+$ ($Y \approx 1.5$ atoms/ion) and slightly increased for $\text{Xe}^+$ (from $\sim 1$ to 1.7) at normal incidence [4.53, 54]. Aligning the beam with the [111] and [100] crystal axes (which was accomplished by tilting the beam direction by 35° and 45°, respectively, off the surface normal) generated a greater number of larger defects and the average crater size increased to 4.6 unit cells for 3 keV ion impact.

Ion irradiation of GaAs at elevated temperatures (625 to 775 K) shows [4.54, 55] that adatom diffusion and adatom–vacancy annihilation processes occur, and vacancy migration and coalescence become apparent. Figure 4.5b shows an STM image of a GaAs(110) surface bombarded at 625 K by 300 eV $\text{Ar}^+$ ions [4.55]; the ion fluence amounted to $3 \times 10^{11} \, \text{cm}^{-2}$, which corresponds to the removal of $\sim 0.05$ ML. The images reveal that monolayer-deep vacancy

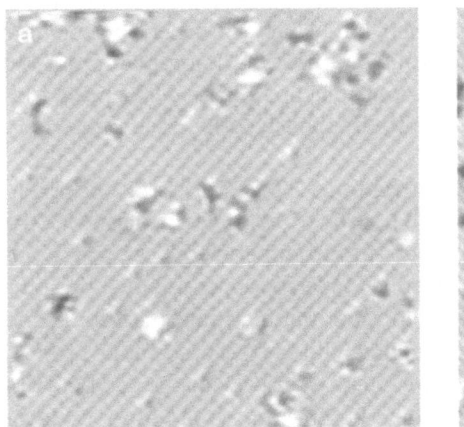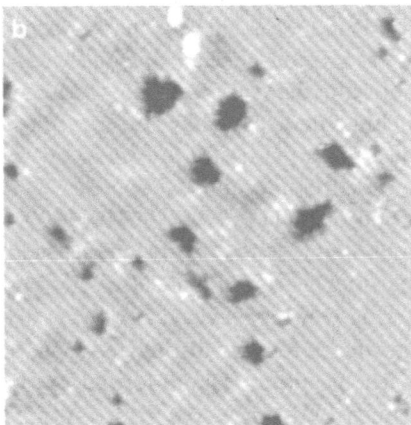

**Fig. 4.5a,b.** Filled-state STM images of GaAs(110) after ion bombardment with (a) 3 keV $\text{Ar}^+$ ions at a sample temperature of 300 K (ion fluence $5.4 \times 10^{12} \, \text{ions/cm}^2$, scan area 20 nm × 20 nm) and (b) 300 eV $\text{Ar}^+$ at a sample temperature of 625 K (ion flux $3 \times 10^{11} \, \text{ions/cm}^2\text{s}^1$, $\sim 0.05$ ML removed, scan area 25 nm × 25 nm). The [1$\bar{1}$0] direction points from the *lower left* to the *upper right* in (a) and from the *upper left* to the *lower right* in (b). In the *upper right* of (a), defects extending over up to 20 unit cells are seen, whereas in (b), single-layer-deep vacancy and adatom islands (*black* and *white regions*, respectively) are visible. From [4.54, 55]

islands are the dominant structures, but with very few adatoms ejected onto the surface (the bright features in Fig. 4.5b). The lateral dimensions of these islands increased with temperature and their density decreased. (Sputtering below $\sim 600$ K produced surface structures comparable to those created by single-ion impacts at room temperature.) Weaver and coworkers [4.55] studied also the vacancy kinetics in great detail. For the temperature range $625 \leq T \leq 775$ K, these authors derive an activation energy $E_d = 1.3 \pm 0.2$ eV. This value refers to divacancy motion which was found to be highly anisotropic along the Ga–As zigzag row of $[1\bar{1}0]$ orientation and is slightly smaller than the typical value for bulk diffusion ($\sim 1.7$ eV) of vacancies.

### 4.1.2  Multilayer Removal and Surface Roughening

The low-fluence ion bombardment regime discussed in the previous section produces isolated surface defects (vacancies, adatoms), which, at room temperature, are frozen out and do not interact. At elevated temperatures, however, these defects tend to form adatom and vacancy islands. The surface morphology that develops during ion bombardment at fluences beyond these single-ion impact regime might depend in a complex way on irradiation- and specimen-related parameters. As noted in Sect. 2.2.2, these interactions may lead, in the case of multilayer removal, to pronounced surface topographic features or, in favorable cases, to layer-by-layer sputtering. The similarity of these distinct removal processes to the corresponding growth processes has been noted by several authors. Because of their technological importance, semiconductor surfaces have been studied intensively in order to elucidate these erosion mechanisms and their possible application in ion-beam-assisted thin-film deposition. While the atomic-resolution capabilities of scanning tunneling microscopy have been employed quite frequently, other techniques have also provided useful information.

Bedrossian and Klitsner [4.57–59] carried out STM studies of ion irradiation of Si(111) (7×7) and Si(100) (2×1) surfaces at high fluences ($> 10^{15}$ ions/cm$^2$), with the specimens held at elevated temperatures. At low bombarding energies (200–250 eV Xe$^+$) controlled layer-by-layer removal by sputtering can be achieved: Above $\sim 640$ K, mutual annihilation of adatoms and vacancies formed during ion bombardment enables layerwise removal of Si without adatom island formation. In agreement with previous data [4.60] obtained by high-energy electron diffraction, the authors report layer-by-layer removal from Si(100) (2×1) surfaces under these conditions; they imply that this process may be described as the inverse of silicon homoepitaxial growth [4.61]. Whereas the evolution of surface morphology during growth is mediated by mobile adatoms (the activation energy for surface diffusion of Si atoms on Si(100) amounts to $0.67 \pm 0.08$ eV [4.62]), which can either nucleate adatom islands or attach at step edges [4.61], layer-by-layer sputtering was found to be mediated by mobile surface vacancies, which are created during sputtering and which can either nucleate monolayer-deep vacancy islands or annihilate at step edges [4.58, 59]. In the temperature range in which layer-by-layer sputtering of Si(100) is operative the $2 \times 1$ periodicity of this surface was found to be preserved.

On Si(100), Bedrossian et al. [4.58, 59] observe with increasing temperature both a lower density and a larger average size of vacancy islands for a given fluence, a consequence of an increased vacancy mobility. Mobile surface vacancies were found to annihilate preferentially at the ends of dimer rows, analogously to the growth situation where adatoms preferentially stick at the ends of dimer rows. At still higher temperatures (725 K), these authors [4.58] observe a transition from vacancy island nucleation to step retraction, which is a consequence of anisotropy in the kinetics of mobile surface vacancies on Si(100): preferential annihilation of vacancies at the ends, rather than the sides, of dimer rows leads to the more rapid retraction of the "ragged" $S_B$ step relative to the "straight" $S_A$ step; this enables, as the authors noted [4.58], the realization of a new, single-A-domain surface which is not an equilibrium structure (and hence not accessible by epitaxial growth alone), but is stabilized at a moderate temperature ($\sim 725$ K) by kinetic arguments. (In this context, A and B refer to the different domain orientations usually observed on adjacent terraces of the Si(100) ($2 \times 1$) surfaces, with the dimer rows running parallel to the step edges for type A and perpendicular to the step edges for type B. The downward steps from B to A domains are comparatively "ragged" whereas the boundary from A to B is rather "straight" [4.63].) These features are illustrated in Fig. 4.6, taken from [4.58]. An image of the pristine surface (offcut by 0.2° from (100)) depicts (Fig. 4.6a) the A and B terraces and the corresponding $S_A$ and $S_B$ steps (scan width 180 nm × 180 nm). Figure 4.6b is a high-resolution image (36 nm × 36 nm) of a 1°-offcut Si(100) ($2 \times 1$) surface, showing the alternating dimer orientation on adjacent terraces. Bombardment of the surface shown in Fig. 4.6a with 225 eV Xe$^+$ ions at a specimen temperature of 725 K after the removal of about 0.5 ML, shows depletion of the B domains (Fig. 4.6c). After removal of $\sim 1$ ML, the B terraces are further reduced and depressions in the A domains show up (Fig. 4.6d). Annealing this surface for 6 min at 725 K results in the virtual disappearance of the B domains, leading to a double-stepped single-A-domain surface [4.58]. Similar STM observations [4.64] reported step-flow kinetics during Si(100) surface erosion by low-energy ion bombardment. These studies generally stress the similarity of these removal processes to the adatom kinetics observed in Si/Si(100) homoepitaxial growth [4.65–67].

Layer-by-layer sputtering of Si(111) proceeds via nucleation of vacancy islands, similarly to the mechanisms that are operative on Si(100). In addition, Bedrossian and Klitsner [4.57] observed on this surface a transition from destruction to preservation of the ($7 \times 7$) reconstruction with increasing specimen temperature. At high temperature (850 K) they found monolayer-deep vacancy islands which exhibit near-perfect ($7 \times 7$) reconstruction. Conversely, at 670 K (or lower), exposed atomic layers with no long-range reconstruction

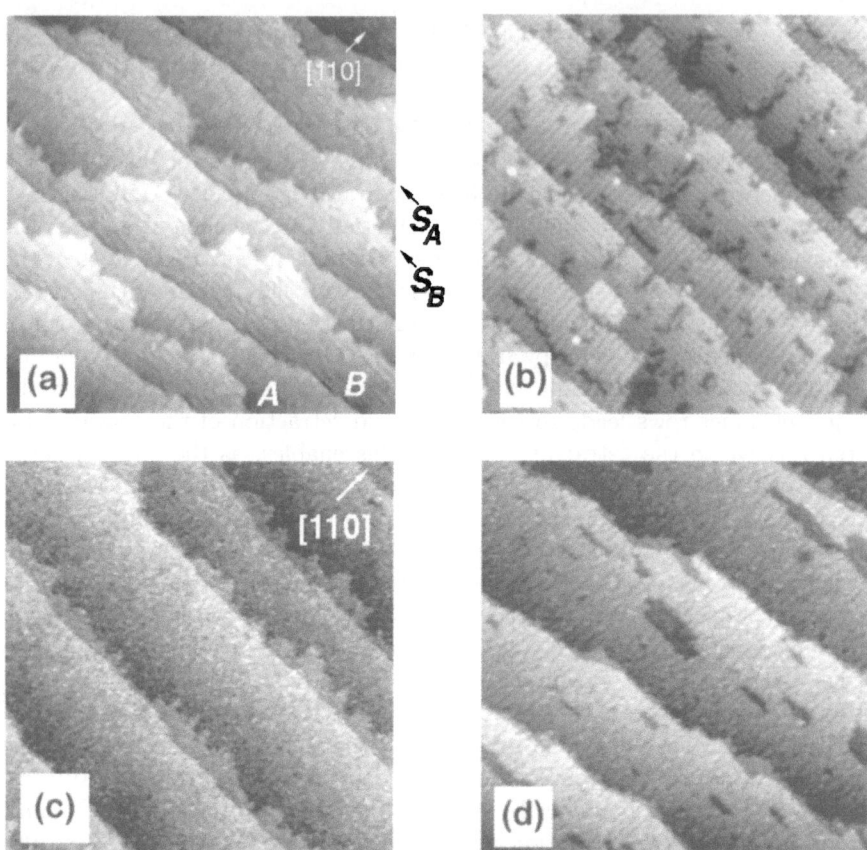

**Fig. 4.6a–d.** Filled-state images of (**a**) a 0.2°-offcut (scan size 180 nm × 180 nm) and (**b**) a 1°-offcut (scan size 36 nm × 36 nm) pristine Si(100) (2 × 1) surface, showing the A and B terraces, the $S_A$ and $S_B$ steps and the alternating dimer orientation on adjacent terraces. Images (**c**) and (**d**) were taken from the surface shown in (a) after 225 eV Xe$^+$ ion bombardment at 725 K with ∼ 0.5 ML (c) and ∼ 1 ML (d) removed. Increasing surface erosion results in depletion of the B domain. After further annealing for 6 min at 725 K the B domain has virtually disappeared. From [4.58]

were observed adjacent to step edges. High-fluence bombardment of Si surfaces at room temperature generally results in a highly disordered surface and undulating surface topography.

The surface morphology of Ge(100) formed during bombardment by low-energy ions was investigated in great detail by Chey et al. [4.49, 50] and by others [4.68, 69]. The former authors used STM to monitor the surface after bombardment with 240 eV Xe$^+$ ions for a wide range of exposure times (fluences) and temperatures. For high fluences (corresponding to ∼ 130 ML of Ge removed), the characteristic in-plane length of the roughness increases with increasing sample temperature from 440 K to 570 K. The authors [4.49] report a transition of the morphology from a relatively disordered arrangement of "mounds" to a more regular pattern of "pits" at $T \approx 545$ K. This multilayer pit formation is, according to Chey et al. [4.49], reminiscent of related observations on Pt(111); cf. Fig. 2.6. They suggested that this roughening is driven by the asymmetric attachment of surface (dimer) vacancies at ascending steps compared to descending steps.

Multilayer erosion of a GaAs(110) surface exhibits distinct differences from what is found for Si surfaces. Weaver and coworkers [4.54–56] employed STM imaging to investigate multilayer removal by 0.3–3 keV $Ar^+$ and $Xe^+$ ions at temperatures between 625 K and 775 K. They found erosion due to sputtering to proceed in a simultaneous-multilayer fashion, with the probability of sputter removal from a particular layer governed by the fractional area exposed. In general, these surfaces are composed of a main layer, remnant islands and vacancy islands. The total number of exposed layers reflects the effectiveness of *interlayer* transfer, a temperature-dependent quantity. The remnant structures and the single-layer vacancy islands appear in stacks as more material is removed; the step height between layers is generally one layer. The data of Weaver and coworkers indicate that there is no preferred alignment of the vacancy islands relative to the specimen's crystallographic directions.

The authors characterize the morphology by the surface height as a function of the lateral coordinates $x$, and measure roughness by the surface width $w$ [4.56, 70],

$$w\left(L, \tau\right) = \left[\left\langle h^2\left(x\right)\right\rangle - \left\langle h\left(x\right)\right\rangle^2\right]^{1/2}, \tag{4.1}$$

where $h(x)$ is the height at position $x$, $\langle\rangle$ denotes an average over the sampling area $L$ and $\tau$ is the duration of sputtering (i.e. a measure of the material removed). The surface width may exhibit the general scaling behavior $w \propto L^\alpha$ for large $\tau$ and small $L$, and $w \propto \tau^\beta$ for large $L$ and small $\tau$, where $\alpha$ and $\beta$ are related to the surface transport mechanisms. (For a review see e.g. [4.71, 72].)

Figure 4.7 shows STM images [4.56] from a GaAs(110) surface irradiated by 2 keV $Xe^+$ ions at 725 K (Fig. 4.7a,b) and at 625 K (Fig. 4.7c,d). The amount of material removed corresponds to 1.4 ML in Fig. 4.7a, with about 3% of the original top layer remaining as isolated monolayer islands. Vacancy islands as large as 30 nm appear in the second layer. The surface width $w$ is 0.6 ML. (The authors derive values of $w$ from the coverage $\theta_i$ of the $i$th layer by a procedure described in detail in [4.54, 56].) Removing 10 ML at 725 K (Fig. 4.7b) results in $w = 1.8$ ML, with up to ten layers exposed. At 625 K, the surface roughness evolves in a different way. The surface width $w$ grows rapidly and the step density increases quickly by the formation of many small-sized and rough-edged vacancy islands (Fig. 4.7c,d). The values of $w$ are 0.9 and 1.8 ML at 1.8 and 10 ML removed, respectively.

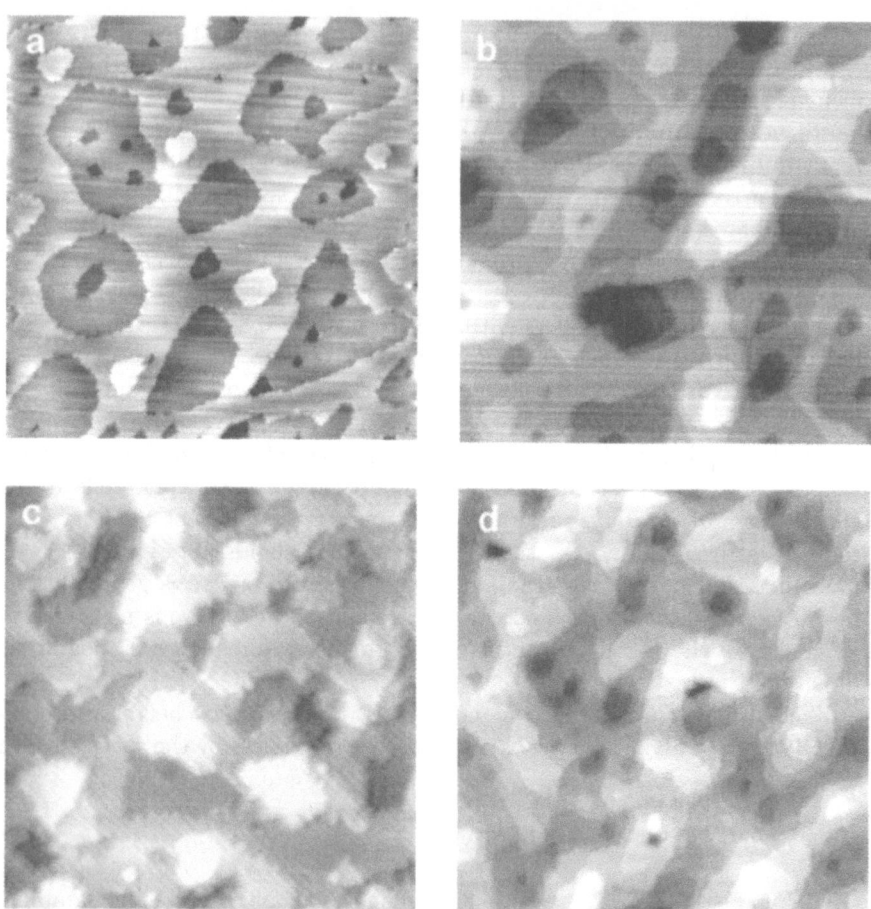

**Fig. 4.7a–d.** Filled-state STM images of GaAs(110) after 2 keV Xe$^+$ ion irradiation at 725 K ((**a**) and (**b**)), and 625 K ((**c**) and (**d**)). The image sizes are 90 nm×90 nm (a), 45 nm×45 nm (c) and 108 nm×108 nm ((b) and (d)). The material removed is 1.4 ML (a), 1.8 ML (c) and 10 ML ((b) and (d)). From [4.56]

This group [4.56] determined for each temperature and $\tau$ value the values of $w$ as a function of sampling area $L$. These data are shown in Fig. 4.8. Generally, $w$ increases with $L$ initially and reaches saturation at some length $L = L_c$. For $L < L_c$ the data can be fitted by $L^\alpha$, where $\alpha = 0.38$. The values of $L_c$ increase linearly with increasing $\tau$ in the range studied in the experiment (1–10 ML); the increase is faster at the higher temperature: from 20 nm to more than 100 nm at 725 K, as compared to 625 K (10 to 25 nm). The authors [4.56] summarize their findings by noting that after removal of ten monolayers at 625 K, the surfaces are rougher on a small scale than those at 725 K, but they are smoother on a large scale. The increased large-scale roughness at high $T$ was ascribed to increased diffusion on terraces and along

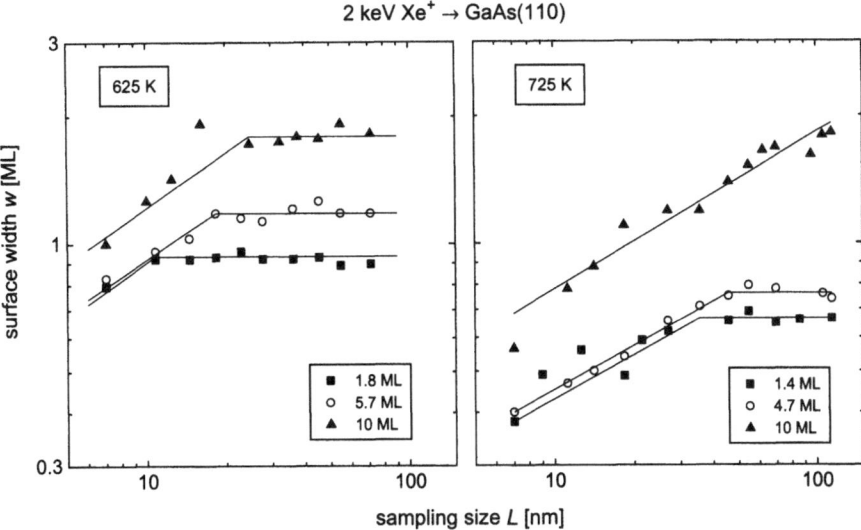

**Fig. 4.8.** Surface width as a function of sampling length for different amounts of material removal at 625 and 725 K. Above the correlation length ($L_c$) the surface width saturates at the large-scale value. Data from [4.56]

step edges, but insufficient cross-step transport. Generally, the short-range surface width is related to step density, which is determined by *intra*layer diffusion, whereas the long-range width depends on *inter*layer mass transport. The high step density created at low $T$ *enhances* cross-step transport, thereby *reducing* large-scale roughness. The surface width at 725 K increases as $w \propto \tau^\beta$, with $\beta = 0.3$. Such a dependence is consistent with a "growth" law [4.72], whereby the interface width increases with deposition time during deposition at a constant flux, the growth exponent $\beta$ depending on temperature. In multilayer erosion, $\beta$ is a measure of the effectiveness of interlayer transport. (If there is no such transport, then $\beta = 0.5$. For layer-by-layer sputtering, $\beta = 0$.) Conversely, at 625 K no such simple relationship was valid, and $\beta > 0.3$ for $\tau < 1$ ML and $\beta < 0.3$ for $\tau > 1$ ML was observed. The authors [4.56] emphasize the congruence of their observations with the mechanism of reentrant layer-by-layer growth of Pt on Pt(111) [4.73]. On GaAs(110) this reduces the rate of increase of $w$, but does not lead to layer-by-layer removal in the temperature range accessible, $T \leq 775$ K. (The surface starts to decompose at about 800 K because of thermal desorption of As as $As_x$ clusters.)

## 4.2 Near-Surface Structural Modifications

The amorphization of silicon and, to a lesser degree, of other semiconductor materials under ion bombardment has aroused considerable interest for over three decades [4.3]. Early experimental and theoretical studies [4.8–10] indicated that the ion energy, the ion mass, the fluence and the substrate temperature were the key parameters influencing the production of stable defects and ultimately the formation of amorphous layers. Various amorphization models have been proposed to explain the experimental observations. Basically, they fall into two categories: (i) direct-amorphization models [4.16] in which each ion creates an amorphous zone whose size depends on ion mass, energy and substrate temperature, and (ii) critical-energy-density models [4.15] in which defects build up with ion fluence until a critical defect density is achieved and the lattice collapses to an amorphous phase.

More recent studies [4.74] of MeV ion-induced amorphization of silicon at elevated temperatures have indicated strong flux-density (ions/area and time) effects for heavy ions. Such behavior was not anticipated by the above models, which were devised for low-temperature irradiations, but provides evidence that amorphization is produced in cases where the defect production rate exceeds the defect annihilation rate during ion bombardment. Thus, the rate of energy deposition has a dominant influence on amorphization at elevated temperatures. It has been proposed that the competition between formation and dissociation of silicon divacancies controls the amorphization process. This proposal was supported by the observed temperature and flux-density dependence of amorphization: the higher the specimen temperature, the higher the flux density has to be in order to amorphize the sample. From an Arrhenius-type correlation of this critical flux density with the corresponding temperature an activation energy of 1.2 eV was derived [4.74], a value close to the dissociation energy of divacancies.

### 4.2.1 Amorphized Surface Layer

As outlined above, the modifications induced by ion irradiation in semiconductors at low energies (a few keV) are of great importance in a variety of applications. In this regime, bombardment effects are typically confined to the topmost surface layers. To investigate these effects, *surface-sensitive* techniques have to be employed. This section will describe some of these studies, emphasizing the formation and the characteristics of amorphized layers at the surface, and the dependence on various irradiation parameters (e.g. ion energy, ion fluence and, to some extent, sample temperature).

In situ X-ray photoelectron spectroscopy has been used by Lu et al. [4.75] to analyze the nature and extent of damage of Si(100) surfaces bombarded with Xe$^+$ ions in the energy range 0.25–2 keV. Dramatic changes in the Si 2p core levels were found upon ion irradiation, and analysis of the peak shape showed that an amorphous silicon overlayer was formed on the Xe$^+$ bombarded surface. Fluctuations in the valence electron charge distribution are known to exist in the amorphous-Si (*a*-Si) continuous random network, as a consequence of deviations in the bond length and bond angle distributions from their crystalline values. These valence charge fluctuations therefore cause the observed broadening in the Si 2p core levels; furthermore, a shift by 0.2 eV towards lower binding energy for the damaged layer was reported, indicative of the existence of a valence energy-band offset at the $a$-Si/$c$-Si interface. The results provide evidence that the depth of damage increases with ion fluence, reaching a saturation for all impact energies at about $1 \times 10^{15}$ ions/cm$^2$. From the attenuation of the photoelectron signal, the saturation damage depth was found to increase linearly with ion energy and amounted to $\sim 4.4$ nm for 2 keV irradiation. These data are included in Fig. 4.9 (see below).

A similar experiment was carried out by Ishii et al. [4.76], who bombarded Si(100) surfaces with rare-gas (Ne, Ar and Xe) ions at energies between 0.3 and 0.6 keV. Amorphous layers with a thickness of 1–3 nm were formed and the amorphization saturated for a fluence of $\sim 10^{15}$ ions/cm$^2$. Specifically, the thickness of the amorphous layer for 0.4 keV bombardment amounts to 2.5 nm for Ne$^+$ ions, 1.5 nm for Ar$^+$ and 1.6 nm for Xe$^+$. Cross-sectional TEM investigations of these specimens revealed amorphous-layer thicknesses that were larger by roughly 50%. Contrary to the notion of Lu et al. [4.75] discussed above, these authors [4.76] propose that the amorphization of Si at low energies proceeds by the accumulation of $a$-Si regions created by individual ion impacts rather than in a layer-by-layer fashion. From these arguments they estimate the amorphized volume due to a single-ion impact event.

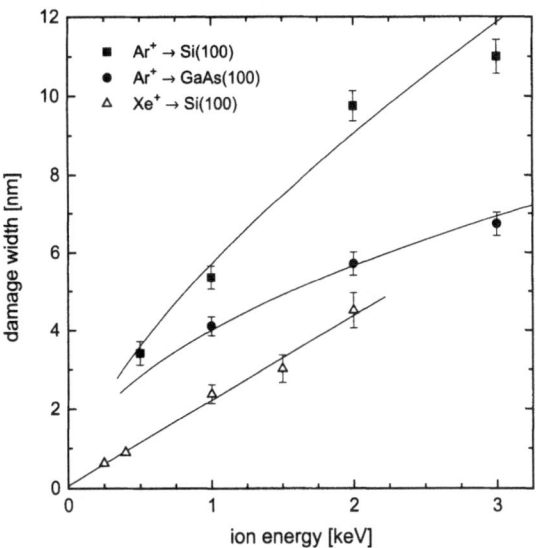

**Fig. 4.9.** Damage width as a function of ion energy for Ar$^+$- and Xe$^+$-irradiated Si(100) and GaAs(100) surfaces. The data were derived from MEIS measurements (*closed symbols*) [4.77] and from XPS analysis [4.75]. Both data sets refer to ion fluences in excess of $1 \times 10^{15}$ cm$^{-2}$

Several groups have measured damage and amorphization in low-energy ion irradiation of semiconductors by means of medium-energy ion-scattering (MEIS) spectroscopy. This technique may achieve sub-nanometer depth resolution because of its much improved energy resolution (as compared to conventional Rutherford backscattering) using an electrostatic energy analyzer. Kido and coworkers [4.77, 78] investigated damage profiles in 0.3–3 keV $Ar^+$-irradiated Si(100), GaAs(100) and Ge(111). They found rather sharp interfaces between the damaged (amorphized) layer and the crystalline substrates for an ion fluence of $3 \times 10^{15}$ $Ar^+$ ions/cm$^2$. From an evaluation of the respective ion scattering spectra, these authors derived damage profiles and determined the width of the disordered zone. These data are compiled in Fig. 4.9. For 2 keV $Ar^+$ bombardment the width of the amorphous layer amounts to $\sim 9$ nm in Si and $\sim 6$ nm in GaAs. The damage width was found to increase with the $Ar^+$ energy $E$; the values could be fitted by $5.7 \times E^{2/3}$ for Si(100) and by $4.0 \times E^{1/2}$ for GaAs(100). According to the authors [4.77], this energy dependence is consistent with the relation between the projected range and the ion energy proposed by Kalbitzer [4.79]. For both of these materials and for the energy range investigated, this layer thickness is considerably larger (by about a factor of two) than the projected range of the Ar ions.

For 3 keV $Ar^+$ irradiation of a Ge(111) surface, Kido et al. [4.78] studied the fluence-dependent evolution of the disordered layer in the range from $2 \times 10^{13}$ to $1 \times 10^{16}$ $Ar^+$/cm$^2$. At a fluence of $\sim 1 \times 10^{14}$ cm$^{-2}$ an amorphous layer is formed and the thickness of this layer essentially saturates at $1 \times 10^{15}$ cm$^{-2}$. Then, it has a thickness of $\sim 12$ nm and the total number of defects produced amounts to $\sim 5 \times 10^{16}$ Ge atoms/cm$^2$. The former result was corroborated by cross-sectional transmission electron microscopy.

Using medium-energy ion scattering, Al-Bayati et al. [4.80, 81] investigated radiation damage induced by low-energy (60–510 eV) $Ar^+$ bombardment of Si(100) at temperatures between 300 K and 673 K. The ion fluence was varied from $1 \times 10^{15}$ to $1 \times 10^{16}$ $Ar^+$ cm$^{-2}$. The residual disorder decreases as the bombardment temperature increases. This finding illustrates the dynamic annealing of the damage created during irradiation. The depth resolution of this experiment ($\sim 0.3$ nm) was sufficient to show that the reduction in damage depth for the higher-temperature bombardment is due to annealing of the regions at the *interface* with the crystalline material rather than of those near the surface, where completely amorphous layers were produced. This competition between damage creation and annealing was found also when irradiating the Si(100) specimen at fixed temperature with different ion energies: for a given total fluence, ions of higher energies produce greater disorder and the width of the damage peak in the ion-scattering spectra increases from 2 to 4 nm as the energy increases from 60 to 510 eV.

On the basis of studies that investigated post-annealing damage in Si, the authors [4.80] argued that annealing at elevated temperatures occurs at the edge of the damage distribution, where cascades are dilute and isolated defects are created. As the activation energy for monovacancy migration in Si is about 0.33 eV [4.15], which is greater than that associated with interstitial migration, monovacancies and interstitials are expected to anneal out at the end of the damage range. With increasing temperature, the thickness of the disorder layer would therefore shrink. By contrast, the 300 K data indicate that the outer $\sim 4$ nm of the bombarded sample (at 510 eV) is largely amorphous. Apparently, the buildup of small defect clusters (e.g. divacancies) and the overlap of the damaged regions leads to the c–a phase change at sufficiently high fluences. Vook and Stein [4.15] have shown that divacancies in silicon anneal out with an activation energy of 1.2 eV, and hence a significant concentration of these defects could be built up at room temperature.

Following a previous proposal, Al-Bayati et al. [4.80] estimated the energy density from their low-energy bombardment of Si(100) and derived values ranging from $9.4 \times 10^{22}$ eV/cm$^3$ for 110 eV Ar$^+$ impact to $3.5 \times 10^{22}$ eV/cm$^3$ for 510 eV irradiation (using the atomic density of Si $N_{Si} = 5 \times 10^{22}$ atoms/cm$^3$). They noted that these values are comparable to those for heavier ions at higher energies. These authors, furthermore, determined the total number of displaced Si atoms $N_d$: this number increases linearly with ion energy in the range from 60 to 510 eV. As can be seen from Fig. 4.10, $N_d$ amounts to $\sim 2 \times 10^{15}$ Si atoms cm$^{-2}$ at 60 eV and to $\sim 10 \times 10^{15}$ Si cm$^{-2}$ at 510 eV for an ion fluence of $1 \times 10^{15}$ Ar$^+$ cm$^{-2}$ and a sample temperature of 473 K. For the higher fluence, $N_d$ is larger by a factor of about 2–3. This increase in thickness was thought to be due to the migration and accumulation of defects at the amorphous/crystalline interface (i.e. layer-by-layer amorphization). The authors [4.80, 81] compared the number of displaced Si atoms with

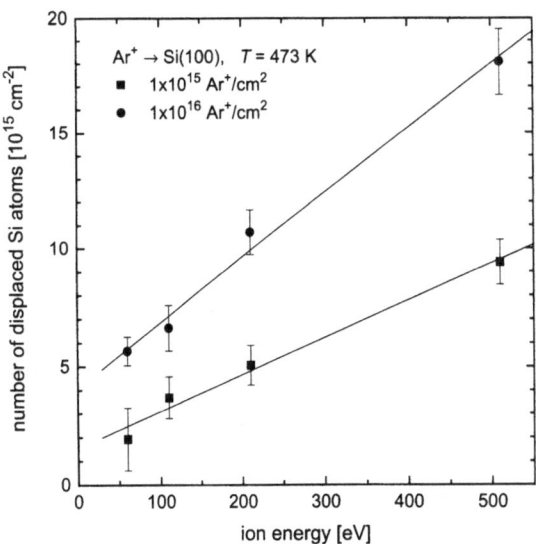

**Fig. 4.10.** Number of displaced Si atoms as a function of Ar$^+$ ion energy for Si(100) bombarded at a sample temperature of 473 K. The parameter is the ion fluence. Data from [4.80]

the values expected from the modified Kinchin–Pease expression; while for the lower fluence ($1 \times 10^{15}$ cm$^{-2}$) the data sets are comparable, they diverge by the aforementioned factor for the high fluence ($1 \times 10^{16}$ cm$^{-2}$), indicating that most of the ions at this fluence interact with an already damaged material. The experimental FWHM damage range increases from 1.2 nm or 2 nm at 60 eV ion energy to 3.1 nm or 4.7 nm at 510 eV for an ion fluence of $1 \times 10^{15}$ cm$^{-2}$ or $1 \times 10^{16}$ cm$^{-2}$, respectively. This extension of the amorphous layer with fluence suggests that more defects accumulate at the amorphous/crystalline interface, leading to an increase in the depth of the damage layer (the layer-by-layer scenario discussed above).

Using 1.8 MeV He$^+$ RBS, Williams [4.82] determined the ion-induced damage on Si(100), GaAs(100) and InP(100) for 1–3 keV Ar$^+$ ion impact. The numbers of displaced atoms were $1.8 \times 10^{16}$, $3.4 \times 10^{16}$ and $5.0 \times 10^{16}$ Si atoms cm$^{-2}$ for the 1, 2 and 3 keV bombardments, respectively, of Si(100). For these energies, the respective values for GaAs(100) were 1.1, 2.2 and $4 \times 10^{16}$ and for InP(100) 1.0, 2.1 and $2.9 \times 10^{16}$ cm$^{-2}$.

The near-surface region of low-energy (0.5–1.5 keV) Ar$^+$-bombarded Si(100) has been studied by Huang et al. [4.83] using high-resolution X-ray absorption near-edge structure spectroscopy (XANES), extended X-ray absorption fine structure spectroscopy (EXAFS), medium-energy ion scattering (MEIS), variable-energy positron annihilation spectroscopy and angle-resolved XPS. The XANES data, which are a direct measurement of the near-surface structure, indicate a disordered layer after ion bombardment: the peaks corresponding to the $L_{III}$ and $L_{II}$ absorption edges for the irradiated samples were much broader ($\sim 2.5$ times) than those for the c-Si. The peak broadening resulted in a tail towards the valence band maximum. For a sample bombarded to an ion fluence of $10^{16}$ Ar$^+$/cm$^2$, this tail extended about 0.5 eV below the conduction band minimum, to a position approximately in the middle of the band gap. This finding suggests a high density of surface states across the band gap, characteristic of amorphous silicon.

Employing MEIS, the authors [4.83] determined the silicon-displacement areal densities. For normal-incidence Ar$^+$ ions of energies 0.5 and 1 keV, these values were $5.9 \times 10^{15}$ and $2.7 \times 10^{16}$ Si atoms/cm$^2$, respectively, for an irradiation fluence of $1 \times 10^{16}$ Ar$^+$/cm$^2$ (see Fig. 4.10). The projected ranges of the disordered Si atoms derived from the spectra amounted to 1.7 nm and 3.1 nm for these energies. The Ar retention ratio (total incorporated Ar over total ion fluence), established again from the MEIS results, was found to decrease from $\sim 0.4$ at $1 \times 10^{15}$ Ar$^+$ cm$^{-2}$ to $\sim 0.1$ at $\geq 1 \times 10^{16}$ Ar$^+$ cm$^{-2}$. The projected depths for Ar$^+$ ions of energies 0.5, 1 and 1.5 keV were, respectively, 1.8, 3.1 and 4.5 nm. These values are essentially identical to those of the displaced Si atoms. The depth distributions of the implanted Ar could be approximated with Gaussian distributions. Although no definite statements as to the defect structure of the retained argon could be made, the authors concluded that no Ar bubbles had been formed in the bombarded specimens.

To address the distinct discrepancies in the literature as to the existence of extended defects much beyond the disordered (amorphous) zone, Huang et al. [4.83] utilized positron annihilation spectroscopy; to fit their spectra, they postulated an interface layer (of thickness 15 nm) between the amorphous near-surface region and the crystalline substrate, with an atomic defect density approximately two orders of magnitude less than that in the bombardment-induced disordered layer.

Annealing of bombarded silicon and epitaxial recrystallization are important issues in semiconductor device manufacture. The incorporated Ar was thought to have a strong inhibiting effect on this recrystallization [4.84]. Huang et al. [4.83] observe that the amount of retained Ar in a low-energy-bombarded Si sample decreases very gradually with increasing annealing temperature: a 773 K anneal for 30 min reduced the retained Ar by only about 35%. Raising the temperature to 973 K, however, decreases the Ar content below the MEIS detection limit of $\sim 1 \times 10^{13} \, \mathrm{cm}^{-2}$. The structural changes upon annealing at that temperature, monitored by XANES, reveal that all the crystalline features are recovered. Evidence for epitaxial recrystallization was provided also by MEIS data [4.83] taken after annealing at 973 K: the c–a interface moves towards the surface and the areal concentration of displaced Si atoms is reduced to $\sim 7.8 \times 10^{15} \, \mathrm{cm}^{-2}$, which was identical to that obtained from a reference $c$-Si sample. Conversely, annealing at the lower temperature of 573 K only slightly decreases the number of displaced atoms. Finally, positron annihilation spectroscopy provides strong evidence that the epitaxially recrystallized layer is comparable to defect-free $c$-Si.

### 4.2.2 Ion-Energy and Fluence Dependence of Amorphization

The amorphization of Si(111), Ge(100) and GaAs(110) surfaces due to low-energy ion bombardment at room temperature was investigated by Bock et al. [4.85–87]. These authors employed low-energy electron diffraction (LEED) and electron-energy loss spectroscopy (EELS) to determine the c–a transition in those materials and its dependence on $Ar^+$ ion energy (0.1–3 keV) and ion fluence ($1 \times 10^{12}$ to $1 \times 10^{16} \, \mathrm{cm}^{-2}$). Changes of the surface composition in GaAs(110) due to ion irradiation were monitored by AES (see Sect. 4.4). The LEED images obtained from the pristine surfaces were representative of the reconstructions usually observed [4.88, 89], namely Si(111) (7×7) and Ge(100) (2×1). These crystalline surfaces were exposed to ion bombardment in an incremental fashion, and EELS spectra and LEED images were recorded after each step; this procedure was repeated until the accumulated ion fluence was sufficient for complete amorphization of the near-surface region, as evident from the disappearance of all LEED spots and an equilibrium state of EELS peak heights and shapes. The crystalline state was restored by annealing of the samples at temperatures beyond about 600 K for Ge and 800 K for Si.

In this section the impact-energy dependence of the ion fluence required to amorphize the near-surface regions of Si, Ge and GaAs single crystals is

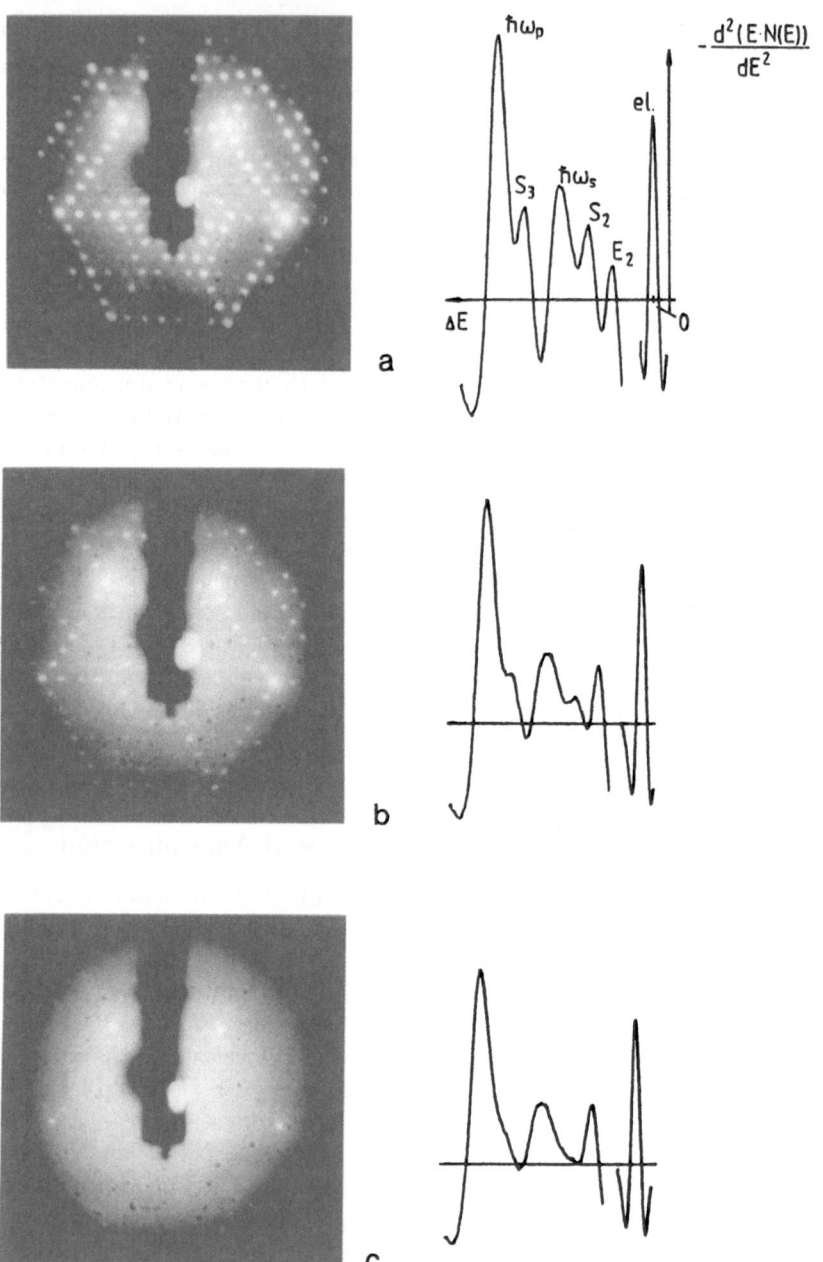

**Fig. 4.11a–c.** LEED patterns and EELS spectra obtained from a virgin Si(111) surface (**a**) and after bombardment with 250-eV Ar$^+$ ions at fluences of $7 \times 10^{13}$ cm$^{-2}$ (**b**) and of $1.5 \times 10^{14}$ cm$^{-2}$ (**c**). EELS spectra are given as the negative second derivative versus the energy relative to the elastic peak ("el"). The characteristic peaks are labeled (see text). Electron energies were 38 eV and 100 eV for LEED and EELS, respectively. Data from [4.85]

compiled [4.85, 86]. These results will be complemented, in Sect. 4.2.3, by data deduced from a theoretical model of ion-induced damage production in solids which closely follows the approach of Brice et al. [4.36]. Figure 4.11 shows a series of LEED patterns and EELS spectra obtained from a Si(111) surface bombarded with 250 eV Ar$^+$ ions at different fluences. Clearly, with increasing ion fluence the LEED spot intensities diminish and the diffuse background increases, until all spots have vanished at fluences of $\sim 2 \times 10^{14}$ Ar$^+$ ions/cm$^2$. Furthermore, it is observed that the spots which are due to the (7×7) reconstruction, and thus contain information about the very first surface layers, disappear at a somewhat lower fluence. In the following, the fluences required for a complete disappearance of bulk- and surface-related LEED spots will be called *amorphization fluences* and denoted $\Phi_b$ and $\Phi_s$, respectively. Results similar to those for Si(111) have been obtained for Ge(100). Again, for fluences of about $2 \times 10^{14}$ cm$^{-2}$, the diffraction spots disappear and the electron energy loss peaks attain an equilibrium state.

The EELS spectra in Fig. 4.11 are characterized by a number of peaks, which for the clean unbombarded surface agree with previous reports [4.90–92]. These peaks are due to the following transitions (their measured energy relative to the peak of the elastically scattered electrons, labeled "el" in Fig. 4.11 is given in parentheses): $E_2$ (4.5 eV), transition between bulk valence and conduction band; $S_2$ (7.4 eV), transition between back-bond surface states and either the bulk conduction band [4.91] or a dangling bond [4.92]; $\hbar\omega_s$ (10.3 eV), surface plasmon with possible contributions from other surface states [4.92]; $S_3$ (14.2 eV), similar to $S_2$; $\hbar\omega_p$ (17.3 eV), bulk plasmon. Upon ion bombardment, these peaks exhibit changes which are correlated with the ion fluence, and a steady-state condition is reached for a fluence value roughly identical to that causing the disappearance of the LEED spots. In particular, peaks $S_2$ and $S_3$, which are both surface related, show a drastic reduction in height for low fluences (some $1 \times 10^{13}$ cm$^{-2}$). These features are more clearly seen when plotting the peak heights normalized to the respective values for the unbombarded surface as a function of the fluence. The results for 250 eV Ar$^+$ incident on Si(111) are displayed in Fig. 4.12. The fluences for bulk and surface amorphization derived from LEED, $\Phi_b$ and $\Phi_s$ (see Fig. 4.14 and below), are also marked.

For Ge(100), EELS peaks due to the following excitation processes (energies with respect to the elastic peak are given in parentheses) were observed: $E_2$ (4.7 eV), transition from the valence band to the conduction band; $S$, a convolution of three distinct peaks [4.93, 94] not resolvable in this work, namely the surface plasmon (11 eV) and two back-bond to dangling bond transitions (8 and 9.4 eV); $\hbar\omega_p$ (16.3 eV), bulk plasmon. Apart from the minimum between the $S$ and $E_2$ peaks (labeled "M"), the bombardment-induced changes in the EELS spectrum were found to be more moderate for Ge than for Si. Nevertheless, plotting the respective normalized intensities versus ion fluence (see Fig. 4.13) yields a distinct and reproducible dependence, char-

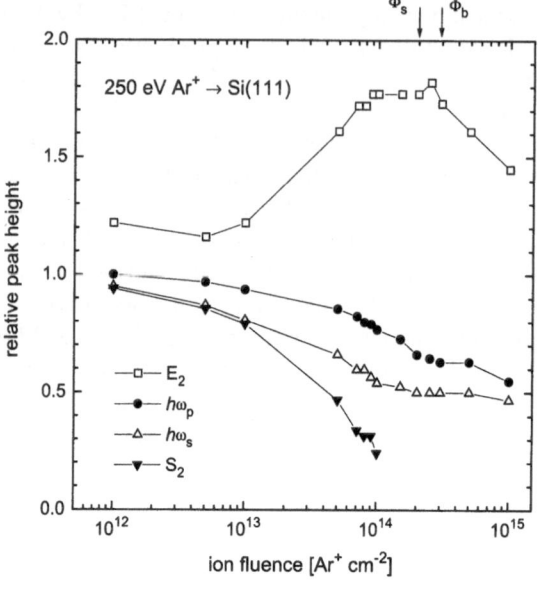

**Fig. 4.12.** EELS peak heights normalized to the values for the unbombarded surface as a function of ion fluence for 250 eV Ar$^+$ ion impact on Si(111). $\Phi_b$ and $\Phi_s$ characterize the amorphization fluences for the surface and the bulk as derived from LEED. Data from [4.85]

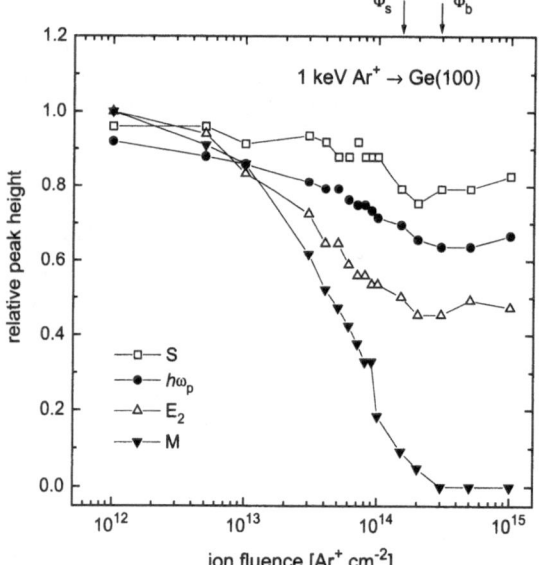

**Fig. 4.13.** EELS peak heights normalized to the values for the unbombarded surface as a function of ion fluence for 1 keV Ar$^+$ ion impact on Ge(100). $\Phi_b$ and $\Phi_s$ characterize the amorphization fluences for the surface and the bulk as derived from LEED. Data from [4.85]

acteristic of the modifications caused by ion impact. Similarly to Si (see Fig. 4.12), the amorphization fluences based on the LEED data are shown in Fig. 4.13 and are found to correspond closely to the leveling off in the EELS intensities at high fluences. Note that the fluences applied here cause negligible sputter erosion of the specimen surfaces (one monolayer or less).

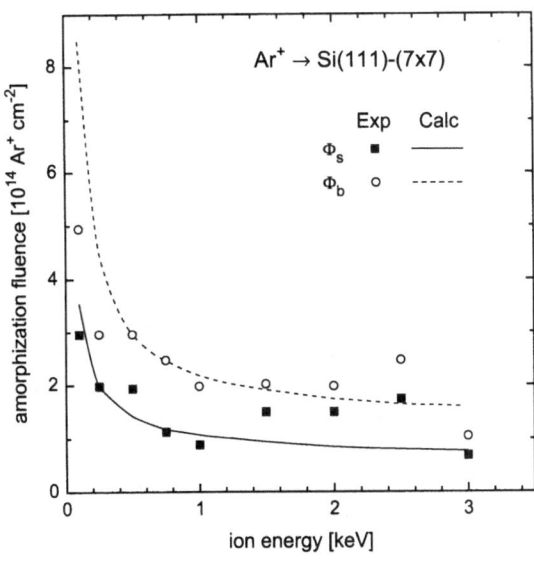

**Fig. 4.14.** Amorphization fluence derived from LEED versus the impact energy for Si(111). The *squares* and *circles* represent the values for the surface and bulk layers, respectively. The *solid* (surface) and *dashed* (bulk) *lines* are the results of a model of defect production outlined in Sect. 4.2.3. Data from [4.85]

The results obtained [4.85] demonstrated that with increasing ion-bombarding fluence, distinct changes are observed both in the LEED patterns and in the EELS spectra, indicative of a (gradual) transition from the crystalline to the amorphous state. The amorphization fluences $\Phi_b$ and $\Phi_s$ exhibit a dependence on the impact energy and, for a given energy, $\Phi_s$ is lower than $\Phi_b$. $\Phi_b$ and $\Phi_s$ are plotted in Fig. 4.14 for the Si(111) (7×7) surface and in Fig. 4.15 for Ge(100) (2×1). The solid and dashed lines in these figures are the

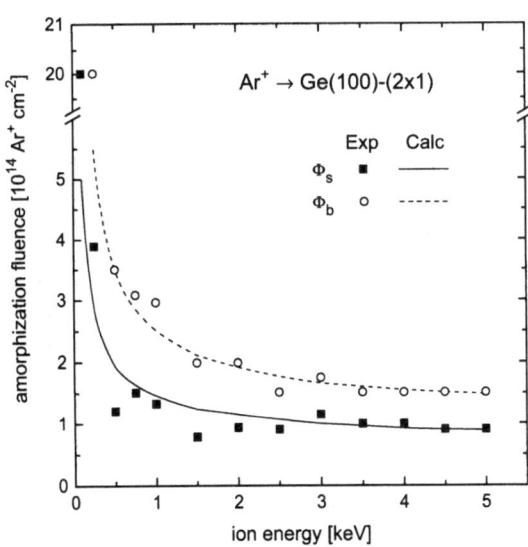

**Fig. 4.15.** Amorphization fluence derived from LEED versus the impact energy for Ge(100). The *squares* and *circles* represent the values for the surface and bulk layers, respectively. The *solid* (surface) and *dashed* (bulk) *lines* are the results of a model of defect production outlined in Sect. 4.2.3. Data from [4.85]

results of theoretical calculations, which will be presented below. As expected, $\Phi_b$ and $\Phi_s$ decrease with increasing impact energy although the variations are rather moderate above 1 keV. Below that value the amorphization fluences increase pronouncedly, in particular for Ge, where at up to $5 \times 10^{15}\,\mathrm{cm}^{-2}$ no bulk amorphization could be established at 100 eV energy. Apart from these discrepancies at low energies, the energy dependence is quite similar for Si and Ge and in good agreement with the theoretical curve.

Though data for the amorphization fluence are rare in the low-ion-energy regime investigated here, the few previous results agree fairly well with the present ones. For 750 eV $Ar^+$ bombardment of Si(111) a value $\Phi_b = 4 \times 10^{14}\,\mathrm{cm}^{-2}$ was reported [4.95]. Jacobson and Wehner [4.96] employing LEED to study the amorphization of Ge(111), found $\Phi_b \approx 10^{14}\,\mathrm{cm}^{-2}$ at 1000 eV, and $10^{15}\,\mathrm{cm}^{-2}$ and $10^{16}\,\mathrm{cm}^{-2}$ at 250 eV and 100 eV, respectively, i.e. a drastic increase at low energies, which is quite comparable to the present findings. High-energy (some ten keV and more) data [4.8, 16] are by far more abundant and yield amorphization fluences of the order of $\leq 1 \times 10^{14}\,\mathrm{cm}^{-2}$ which is in agreement with an extrapolation to higher energies of the present bulk values (see Figs. 4.14 and 4.15). Note, however, that these high-energy studies usually employ techniques (e.g. Rutherford backscattering) which are not surface sensitive, but, rather, probe a region extending (more or less) deeply into the bulk of the solid. By contrast, in the present study, emphasis was laid on the modifications induced in the topmost (one to three) surface layers.

It can be seen from Figs. 4.14 and 4.15 that, at a given energy, the amorphization fluences for the bulk are higher than for the surface. For Si the ratio $\Phi_b/\Phi_s \sim 1.6$ is essentially independent of the bombarding energy. The same statement holds for Ge for energies down to $\sim 1$ keV, while $\Phi_b/\Phi_s$ rises steeply below that value; this observation appears associated with the afore-mentioned increase of $\Phi_b$ at low energies. The reason for this pronounced increase of $\Phi_b$ for Ge and the specific magnitude of these ratios could not be identified unambiguously [4.85]. The ion reflection coefficient for 3 keV $Ar^+$ on Ge is $\sim 0.08$ [4.97] and is thus more than an order of magnitude higher than for Ar on Si (at 3 keV, about $5 \times 10^{-3}$ [4.97]). Thus the enhanced ion reflection in Ge may account for higher amorphization fluences, but probably cannot explain the steep rise of $\Phi_b$ at low energies, as the reflection coefficient rises only moderately (by a factor of two) if the bombarding energy is reduced from 3 keV to 100 eV. Another possibility would be the more open Ge(100) surface as compared to Si(111). An increased ion penetration should lead to a higher amorphization fluence, but it is not expected that this kind of transparency effect is pronounced in this low-energy regime. Furthermore, Jacobson and Wehner [4.96] investigated the (111) surface of Ge and found comparably high amorphization fluences at the same low energies. A third explanation [4.85] may be found in comparing the different ranges of recoiling surface atoms in Si and Ge with the probing depth of LEED. For a 100 eV projectile the mean range of a Ge recoil is about 0.75 nm, which is comparable to the depth that LEED probes ($\sim 0.58\,\mathrm{nm}$). Therefore, at low energies recoiling atoms remain in the near-surface region and may be captured by an already existing vacancy, causing continuous annealing and a reduced defect production rate in the surface layer examined by LEED. By contrast, for the same projectile energy, Si recoils have a slightly higher average range, of about $1\,\mathrm{nm}$, which is twice the probing depth of the diffracted electrons. This leads to the most plausible explanation for the low-energy differences: a continuous annealing which may take place in semiconductors even at room temperature (see the data on GaAs below). Ion-bombardment-induced amorphization is only possible if the damage production rate exceeds the annealing rate. This is not the case for $100\,\mathrm{eV}$ $\mathrm{Ar}^+$ bombardment of Ge [4.35], but is apparently for Si because of its higher annealing temperature.

The GaAs(110) surface is different from the Si and Ge surfaces discussed in the previous section in that it does *not* exhibit a reconstruction. In contrast to other low-index surfaces of GaAs, the (110) surface is nonpolar and thus contains an equal number of Ga and As atoms. Detailed LEED investigations [4.98, 99] revealed, however, that the atoms of the topmost layers are slightly shifted relative to the underlying bulk layers, i.e. the surface layer is relaxed which results in an energetically more favorable condition [4.99]. This vertical cation–anion shear represents about 30% of the original interlayer spacing ($d = 0.1999\,\mathrm{nm}$) and causes a rotation of the Ga–As bond angle by about $27°$ from the ideal surface plane [4.100]. The atoms at the surface experience, because of the missing bonding partners, a rearrangement of their binding-orbital configuration. The As atoms (lying slightly above the Ga atoms on the relaxed surface) now have three p-like orbitals (roughly at angles of $90°$ to each other) which form the bonds to their neighbors and a doubly-occupied dangling-bond state which lies in the valence band. The Ga atoms have three binding $\mathrm{sp}^2$ orbitals and an empty dangling bond in the conduction band of the crystal [4.101].

In the following, the energy dependence of the ion fluence necessary to amorphize the near-surface region of GaAs(110) single crystals [4.86, 87] is presented. These measurements were carried out at room temperature. Owing to the absence of a surface reconstruction, no distinction between surface- and bulk-related diffraction spots as in the case of Si and Ge is possible here. The fluence required for complete extinction of the LEED pattern will thus be called the amorphization fluence $\Phi_\mathrm{a}$ in the following. LEED patterns of GaAs bombarded by $\mathrm{Ar}^+$ ions were recorded from the pristine surface and from surfaces exposed to an increasing total fluence. From such exposure series the values of $\Phi_\mathrm{a}$ were derived for different bombarding energies. Apart from the impact energy, a dependence of $\Phi_\mathrm{a}$ on the ion flux density was also observed. The latter is shown in Fig. 4.16 which plots the scaling of $\Phi_\mathrm{a}$ with impact energy.

Above an impact energy of $\sim 1.5\,\mathrm{keV}$, $\Phi_\mathrm{a}$ is roughly constant, but rises steeply below that value. Two major differences are apparent for GaAs as

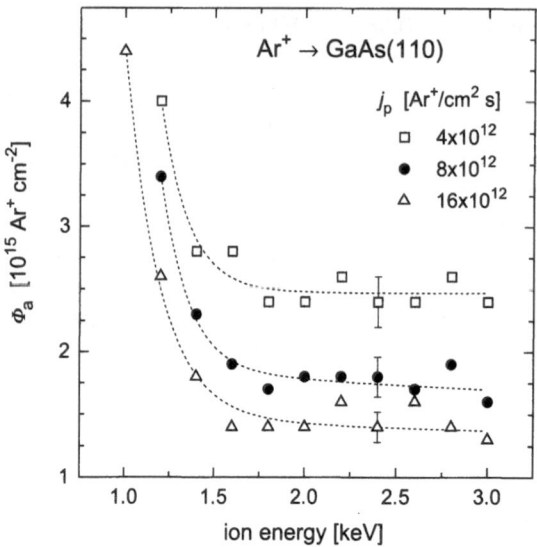

**Fig. 4.16.** Amorphization fluence $\Phi_a$ derived from LEED versus the bombarding energy for $Ar^+$ ion bombardment of GaAs(110). The parameter is the ion flux density. Data from [4.86]

compared to the elemental semiconductors Si and Ge. First, $\Phi_a$ is strongly dependent on the ion flux density [4.102], in such a way that decreasing the latter increases the value of $\Phi_a$ at a given bombarding energy. In these room-temperature experiments, no amorphization occurred below an ion energy $E = 1\,\text{keV}$ for the current densities accessible [4.86]. Similarly, for a flux density of $1 \times 10^{12}$ ions/cm$^2$ s (or less), amorphization of the GaAs(110) surface was not observed for the impact energies employed, even for total fluences in excess of $10^{16}$ ions/cm$^2$. Second, the values of $\Phi_a$ (Fig. 4.16) for GaAs(110) are about an order of magnitude higher than the corresponding ones reported for Si and Ge single crystals bombarded under similar conditions. As will be shown in Sect. 4.2.3, for the latter specimens a correlation between the defect density and the amorphization fluence, largely based on the Kinchin–Pease formula, could be established. Since a flux-density dependence does not enter this model, its application to the results obtained for GaAs appears not possible.

Experiments by Weaver et al. [4.103] on defect production in GaAs(110) using thermal He scattering demonstrated that apart from thermal annealing, another process can reduce the number of defects. In this picture, bombardment-induced recombination of defects with weakly bonded adatoms at the surface can occur. Owing to ion impact, the latter gain sufficient mobility for such a process to take place. The data in Fig. 4.16 indicate, therefore, that annealing processes can occur sufficiently rapidly to counterbalance the production of ion-bombardment-created defects. If, in addition, the creation rate is too low (at low current densities), the crystalline structure remains largely intact (i.e. at least on a scale sufficient to produce LEED images). To verify this possibility additional experiments at different sample temperatures during bombardment would be desirable.

The local density of states at the surface is modified by any distortion or destruction of the crystalline lattice [4.104]. Thus ion-bombardment-induced surface modifications can be monitored using low-energy electrons, which, in the energy range from 10 eV to about 200 eV, are very surface-sensitive. Similarly to the Si and Ge surfaces described above, EELS was also employed [4.86] for GaAs(110) to investigate these irradiation effects. Inelastically scattered electrons provide information on characteristic energy losses due to collective excitations or inter- and intraband transitions. The collective excitations result from the excitation of valence band electrons (bulk and surface plasmons); the inter- and intraband are a consequence of electron transitions between different energy states both within a band and among different bands. The surface relaxation causes the development of additional dangling- and back-bond states (see above), which can take part in electronic transitions.

Figure 4.17a shows a typical energy loss spectrum obtained from a crystalline GaAs(110) surface using 80 eV electrons for excitation. The data were

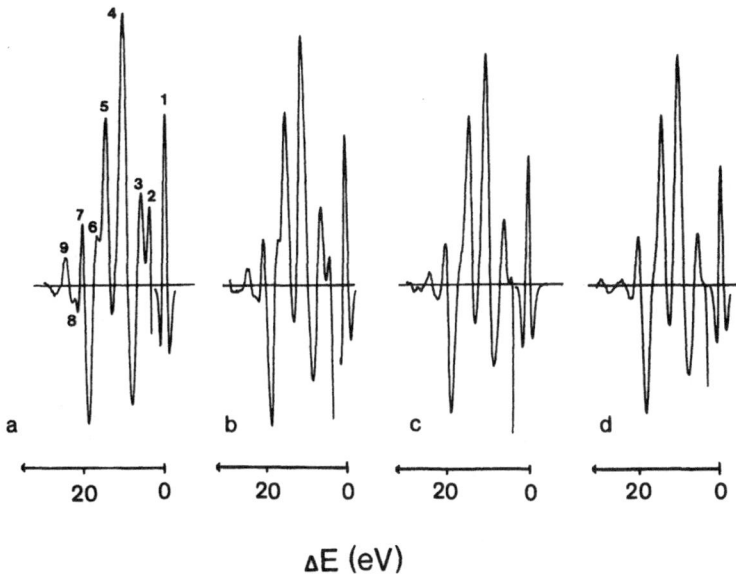

ΔE (eV)

**Fig. 4.17a–d.** EELS spectra of crystalline GaAs(110) recorded as the negative second derivative versus the energy loss relative to the elastic peak ("1"). The energy losses and the respective transitions of the designated peaks are compiled in Table 4.1. The electron energy was 80 eV and the current 1.7 μA. The spectra are from a pristine GaAs(110) surface (**a**) and after bombardment with 3 keV Ar$^+$ at fluences of $5 \times 10^{13}$ cm$^{-2}$ (**b**), $3 \times 10^{14}$ cm$^{-2}$ (**c**) and $2 \times 10^{15}$ cm$^{-2}$ (**d**). Data from [4.86]

recorded as the negative second derivative by means of lock-in techniques and the energy loss is plotted relative to the peak of the elastically scattered electrons. Table 4.1 lists the energy loss of the most prominent peaks (labeled in Fig. 4.17a) and the associated transitions, which are well documented in the literature.

**Table 4.1.** Energy loss peaks shown in Fig. 4.17 with their position ($\Delta E$) relative to the elastic peak, and the electron transitions

| Peak no. | $\Delta E$/eV | Transition |
|---|---|---|
| 1 | 0 | Elastic peak |
| 2 | 3.5 | As dangling bond $\rightarrow$ Ga dangling bond [4.105] and bulk valence to conduction band [4.106] |
| 3 | 5.8 | Bulk valence band to conduction band [4.105, 107] |
| 4 | 10.6 | Surface plasmon [4.105] |
| 5 | 15.5 | Bulk plasmon [4.105, 107] |
| 6 | 18.0 | As back bond $\rightarrow$ Ga dangling bond [4.106] |
| 7 | 20.0 | 3d core electron $\rightarrow$ excitonic surface state [4.108, 109] |
| 8 | 21.6 | 3d core electron $\rightarrow$ Ga dangling bond [4.106] |
| 9 | 24.1 | 3d core electron $\rightarrow$ conduction band [4.105] |

The changes of the EELS spectra due to ion bombardment are depicted in Fig. 4.17b–d. Since variations in peak heights have already been observed for rather low fluences ($\sim 5 \times 10^{13}$ ions/cm$^2$), the ion irradiation was done at a flux density of $1 \times 10^{12}$ Ar$^+$ ions/cm$^2$ s. Some data taken at higher densities produced similar fluence-dependent changes of the EELS spectra, but low-fluence features could not be followed as accurately.

Since no changes in peak shape and width were observed with ion bombardment, the peak heights in the doubly differentiated spectra can be assumed to represent the actual intensities of the backscattered electrons. The normalized peak heights of selected peaks are plotted in Fig. 4.18 as a function of bombarding fluence. For several transitions drastic variations are found at comparatively low fluences ($< 10^{14}$ cm$^{-2}$), and the signals level off to a constant value at about $1 \times 10^{15}$ ions/cm$^2$. It appears that the qualitative shape of these fluence dependence curves are essentially *independent* of impact energy, even for bombardment energies as low as 200 eV (the lower limit accessible experimentally). By contrast, LEED patterns did not show any irradiation-related changes below $E \sim 1$ keV. The EELS data nevertheless demonstrate that some defect production in the near-surface layers does occur and exemplify the sensitivity of this technique.

The most pronounced variations in signal height are found for peaks no. 2 and 9 (see Fig. 4.17). Two processes can contribute to the former peak (Table 4.1). The complete extinction upon ion bombardment indicates, however, that the transition between Ga and As dangling-bond states dominates here. This is because, as discussed in [4.110, 111], the Ga surface state responds very

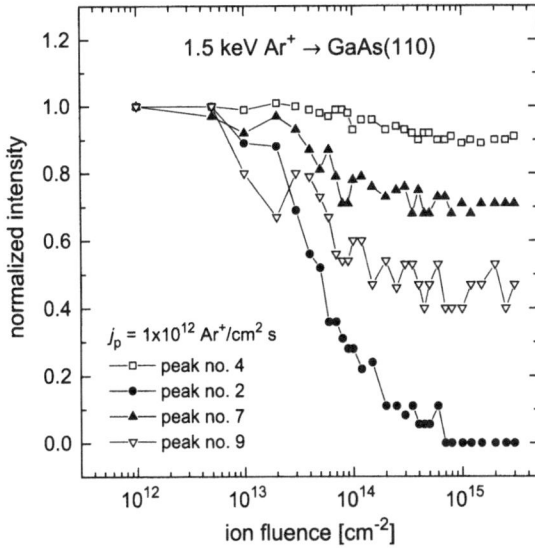

**Fig. 4.18.** EELS peak heights normalized to the values for the unbombarded surface versus ion fluence for 1.5 keV $Ar^+$ ion impact on GaAs(110). Data from [4.86]

sensitively to changes of the Ga–As bond angle at the relaxed GaAs(110) surface; very likely, this angle is readily modified by ion impact and the associated relocation of (surface) atoms. As peak no. 2 disappears at fluences ranging from $6 \times 10^{14}$ to $10 \times 10^{14}$ $Ar^+$ $cm^{-2}$ (which is slightly less than the number of atoms per monolayer), it may be concluded that each impinging Ar ion creates at least one (and possibly more) defect sites in the surface. Peaks no. 6 and 8 also involve the excitation into a Ga dangling-bond state and, not surprisingly, react very prominently to ion bombardment (see Fig. 4.18). Owing to their intrinsic weakness even in the crystalline state, their signal evolution with increasing fluence could not be recorded quantitatively.

Apart from the work outlined in the foregoing paragraphs, the amorphization of GaAs surfaces by low-energy ions is well established through many other investigations [4.112–120]. These studies frequently indicate the existence of a transition temperature: for a given ion current density, above this transition temperature defects created by ion bombardment are annealed out faster than they are produced by ion impact, leaving the specimen largely crystalline. Such conclusions were also drawn from sputtering investigations of GaAs samples: above the transition temperature distinct spot patterns (see Sect. 2.3.3) were observed in the angular emission distribution; these transformed into a featureless, radially symmetric distribution below this temperature (see [4.14] for a review of pertinent data). The occurrence of such a critical temperature will reemerge in the discussion of bombardment-induced composition changes in compound semiconductors (Sect. 4.4). The depth of the damaged (amorphized) zone is comparable to or slightly exceeds [4.77, 112, 116] the ion's range in GaAs (cf. Fig. 4.9 and the associated discussion), an observation that holds also for energies in the ten to hundred keV range [4.121, 122].

Most of what was said about damage in GaAs appears valid also for InP and, possibly, other compound semiconductors, although some distinct differences have been noted [4.14]. For example, InP has a lower threshold damage density (i.e. energy/volume) for amorphization than does GaAs. A systematic study [4.123], using TEM, determined these threshold densities for a number of III–V compound semiconductors. For 20 keV $Si^+$ irradiation at 77 K and a fluence of $1 \times 10^{15}$ cm$^{-2}$, the threshold densities for amorphization fall in the range from $\sim 3 \times 10^{22}$ eV/cm$^3$ to $\sim 7 \times 10^{23}$ eV/cm$^3$, with AlAs and GaAs at the upper and InP, InAs and GaSb at the lower end of this range (see also Sect. 4.2.3). Similar trends can be inferred from related studies [4.18]. The reason for these pronounced differences in the damage production of III–V semiconductors appears to be still unresolved, although different amorphization models were invoked [4.14, 18] as an explanation.

EELS was utilized also to determine the state of the polar InP(100) and InP(111) and the nonpolar InP(110) surfaces [4.124, 125]. Owing to preferential P loss upon ion bombardment, the resulting small In clusters and islands (see below) give rise to bulk and surface plasmon excitations at 11.7 and 8.7 eV, respectively. With this signature at hand, the procedures to produce clean and well-defined surfaces were optimized [4.124]: oblique ion incidence reduces the extent of preferential P sputtering (see Sect. 4.4), and repeated cycles of P deposition and annealing eventually produce an InP(111) (2×2) reconstructed surface and, on InP(100), both In-terminated (4×2) and P-terminated (1×4) reconstructions.

The amorphization and the formation of near-surface damage due to ion irradiation has been studied for InP in more detail [4.14]. This material shows a (near-surface) depletion of P upon ion bombardment (see Sect. 4.4) and, possibly related, the development of a pronounced surface morphology at higher irradiation fluences [4.126–130]. Submicroscopic metallic islands of pure In and monocrystalline conical protrusions have been reported in a number of publications. (Malherbe [4.14] discussed thoroughly the development of gross surface topography such as cone and whisker formation and ripple growth induced by prolonged ion irradiation, and outlined the major models devised to understand this surface morphology.) The lower resistance of InP to beam-induced destructive effects as compared to GaAs has been noted above. The changes in the short-range order effected by ion bombardment in the near-surface region of InP single crystals was investigated by Valeri et al. [4.131]. They bombarded InP(110) surfaces with 0.6 and 1 keV $Ar^+$ ions at normal and oblique incidence and monitored the early stages of defect generation by means of primary-beam diffraction-modulated electron emission (PDMEE). This technique samples the structural arrangement of the surface over a depth determined by the inelastic mean free path of the detected electrons (typically Auger electrons to yield also chemical information) and exploits the modulation of the primary electron wave amplitude due to scattering-interference processes that these electrons experience in *ordered* solids. If the primary-beam incidence angle is scanned, the electron yield reflects this modulation, showing forward focusing maxima for beam alignment with major crystal axes. The anisotropy thus created is directly related to the near-surface structure; in amorphous specimens, such features are averaged out because of the random atom arrangement.

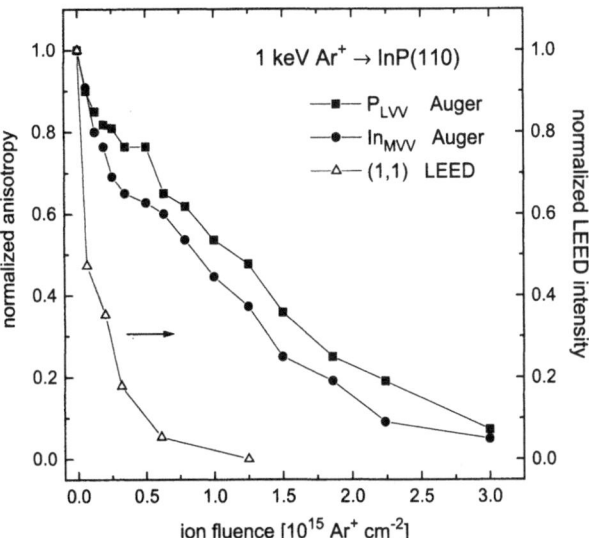

**Fig. 4.19.** Normalized anisotropy observed in primary-beam diffraction modulated electron emission spectra monitoring $P_{LVV}$ and $In_{MVV}$ Auger electrons as a function of ion fluence for 1 keV $Ar^+$ irradiation of InP(110). Also depicted is the normalized intensity of the (1,1) LEED spot. Data from [4.131]

Valeri and coworkers [4.131] monitored these anisotropic features as a function of the ion fluence for 0.6 and 1 keV $Ar^+$ bombardment of InP(110). Figure 4.19 depicts such data for 1 keV $Ar^+$ impact. The normalized anisotropy, derived from the incidence-angle-dependent $In_{MVV}$ and $P_{LVV}$ Auger electron intensities, constitutes a measure of the crystalline order of the specimen. With increasing ion fluence, the anisotropy rapidly decreases and, upon further irradiation, levels off to a value close to zero at a fluence of $\sim 3 \times 10^{15}\,cm^{-2}$. This amorphization fluence was found to be slightly higher for 0.6 keV $Ar^+$ ions, but smaller by a factor of about two for 1 keV ions at an incidence angle of 55°. (In agreement with the data of Gnaser et al. [4.86], these authors found for a GaAs(110) specimen that no complete amorphization is reached for 1 keV $Ar^+$ bombardment, with the residual anisotropy at a fluence of $\sim 3 \times 10^{15}\,cm^{-2}$ amounting to roughly 10%.)

Apart from PDMEE, Valeri et al. [4.131] also employed LEED to monitor the gradual amorphization of the InP(110) surface upon bombardment. The spot intensities diminish with increasing ion fluence and no spot pattern is visible for values larger than $\sim 5 \times 10^{14}$ $Ar^+$ $cm^{-2}$. The decrease of the normalized (1,1) spot intensity is depicted also in Fig. 4.19. On the basis of modeling of their experimental results, the authors proposed that structural damage in crystalline InP surfaces nucleates in a *subsurface* region, at a depth that slightly exceeds the ion's range. Crystalline order is completely removed in this subsurface range at an ion fluence of $\sim 5 \times 10^{14}$ $cm^{-2}$ for 1 keV $Ar^+$ ions. Further ion bombardment and concurrent sample erosion move the disordered layer towards the surface, resulting in complete amorphization over a depth larger than 6 nm at a fluence of $\sim 3 \times 10^{15}$ $cm^{-2}$. From the LEED data it was concluded that long-range order disappears on InP(110) surfaces at a somewhat lower fluence.

Whereas in III–V semiconductors, owing to their strong covalent bonding, radiation-induced defects are usually localized within the region of the ion's nuclear energy deposition, II–VI compound semiconductors are characterized by lower binding energies and higher ionicity; extended defects (e.g. dislocation loops) are often observed which propagate beyond the ion's penetration range [4.132]. Owing to ionic forces, recombination of point defects is more readily effected and amorphization is less likely to occur even for high fluences [4.14]. Deep radiation damage (for high-energy ion bombardment) is, however, a very common feature for II–VI semiconductors [4.133–136]. By contrast, low-energy experiments performed at low ion fluences in order to monitor the possible transition to an amorphization of the near-surface region are still limited [4.136, 137].

### 4.2.3 Bulk Versus Surface Defects: Computations of Defect Production

Epitaxial growth at the surface may be modified by ion-beam-generated defects. Several different phenomena may occur as a result of low-energy ion bombardment during (epitaxial) growth, including enhanced adatom diffusion, sputtering, dissociation of small islands and formation of alternate nucleation sites. Hence, the modification of the surface kinetics is of interest but it is usually necessary to limit the bulk damage caused by the impinging ions. The boundary between these two regimes (surface damage *with* or *without* concurrent bulk damage) in terms of irradiation parameters (energy, ion species) is rather poorly known even for the best-studied semiconductor, Si. Computer simulations [4.21, 22, 27, 138–140] have provided some insight into the optimal bombardment conditions.

Atomically rough Si(100) surfaces have been studied by MD simulations [4.21] using Ar ions of 10–50 eV energy at an incidence angle of 45° as the bombarding species. The numbers of displacements per ion for each layer are depicted in Fig. 4.20 for different irradiation energies. (The criterion for a

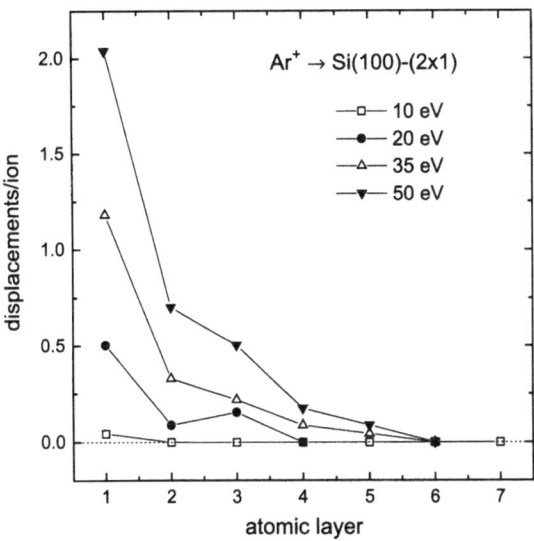

**Fig. 4.20.** The distribution of displacements in individual layers (layer 1 corresponds to the surface) for Ar$^+$-bombarded Si(100) (2×1) at an incidence angle of 45° as derived from MD computer simulations. The parameter is the Ar energy. Data from [4.21]

displacement was such that no Si atom was located within a sphere of a radius equal to half the equilibrium Si–Si bond length.) For the lowest energy, the defects are almost exclusively confined to the top layer, whereas deeper-lying layers are affected with higher energies, up to the fifth layer at 50 eV. The yield data indicated, furthermore, that the surface displacements are site-specific. The ratio of displacements in the topmost reconstructed layer to all deeper layers decreases from a value of 2–5 for 15 eV ion impact to 0.6–0.8 for 50 eV. The total number of displacements (i.e. the sum over all layers in Fig. 4.20) increases with the ion energy from $\sim 0.05$ at 10 eV Ar impact to $\sim 4$ displacements/ion at 50 eV bombardment. The authors conclude that ions with energies less than 20 eV can alter the surface kinetics without causing bulk damage. This is in agreement with successful epitaxy in low-energy bias sputtering experiments carried out at 570 K. The simulations implied that the most important effect of ion bombardment on surface self-diffusion and morphology is an *increased formation rate of single adatoms*.

Hensel and Urbassek [4.27] performed a very detailed MD simulation study of implantation and damage under low-energy (50 and 100 eV) Si self-bombardment, utilizing (100) (2×1), (110) and (111) oriented crystals. For these surfaces and the two Si impact energies, they derived the range distributions and the depth of disordered atoms and of empty sites. (These authors define an atom as disordered if it is located outside the Lindemann radius, i.e. the atom's vibration amplitude at the melting point, and, in analogy, they define an empty site as one with no atom within that sphere.) Furthermore, Hensel and Urbassek [4.27] conclude from their data that low-energy irradiations produce near-surface disordered zones. For 100 eV Si bombardment, these disordered zones contain about 15 displaced atoms, largely independent of crystal orientation. This number can be derived by dividing the projectile energy by an amorphization energy of 6–7 eV. The magnitude of such an energy is comparable with values from other simulations. The authors view target amorphization to proceed via the overlap of single-ion-impact-induced amorphous regions. These amorphous zones are metastable and recrystallize at sufficiently high temperatures ($> 900$ K), leaving only point defects at the surface. With the exception of the fairly open (110) surface, they found the damage to extend to a considerably larger depth into the target than the ion's range, indicative of a massless energy and momentum transport.

An approximate relationship between surface and bulk defect generation was computed by Brice, Picraux and coworkers [4.36]. They determined isocontours for constant ratios of the number of displaced atoms at the surface to that in the bulk. These were calculated by analytical evaluation of a displacement energy loss function $S_d$. This was obtained from the usual stopping-power expression, but with integration over only those energy loss values $T$ higher than the displacement threshold energy $E_d$. The energy into displacements was derived by integrating $S_d$ along the path of the projectile, first in the surface layer and then in the underlying bulk (see below). For the computations the surface displacement threshold energies were taken to be half the bulk displacement threshold energies.

Obviously, neither surface nor bulk displacements are possible if the incident energy falls below a limiting value; the latter is mass-dependent and its minimum (exactly equal to the surface displacement threshold) occurs when the incident and target masses are equal and momentum transfer is most efficient. Above this lower limit but below the bulk displacement threshold, there exists an energy/mass range where only surface displacements and no bulk displacements can occur. Owing primarily to the increased energy loss per unit length for heavier ions, the computations [4.36] show that this window widens significantly with increasing ion mass. For energies beyond the bulk displacement threshold, defects would be created both at the surface and in the bulk. In this range, the higher the energy, the deeper the ion penetration and the higher the number of bulk to surface displacements. Clearly, the calculations depend sensitively on the values of the displacement energies; the corresponding data for surface displacement thresholds are not known precisely. For their computations, these authors [4.36] used, for example, values of 11 eV and 7.5 eV for the surface displacement energies of Si and Ge, respectively. With these values, the calculated window for the creation of displacements at the surface but not in the bulk of Si spans an energy range from 12 eV to 45 eV for a light ion (mass 10 amu) and widens to a range from 20 eV to 100 eV for a heavy ion of mass 100 amu.

Following the procedure of Brice et al. [4.36], the defect and energy densities necessary for the amorphization of the surface were calculated by Bock et al. [4.85] and compared with their LEED data on Si(111) and Ge(100) (see Sect. 4.2.2). The number of defects (vacancies) created per incoming ion was calculated according to Sigmund's version [4.141] of the Kinchin–Pease formula (see (2.26) in Sect. 2.2):

$$\nu\left(E, E_d\right) = \alpha \frac{E_n\left(E, E_d\right)}{2E_d} ,$$ (4.2)

where $E_d$ is the displacement energy for removing a target atom from its lattice site, $E_n(E, E_d)$ is the fraction of the projectile's energy $E$ spent in elastic collisions and $\alpha$ is a constant ($\sim$0.8 for semiconductors). To obtain an estimate of the number of defects created in a depth interval $z_1 \leq z \leq z_2$ of the solid (i.e. in the surface or near-surface layer accessible by the experimental technique used), the energy $\varepsilon_{dis}$ lost by the primaries in defect (displacement) generation within this depth has to be calculated. For this purpose Brice et al. [4.36] defined, in analogy to the nuclear (elastic) stopping cross section $S_n(E)$, a partial stopping cross section for displacements

$$S_{dis}\left(E\right) = \begin{cases} \displaystyle\int_{E_d}^{T_{max}} T d\sigma\left(E, T\right) , & T_{max} > E_d , \\ 0 , & T_{max} \leq E_d . \end{cases}$$ (4.3)

As in Sect. 2.1.1, $T$ is the recoil energy transferred to a target atom in a collision, $d\sigma(E, T)$ is the differential elastic scattering cross section for recoil energy in the interval $(T, T + dT)$, and $T_{max} = 4M_1 M_2 E / (M_1 + M_2)^2$ is the maximum energy that a projectile with mass $M_1$ and energy $E$ can transfer to a target atom with mass $M_2$. Then, the spatial rate of energy deposition into atom displacement is

$$\frac{d\varepsilon_{dis}}{dz} = -N S_{dis}\left(E, E_d\right) ,$$ (4.4)

where $N$ is the target atom density; the total energy deposition into displacements in a surface layer of thickness $\Delta z$ is given by

$$\varepsilon_{dis}\left(E, E_d\right) = -N \int_0^{\Delta z} S_{dis}\left[E\left(z\right), E_d\right] \langle \cos\varphi \rangle \, dz .$$ (4.5)

Here $E(z)$ is the projectile's energy at depth $z$, and $\langle \cos\varphi \rangle$ is the mean direction cosine of the ion motion ($\varphi$ is the angle relative to the ion's initial direction) and accounts for the scattering the projectile experiences during slowing down. According to Biersack [4.142] this influence is described by

$$\langle \cos\varphi \rangle = \exp\left[-2\tau\left(E_0, E\right)\right] ,$$ (4.6)

where $E$ is the initial energy of the projectile, $E_0$ is its energy at a given stage of slowing down and

$$\tau\left(E_0, E\right) = -\frac{\mu}{4} \int_E^{E_0} \frac{S_n\left(E'\right)}{S_n\left(E'\right) + S_e\left(E'\right)} \frac{dE'}{E'} ,$$ (4.7)

where $\mu = M_2 / M_1$ and $S_n$ and $S_e$ are the stopping cross sections for, respectively, elastic and inelastic energy losses (see Sect. 2.1.1). For normal incidence, the projectile energy $E(z_i)$ at a depth $z_i$ is obtained from

$$z_i = \int_{E(z_i)}^{E} \exp\left[-2\tau\left(E_0, E\right)\right] \frac{\mathrm{d}E'}{N\left[S_{\mathrm{n}}\left(E'\right) + S_{\mathrm{e}}\left(E'\right)\right]} . \tag{4.8}$$

With the total stopping power of the projectile, $\mathrm{d}E/\mathrm{d}z = -N\left[S_{\mathrm{n}}(E) + S_{\mathrm{e}}(E)\right]$, (4.5) and (4.8) yield the energy deposited into displacements in the depth interval $z_1 \le z \le z_2$ (corresponding to particle energies $E_1 \le E \le E_2$) as

$$\varepsilon_{\mathrm{dis}}\left(E_1, E_2, E_{\mathrm{d}}\right) = \int_{E_1}^{E_2} \exp\left[-2\tau\left(E_0, E\right)\right] \frac{S_{\mathrm{dis}}\left(E_1, E_2, E_{\mathrm{d}}\right)}{S_{\mathrm{n}}\left(E\right) + S_{\mathrm{e}}\left(E\right)} \mathrm{d}E, \tag{4.9}$$

employing (4.8) to derive $E_1(z_1)$ and $E_2(z_2)$.

The displacement energy $E_{\mathrm{d}}$ (entering into $S_{\mathrm{dis}}$) was assumed depth-independent here. This approach [4.36] includes some approximations. First, straggling of projectiles is neglected. This causes an underestimation of the deposited energy. Second, energy transport by recoiling target atoms is not taken into account. Since this effect is most important at low energies and in the near-surface region, the calculations will overestimate the energy deposition. Finally, electronic energy losses have been neglected in the numerical computations [4.85]. Since they are at least an order of magnitude smaller than the elastic energy losses in the regime covered by the experiments (see Fig. 2.1), this approximation appears justified. For the numerical evaluation of $S_{\mathrm{n}}$ and $S_{\mathrm{dis}}$ in (4.9) the universal potential of Ziegler et al. [4.143] was used to compute the elastic scattering cross section.

Because of the surface sensitivity of the experimental techniques employed, the deposited energy was calculated according to (4.9) for a surface layer thickness of 0.53 nm for Si and of 0.58 nm for Ge which corresponds to the mean escape depth of the elastically scattered electrons [4.144]. Since the LEED data are also sensitive to the reconstructed surface layer(s), the energy deposition was, in addition, evaluated for these layers, whose thickness amounts to 0.14 nm for Ge(100) [4.89] and 0.26 nm for Si(111) [4.145]. From (4.2) the defect concentration $P$ produced in a layer of thickness $\Delta z$ can be derived:

$$P = \frac{\alpha}{\Delta z \, N_z} \frac{\varepsilon_{\mathrm{dis}}\left(E_1, E_2, E_{\mathrm{d}}\right) \Phi}{2 \, E_{\mathrm{d}}}, \tag{4.10}$$

where $N_z$ is the atom density in $\Delta z$ and $\Phi$ is the ion fluence (ions/cm$^2$). For comparison with the experimental results shown in Figs. 4.13 and 4.14, the fluence $\Phi$ was fitted to the measured data of $\Phi_{\mathrm{b}}$ and $\Phi_{\mathrm{s}}$ using $P$ as a fit parameter. For both Si and Ge, $E_{\mathrm{d}} = 15\,\mathrm{eV}$ was chosen for the bulk and half that value for the surface damage computations [4.36]. The solid and dashed lines in Figs. 4.14 and 4.15 correspond to the fluences calculated according to this procedure. The corresponding critical defect concentration $P$ for amorphization amounts to 0.35 and 0.38 for the surface and bulk layers, respectively, of Si; for Ge the corresponding values are 0.57 for the surface and 0.42 for the bulk. Note that "surface" and "bulk" refer to the above-mentioned layer thicknesses. For Si and Ge, $P$ values ranging from 0.1 to 0.5

have been reported [4.14, 146–148]. These measurements generally probed (by RBS) a much greater depth of the solid than the surface-specific experiments reported here. Obviously, increasing $E_d$ used in the calculations (see (4.10)) would decrease the value of $P$ and vice versa. Still, for reasonable values of $E_d$ (from 10 to 20 eV), $P$ will fall in the range from 0.25 to 0.6, i.e. comparable to previously reported data. The linear theory outlined above is limited to moderately low damage densities (exhibiting no saturation and no appreciable overlap of damaged regions) and to systems showing no annealing of the created damage. Since the present defect densities appear low enough and annealing only plays a role for Ge at very low impact energies (as discussed above), the theoretical treatment appears not to be invalidated by these limitations.

Closely related to a critical defect density is the threshold energy density which causes amorphization. From the energy $\varepsilon_{dis}(E_1, E_2, E_d)$ deposited into the surface and bulk layers, the energy density $\Theta_{b,s}$ can be calculated:

$$\Theta_{b,s} = \frac{\varepsilon_{dis}(E_1, E_2, E_d)}{\Delta z} \Phi_{b,s}(E) , \tag{4.11}$$

where $\Phi_{b,s}(E)$ is the fluence necessary for amorphization at a given projectile energy $E$ and $\Delta z$ is the thickness of the surface and bulk layers considered. From the measured values of $\Phi_{b,s}$, the critical energy density $\Theta_{b,s}$ averaged over all projectile energies is derived. For Si the value is $(4.0 \pm 1.4) \times 10^{23}$ eV/cm$^3$ and $(6.3 \pm 1.9) \times 10^{23}$ eV/cm$^3$ for the surface and the bulk, respectively. The corresponding values for Ge are $(4.3 \pm 0.9) \times 10^{23}$ eV/cm$^3$ and $(6.3 \pm 1.3) \times 10^{23}$ eV/cm$^3$. The bulk result for Si is essentially identical to data reported for the high-energy bombardment of silicon with a variety of ions at different target temperatures [4.17, 149]. The values of $P$ and $\Theta$ are of course averages as, according to the energy distribution profiles, neither $P$ nor $\Theta$ are distributed homogeneously over the depth intervals considered in the calculations. In view of the rather thin layers considered, this effect might, however, be of minor importance.

## 4.3 Modification of Electronic Properties

The structural changes in ion-bombarded semiconductors are closely related to modifications of their electronic and optical properties. Obviously, such effects are enormously important in micro- and optoelectronic device applications and the (degree of) recovery of the required electronic quality following an ion irradiation step has been the subject of long-standing and intensive research efforts. Typically, the threshold fluences for changes of these properties [4.113, 114] are much lower than for structural damage, and the damage regions measured by electrical or optical methods usually extend deeper into the specimen. Some results for low-energy ion bombardment are presented in the following paragraphs. The electrical properties of a semiconductor material in the near-surface region can be modified as a result of Fermi-level pinning caused by the presence of defect states and, for III–V compounds, the preferential removal of Group V elements during ion bombardment. For example, in the case of n-type GaAs, ion irradiation results in a substantial depletion of carriers in the near-surface region [4.150]. In extreme cases (e.g. low-doped p-type InSb) it is possible to type-convert a thin layer of material such that this layer exhibits the properties of n-type material [4.151]. Clearly, the electrical properties of device materials subjected to ion bombardment will be dependent on the modifications induced in such damage layers. Investigations both of the degree and depth of damage and of the subsequent recovery of the electrical properties upon thermal annealing are of considerable importance.

High-resolution electron-energy loss spectroscopy (HREELS) has proved [4.134] to be a sensitive probe of plasmon excitations and is thus a direct means of measuring the free-carrier concentration in semiconductor materials. (The free-carrier concentration $n_c$ is proportional to the square of the surface plasmon frequency $\omega_s$, $n_c \propto \omega_s^2$.) In the dipole scattering process that dominates the small-angle scattering of these long-wavelength excitations [4.152], the *effective* probing depth is not determined by the mean free path of the electrons but is governed by the inverse of the transferred wavevector parallel to the surface. In common situations this depth is of the order of about 100 nm, thus corresponding to typical space-charge layer thicknesses. By changing the kinetic energy of the incident electrons it is possible to probe the free-carrier concentration at different depths within the solid. Not surprisingly, HREELS is widely utilized in experiments that are aimed at determining electrical properties of ion-bombarded semiconductor materials (see [4.152] and references therein).

In attempts to minimize the damage created in ion-etching processes, low energies or light ions are frequently used. HREELS has been applied to assess the near-surface free-carrier concentration in GaAs(100) following low-energy (0.25–1.5 keV) hydrogen ion bombardment [4.150]. For impact energies below 500 eV, the nominal bulk free-carrier density is recovered by annealing at 725 K. At impact energies in excess of this value, damage is retained even after annealing. These defects act to compensate the majority carrier concentration in n-type GaAs and serve to reduce the near-surface carrier concentration.

Photoreflectance studies [4.153] into the effects of ion bombardment on n- and p-type GaAs(100) surfaces show that as a result of 500 eV Ar$^+$ impact the Fermi level moves from a midgap position to near the conduction band for both types of specimen. Subsequent annealing at 620 K restored the Fermi level to its original midgap value. Considerable implications for various models of Schottky barrier formation were anticipated from these observations [4.153].

HREELS was also employed [4.154] to investigate ion-bombardment-induced damage in In-based compound semiconductors. In both n- and p-type InSb(100), 500 eV Ar$^+$ irradiation creates a large excess of free carriers in the near-surface region that persists after annealing to $\sim 500$ K. The damage depth extends approximately 80–100 nm into the bulk of the specimens, corresponding to the formation of an n-type layer near the surface. It is noteworthy that the thickness of this electrically modified layer exceeds the range of the impinging ions by at least an order of magnitude.

InAs is of particular interest for certain device applications because of the natural electron *accumulation* layer formed at the (100) surface; this behavior differs from that of other III–V materials such as GaAs, which forms a natural *depletion* layer at the surface. HREELS measurements [4.155] on a low-doped ($n_c \sim 2 \times 10^{16}$ Si/cm$^2$) InAs(100) sample showed an increased carrier concentration ($\sim 10^{18}$ cm$^{-3}$) after ion bombardment and annealing. Measurements of the plasmon frequency as a function of the incident electron-beam energy indicated that the depth of electronic damage extends at least 40 nm into the material (for 0.5 keV Ar$^+$ ions). Increasing the ion impact energy gives rise to even higher residual carrier concentrations. These results are exemplified in Fig. 4.21 which shows the free-carrier concentration as a function of Ar$^+$ ion energy at an incidence angle of 45° [4.155]. The carrier concentration was derived from the experimentally determined surface plasmon frequencies. The carrier concentration was found to be highest at normal ion incidence and to decrease at oblique impact angles. These and similar studies of ion beam damage in semiconductors indicate that residual damage is often detectable in amounts well below the detectability limits for structural damage; this statement appears to apply also to the near-surface

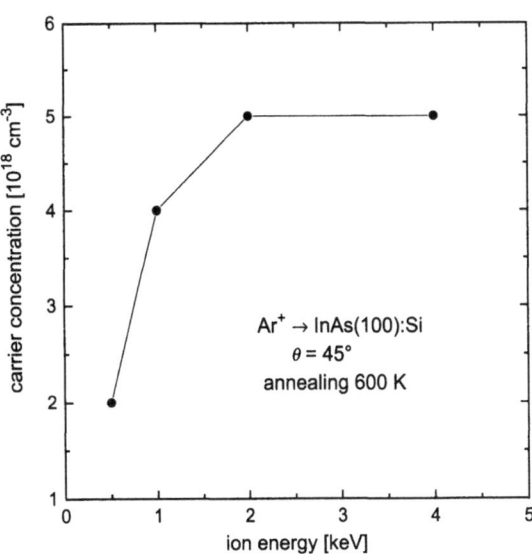

**Fig. 4.21.** The free-carrier concentration plotted as a function of Ar$^+$ ion energy for n-type (Si-doped, $n \approx 2 \times 10^{16}$ cm$^{-3}$) InAs(100) after ion bombardment at 45° and annealing at 600 K. The carrier concentration was obtained from the measured surface plasmon frequency. Data from [4.155]

region of solids. This higher sensitivity might be the reason for the apparently extended depth range of electrical modifications as compared to the ion range noted above.

For an InAs(110) surface bombarded with 0.5–1.5 keV Ar$^+$ ions, no changes in surface stoichiometry were observed [4.156]. However, the electronic properties of the surface are distinctly affected: the surface states of the pristine specimen are quenched by ion irradiation (and do not fully recover on annealing). Extra peaks in the photoemission In 4d core-level spectra indicate partial breaking of In–As bonds with a rearrangement. Furthermore, evidence for a downward bending induced by surface vacancies was reported.

Photoelectron spectroscopy was used [4.157] to study InP single crystals subjected to 3 keV Ar$^+$ ion bombardment. Distinct changes of structural and electrical properties were observed after irradiation fluences as low as $1 \times 10^{12}$ ions/cm$^2$. The surface become In-enriched (P/In ratio $\sim 0.5$) in the topmost two monolayers and the spectra indicate the presence of chemical components such as In–In bonds, P–P bonds and In atoms bonded to P clusters. The shifting of the bulk InP signature in the energy spectra relative to that for the pristine state indicated that the surface band bends down to 0.5 eV from the flat-band condition and pins the Fermi level at $-0.6$ eV from the conduction band minimum. Acceptor defects induced by ion bombardment causes this bending. The authors noted [4.157], in addition, that lowering of the Ar$^+$ ion energy to 1 keV does not modify their findings. Annealing of the irradiated samples at 520 K shifts the Fermi level from $-0.6$ to $-0.4$ eV below the conduction band minimum and causes some of the metallic In to diffuse into the bulk of the crystal and/or to disperse into smaller clusters.

The effect of 350 eV He$^+$, Ar$^+$ and Xe$^+$ bombardment on the induced carrier density in PbTe thin films was investigated by Kubiak et al. [4.158]. (The areal density of carriers was derived in this work from Hall measurements.) With increasing ion fluence, the induced carrier density was reported to increase and to saturate for fluences of about $5 \times 10^{15}$ ions/cm$^2$. Then, the areal density of carriers amounted to $\sim 6 \times 10^{13}$ cm$^{-2}$ for Ar$^+$ and Xe$^+$ ions and to $\sim 2 \times 10^{13}$ cm$^{-2}$ for He$^+$. The results were discussed in terms of preferential sputtering of Te atoms in combination with relocation by collisional mixing. As an excess Pb component produces n-type doping in bulk PbTe, the authors conclude that n-type carriers produced by ion bombardment were a consequence of a sputter-induced deficiency of Te in the near-surface region of the thin films. Apparently, removal of the weakly bound surface species contributes significantly to the observed changes in carrier concentration.

Owing to its importance as a narrow-gap semiconductor material, the effects of ion irradiation on the electrical properties of $Hg_{1-x}Cd_xTe$ have been investigated extensively [4.133–136]; in particular, the importance of radiation damage for junction formation was recognized. Most of these studies were performed at ion energies that are beyond the energy range considered in this book (for a bibliography see [4.14]). Low-energy investigations, on the other hand, were often in the form of ion-beam milling, employing rather high fluences [4.135, 136]. Ion milling of p-type $Hg_{1-x}Cd_xTe$ ($x \sim 0.21$) with neutralized Ar ions (energy range 200–1000 eV) at ion currents of about $0.5\,mA/cm^2$ produced [4.136] thick n-type layers near the surface. The samples were characterized by electron-beam-induced current, differential Hall effect and conductivity measurements. Typically, the depth of the created p–n junction increases with decreasing hole concentration (at 77 K) before ion milling, ranging from several tens of microns to more than 200 µm. The electron concentration in the n layer is of the order of $10^{15}\,cm^{-3}$ and is essentially constant throughout the layer. According to the authors [4.136], the mechanisms for p–n type conversion are not completely understood. A common explanation is based on the idea [4.135] that a large concentration of extended defects (stacking faults) is created because of stresses during milling, and Hg atoms migrate rapidly along them inside the crystal; this step would be accompanied by annihilation of metal vacancies and recombination of defects.

# 4.4 Composition Changes in Compound Semiconductors

The processes leading to ion-bombardment induced composition changes in multicomponent materials discussed in Chap. 3 may, of course, occur also in compound semiconductors. They will, therefore, not be reiterated here. Contrary to the case for many metallic alloys, studies for these semiconductor materials appear to be much more limited, with the possible exception of GaAs and InP. This somewhat solitary position is probably due to their great technological importance. It might also be partly due to the difficulty of preparing, in situ, well-defined and stoichiometric surfaces of many compound semiconductors. Typically, the Group V components in III–V compounds are fairly volatile and, depending on the crystal orientation, polar surfaces consisting exclusively of one component are encountered. Such a monoelemental topmost surface layer can enhance the difficulties in identifying composition changes near the surface. The following discussion will therefore emphasize results obtained on the (nonpolar) GaAs(110) surface and present only a limited number of data for other compound semiconductors. Malherbe [4.14] has given a rather comprehensive account of composition changes in compound semiconductors. The data presented in this review will not be repeated here; rather, selected examples will be used to illustrate important aspects of these compositional changes in semiconductors and to complement thereby the data and processes discussed in Chap. 3 for other materials.

## 4.4.1 GaAs(110)

Compositional changes induced by low-energy ion bombardment of the GaAs(110) and other GaAs surfaces have been studied extensively by a variety of techniques, most often AES, XPS and ISS [4.14]. Recently, Gnaser et al. [4.86] monitored, using Auger electron spectroscopy, the dependence of the Ga and As surface concentrations on the Ar$^+$ impact energy, the current density and the ion fluence. In addition, some measurements were performed at elevated sample temperatures. Normal-incidence ion bombardment with energies in the range from 200 eV to 3 keV was performed at three different flux densities ($1 \times 10^{12}$, $4 \times 10^{12}$ and $16 \times 10^{12}$ Ar$^+$/cm$^2$ s); the applied fluence was increased in increments of $5 \times 10^{14}$ ions/cm$^2$ until a total fluence of $1 \times 10^{16}$ cm$^{-2}$ was accumulated. AES spectra were recorded after each step; excitation was performed with 3 keV electrons (current $\sim 2\,\mu$A), and the low-energy AES peaks of As (31 eV) and of Ga (55 eV) were monitored as the peak-to-peak amplitude in the (singly) differentiated spectrum. Upon reaching the maximum fluence the sample was recrystallized by annealing at 830 K.

For two impact energies, Fig. 4.22 shows the dependence of the Ga/As surface concentration ratio as a function of fluence, with the flux density as the parameter. The data in Fig. 4.22 represent, therefore, the gradual variation of the surface composition due to ion irradiation. Because of their low

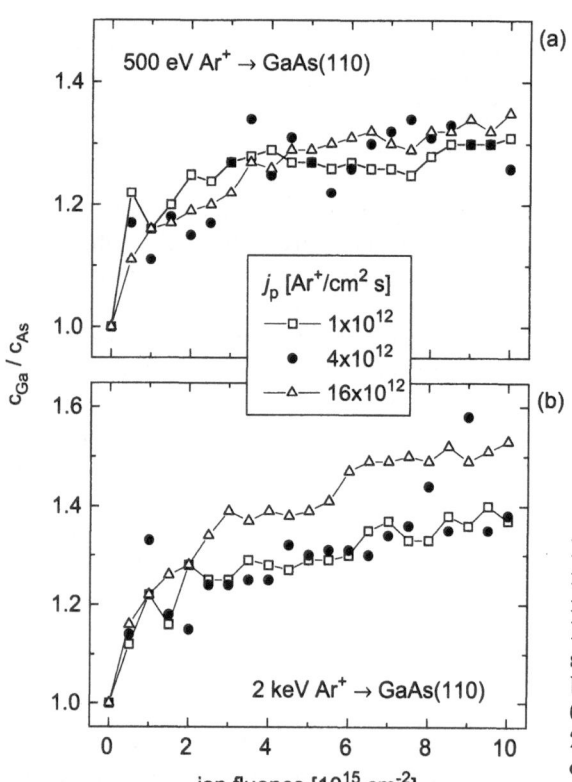

**Fig. 4.22a,b.** Ga/As surface concentration ratios derived from the respective low-energy AES peaks versus the fluence for the bombardment of GaAs(110) by 0.5 keV Ar$^+$ (**a**) and by 2 keV Ar$^+$ (**b**). The parameter is the ion flux density. Data from [4.86]

energy (and the associated short mean free path length), the AES electrons probe a depth of about 2–3 monolayers [4.159, 160]. The penetration depth of $Ar^+$ ions in GaAs amounts to $\sim 1$ nm at $200$ eV and increases to roughly $4$ nm at $3$ keV [4.143]. Thus, at the lowest bombarding energies the modified sample region is comparable to the information depth, but becomes much larger at the higher energies.

Figure 4.22 exemplifies the general tendencies of the results: (i) With increasing bombarding fluence the surface is *enriched* in Ga and an equilibrium is reached for a fluence $\sim 10^{16}$ cm$^{-2}$ for low impact energies and, judging from an extrapolation, at some $10^{16}$ cm$^{-2}$ for $E > 1$ keV. (ii) The near-surface Ga enrichment *increases* with *increasing* impact energy. (iii) While at low bombarding energy the steady-state enrichment is independent of the current density, it appears to increase with flux density above $\sim 1$ keV; this observation is, however, less clear-cut in that the data [4.86] for the intermediate flux density $(4 \times 10^{12}$ cm$^{-2}$ s$^{-1})$ exhibit considerable scatter in this energy regime. It is worthwhile to note that the absolute AES signals of Ga and As produce a different fluence dependence: for all energies and fluxes, the As signal remains roughly constant, but the Ga intensity rises with increasing fluence.

From the data in Fig. 4.22 and similar data for the other impact energies, the steady-state surface concentration ratios $(c_{Ga}/c_{As})_\infty$ were derived [4.86]. For $E \geq 1$ keV an extrapolation to higher fluences was carried out to obtain an estimate of the equilibrium composition. The values of $(c_{Ga}/c_{As})_\infty$ are depicted in Fig. 4.23 versus the $Ar^+$ bombarding energy. As mentioned previously, the equilibrium ratio is found to increase with $E$ and, for higher energies, tends to be higher for higher flux densities. Note that the highest

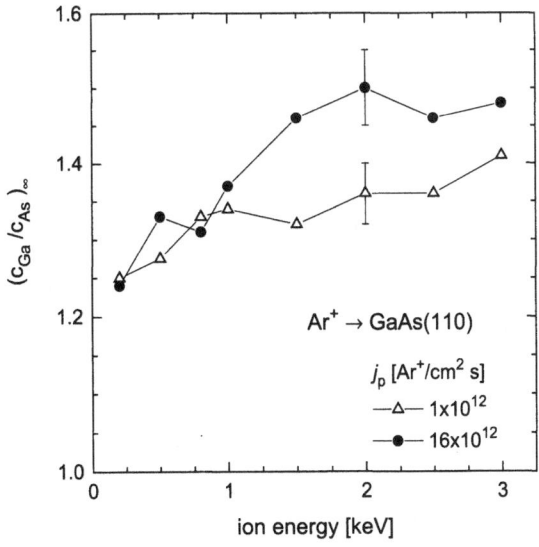

**Fig. 4.23.** Steady-state Ga/As surface concentration ratio $(c_{Ga}/c_{As})_\infty$ as a function of $Ar^+$ impact energy for two ion flux densities. Data from [4.86]

ratio corresponds to Ga and As surface concentrations of 59% and 41%, respectively. To investigate the influence of the AES probing depth, two bombardment series (at 0.2 and 3 keV) were performed, monitoring the high-energy Ga and As Auger electrons; their energies of 1070 eV and 1228 eV, respectively, translate into mean free path lengths of about 2 nm [4.160]. Qualitatively, these experiments produce the same fluence-dependent variation of the Ga/As composition, but the steady-state ratio $(c_{Ga}/c_{As})_\infty$ is lower ($\sim 1.1$ and 1.2 for 0.2 keV and 3 keV Ar$^+$ impact, respectively) than the values observed for the low-energy AES peaks (1.2 and 1.4, see Fig. 4.23).

The steady-state compositions determined in the work of Gnaser et al. [4.86] are in good agreement with previous AES and XPS data [4.161–174] and generally indicate a distinct Ga enrichment over the depth probed by low- and high-energy Auger or photoelectrons; it increases with increasing impact energy and is apparently independent of the incidence angle [4.174] and of the surface crystallographic orientation [4.102, 165, 174]. (An exception is the result of [4.165], which indicates an As enrichment, $(c_{Ga}/c_{As})_\infty \sim 0.77$, for 0.5 keV Ar$^+$ impact.) The reported $(c_{Ga}/c_{As})_\infty$ values exhibit a considerable scatter between different studies at the same ion energy [4.14], and a quantitative comparison is difficult because of the different bombarding flux densities (cf. Fig. 4.23) and the differences in probing depth of the different techniques used in previous work [4.161–174].

A discrepancy in the literature data concerns the depth variation of the As depletion and Ga enrichment. Some results [4.102, 171, 173] indicate an enhanced As concentration in the *outermost* layer of ion-bombarded GaAs. The latter observation strongly points to a surface segregation of As atoms and an As depletion in the subsurface layers, extending at least to the depth probed by (high-energy) Auger electrons (i.e. some 2 nm). Angle-resolved XPS results by Bussing et al. [4.166] for Ar$^+$ ion bombardment (1.5–5 keV) indicate that the top surface layer consists of stoichiometric GaAs, but that there exists a subsurface region with an As depletion. The thickness of this subsurface As-deficient region increases with Ar$^+$ energy. The authors [4.166] attribute this composition profile to the occurrence of Gibbsian surface segregation (see Sect. 3.2.3). This observation is corroborated by data [4.175] obtained from GaAs(100) surfaces bombarded by 3 keV Ar$^+$ ions. According to XPS spectra, the steady-state surface is *depleted* in As, with an As/Ga concentration ratio of $\sim 0.8$. By contrast, low-energy ion-scattering spectroscopy (ISS) indicated that in the top surface layer the As concentration is *enhanced* by 40% [4.175]. This finding strongly suggests the occurrence of Gibbsian segregation, causing an As-enriched outermost layer and a subsurface depletion of arsenic, in agreement with the aforementioned data. Such a situation was also found for other compound semiconductors (see below) and in MBE-grown SiGe(100) specimens [4.176]. In the latter, Ge atoms segregate to the surface and the Si/Ge ratio in the topmost layer is 0.65, whereas this ratio averaged over the XPS information depth is 1.23 under stationary 3 keV Ar$^+$ bombardment. A segregated top layer of As is also in agreement with the more forward-peaked angular emission distribution of sputtered Ga atoms shown in Fig. 3.13: Ga atoms have to penetrate through this outermost As layer.

The apparently high mobility of arsenic implied by such a segregation is reminiscent of the lack of a complete amorphization observed in the LEED data (see Sect. 4.2.2) at low impact energies and/or low current densities. In that context it was argued that thermally activated processes cause a rapid annealing of defects even at room temperature [4.103]. Those authors, using thermal-atom scattering, argue that 600 eV $Ar^+$ ion bombardment of GaAs(110) at $T = 300$ K results in 2.3 to 5 defects, approximately one of which is a target adatom, which can make a few jumps before freezing. They hypothesize, furthermore, that above 600 K adatoms remain mobile until they recombine with vacancies or form adatom clusters, and that at and above 700 K complete thermal annealing of the created defects occurs.

These arguments are in broad agreement with the much earlier data of Anderson and Wehner [4.177] who studied the temperature dependence of sputtered-atom ejection patterns from semiconductors. They observed that below a critical impact energy $E_c$ spot patterns persist, which was taken as an indication of the crystallinity of the specimen. Also, above a threshold temperature $T_a$ the value of $E_c$ steeply increases, to the extent that the sample remains crystalline for all but the very highest bombarding energies. For GaAs $T_a \sim 400$ K and $E_c$ in the range 100 eV to 300 eV were determined [4.178]. Note, however, that these authors employed current densities at least two orders of magnitude larger than those in most other studies. Anderson and Wehner tried to model their findings by a balance of defect production and the concurrent annealing of these defects. Explicitly, an equilibrium defect density $N_0$ (defects/cm$^2$) is given [4.177] by

$$N_0 = \frac{j_p \beta(E)}{\nu \exp(-E_a/kT)} , \qquad (4.12)$$

where $j_p$ is the flux density, $\beta(E)$ is the number of defects produced per ion of energy $E$, $\nu$ is the atom jump frequency and $E_a$ is an activation energy for defect migration. From this approach they derived [4.177, 178] values of $E_a$ $\sim 0.7$–$0.8$ eV for various III–V semiconductors and slightly larger values for Ge (1.1 eV) and Si (1.4 eV).

To study the possible influence of thermal processes, compositional variations were monitored by Gnaser et al. [4.86] at different sample temperatures during ion bombardment. Figure 4.24 shows the gradual change of the Ga/As surface concentration ratio with increasing fluence of 3 keV $Ar^+$ ions at various elevated temperatures. Compared to the room-temperature bombardments, increasing the sample temperature drastically reduces the observed Ga enrichment, until, at about 730 K, essentially no compositional changes are observed upon ion bombardment. These data are in agreement with the tendencies in the other aforementioned investigations [4.103].

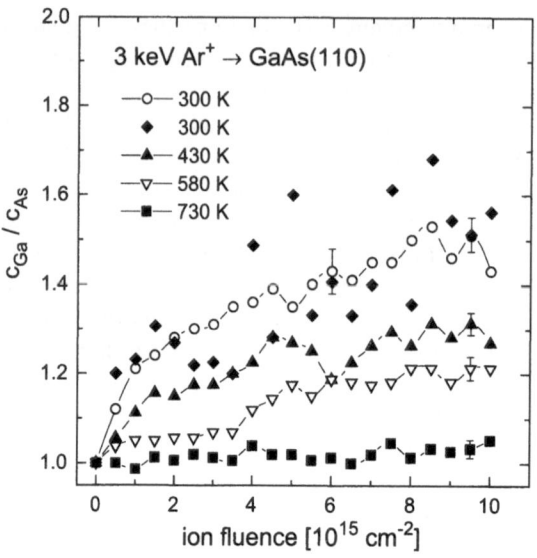

**Fig. 4.24.** Ga/As concentration ratio (derived from low-energy AES) versus fluence for 3 keV Ar$^+$ ion bombardment at the sample temperatures indicated. The ion flux density was $4 \times 10^{12}$ cm$^{-2}$ s$^{-1}$ in all cases with the exception of one of the 300 K data sets (*open circles*), which was taken at $16 \times 10^{12}$ cm$^{-2}$ s$^{-1}$. Data from [4.86]

In a related experiment Singer et al. [4.102] monitored the (high-energy) Ga/As AES intensity ratio as a function of ion current density at various sample temperatures and found that increasing the current density at a given temperature beyond a critical value results in a rapid increase of the Ga/As intensity ratio. This limiting current density rises with the temperature. These authors thus conclude that, depending on $j_p$ and $T$, two regimes may be discerned, one where the Ga/As stoichiometry is largely preserved (at high $T$ and/or low $j_p$) and a second where Ga enrichment occurs for low $T$ or large $j_p$. For 2 keV Ar$^+$ impact and the current density used to obtain the data depicted in Fig. 4.24 ($4 \times 10^{12}$ cm$^{-2}$ s$^{-1}$), they derive a limiting temperature of about 330 K. The results in Fig. 4.24 indicate in fact that in the temperature range between 300 K and 430 K a pronounced change in Ga/As composition takes place. The considerable scatter of the data for this specific value ($4 \times 10^{12}$ Ar$^+$/cm$^2$ s) of the ion flux density (cf. Figs. 4.22b and 4.24) indicates that, at $T = 300$ K, $j_p$ falls into the transition regime (from stoichiometric to nonstoichiometric composition) mentioned above. To assess this possibility, the defect density $N_0$ was evaluated according to the approach of Anderson and Wehner [4.177] (4.12), using $E_a = 1$ eV as determined by Singer et al. [4.102], $\beta = 3$ (see above [4.103]) and $\nu = 10^{13}$ s$^{-1}$ [4.179]. For $j_p = 4 \times 10^{12}$ ions/cm$^2$ s, $N_0 = 8.2 \times 10^{16}$ cm$^{-2}$ at $T = 300$ K, but falls to $3.2 \times 10^{14}$ cm$^{-2}$ and $5.1 \times 10^{12}$ cm$^{-2}$ for $T = 350$ K and $T = 400$ K, respectively. Thus, in agreement with the data shown in Fig. 4.24, a drastic reduction of the number of defects (through annealing) occurs in this temperature interval; this enhanced annealing apparently causes the Ga/As concentration ratio to approach the bulk stoichiometry at elevated temperatures.

In summarizing the ion-bombardment-induced modifications observed on GaAs(110) surfaces [4.86], a prominent feature has to be noticed: a high mobility of target atoms appears to balance the production of defects and the relocation of atoms; this mobility depends sensitively on the sample temperature. Experimentally, this is manifested in a dependence on the flux density (i.e. the defect production rate) of both the amorphization fluence (as derived from electron diffraction) and the Ga/As composition (determined from AES). In the latter case this is also noticed directly from the results of a variation of the specimen temperature. While for GaAs these effects are already evident at room temperature, for the elemental semiconductors Si and Ge they may become apparent only at much higher temperatures (the lack of any amorphization of Ge(100) subjected to 100 eV $Ar^+$ impact might, in fact, be an indication for the occurrence of such processes in this material also). These differences have their origin in the much lower activation energy for atom and defect migration in GaAs ($E_a \sim 1$ eV or less [4.102]) as compared to Ge and Si. This high mobility might also have a decisive influence on the (somewhat surprising) finding that the Ga enrichment *increases* with *increasing* bombardment energy (see Figs. 4.22 and 4.23). If the compositional variations were due solely to preferential sputtering, the reverse dependence would be expected, as discussed in Chap. 3. The probing depth of the low-energy Auger electrons is much shallower ($\sim 0.6$ nm) than the $Ar^+$ range and the thickness of the damaged (altered) layer. The latter was determined (cf. Fig. 4.9) from medium-energy ion scattering to increase from 3 nm for 0.5 keV $Ar^+$ impact to about 6.5 nm at 3 keV and is thus roughly twice the projected range of $Ar^+$ ions in GaAs [4.143]. The results indicate, therefore, that the As relocation out of the surface layer (probed by AES) and the possible replacement by Ga atoms increases in efficiency with enlargement of the damaged volume, that is to say with increasing energy. These transport processes are facilitated by the comparatively high atomic mobility in this kind of specimen.

### 4.4.2 Other Semiconductors

The second compound semiconductor that was investigated more thoroughly in terms of bombardment-induced composition changes is InP [4.14, 128, 173, 180–187]. In this system, P is depleted at the surface because of preferential sputtering, but no segregation is observed at room temperature. The P/In surface concentration ratio as determined from the $In_{MNN}$ (404 eV) and $P_{LMM}$ (120 eV) Auger electron emission amounts to $\sim 0.6$ for $Ar^+$ ion energies of 1–5 keV [4.173]. In addition, the (100) and some vicinal surfaces of InP have been investigated [4.128]. The equilibrium surface composition showed no dependence on the Ar ion energy between 0.5 and 5 keV or on the crystal orientation [4.128]. The steady-state P/In surface concentration ratio as derived from AES data exhibits a dependence on the ion's incidence angle: this ratio is

$0.68 \pm 0.07$ up to angles of $\sim 40°$, but increases sharply for more oblique bombarding directions, amounting to $0.94 \pm 0.09$ at $70°$. This angular dependence might be one of the reasons why different data sets obtained for composition changes in InP surfaces show considerable discrepancies. Another reason might be the development of surface topography frequently observed.

A quantitative comparison of the experimentally derived steady-state surface composition with the predictions of (3.14) is not straightforward because of the uncertainties associated with the correct choice of the surface binding energies. Making several reasonable estimates for $U_i$, Malherbe [4.14] derives values of $(c_P/c_{In})_\infty$ ranging from 0.72 to 0.85, whereas the average of the experimental data is $0.63 \pm 0.19$. It appears that the mass effect alone may be responsible for this pronounced In enrichment at the surface.

The effects of low-energy (5–500 eV) $Ar^+$ bombardment on the polar InP(100) surface were investigated by Rabalais and coworkers [4.130]. For a constant fluence of $\sim 1 \times 10^{16}$ ions/cm$^2$, an increasing surface roughness was observed in atomic-force micrographs for energies above $\sim 40$ eV. Auger spectral information indicates that this topography is due to In-rich clusters or islands at the surface. Similarly, annealing of the specimens at temperatures above $\sim 570$ K produces features that apparently are indium islands, which may have a width of $\sim 500$ nm and a height of $\sim 20$ nm. Conversely, by combining 30 eV ion bombardment and annealing at $\sim 570$ K the authors observed clean and stoichiometric InP(100) surfaces that exhibit the well-known (4×2) reconstruction and no detectable defects on a scale of 1 nm. In other investigations, small nanometer-sized protrusions were found on 0.5 and 5 keV $Ar^+$-bombarded InP(100) at a fluence of $1 \times 10^{15}$ cm$^{-2}$, by employing TEM and AFM [4.129].

The development of a pronounced surface topography (e.g. swelling or ripple formation) under prolonged ion bombardment ($> 10^{16}$ ions/cm$^2$) was reported also for several other III–V semiconductors (GaAs, GaSb, InSb) [4.188–191]. These effects were found to have a rather detrimental influence on depth profiling by secondary-ion mass spectrometry, as their occurrence is usually associated with a (sudden) change in ion intensity. To some extent these features might be generated (or at least enhanced) by the specific primary-ion species ($O_2^+$ and $Cs^+$) employed in this technique (see Chap. 5).

Studies into bombardment-induced composition changes for other compound semiconductors are more limited [4.14]. Part of this restriction might be due to the possible difficulties encountered in preparing clean, stoichiometric surfaces of these materials by the usual procedure of sputtering and annealing. Furthermore, these systems are typically quite sensitive to the loss of one component at elevated temperatures. Very often, large discrepancies show up among the results of different studies on the same material with the same technique, see the pertinent compilations in [4.14]. For many III–V compound semiconductors a surface enrichment of the group III atomic species was found [4.161, 192–195]. However, detailed studies employing techniques with different depth sensitivity reveal surface segregation similar to that discussed for GaAs. Investigations of ion-bombarded GaSb(111), InSb(100) and CdSe were performed by combining XPS and low-energy ISS [4.194]. Both techniques reveal a surface composition prior to ion bombardment close to the samples' stoichiometric values of 1:1. Upon 3 keV Ar$^+$ irradiation, XPS analyses showed that the surface atomic ratios of Sb/Ga, Sb/In and Se/Cd decreased from unity to 0.71, 0.92 and 0.87, respectively, at steady-state conditions. By contrast, the ISS results indicated that these ratios *increased* to 3.33, 1.63 and 1.32, respectively. The authors [4.194] ascribe this variance between the XPS and ISS data to the different sampling depths of the two techniques. Whereas XPS provides an average surface composition over a depth of 7–8 nm, the corresponding information from ion scattering refers to the outermost atomic layer. The results, therefore, give a clear indication of Gibbsian surface segregation of Sb atoms in both GaSb and InSb and of Se atoms in CdSe. Deriving the extent of segregation in terms of an expression that accounts for the differences in surface energies and size of the two constituent atoms, the authors can essentially reproduce the surface composition values in the outermost layer as seen by ISS. The XPS data, on the other hand, point to a pronounced subsurface depletion of Sb (in GaSb and InSb) and Se (in CdSe). These findings are thus similar to what has been reported for GaAs [4.166, 175] (see above). Very recent work [4.195] by this group on ion-bombardment induced composition changes in GaP(111) and InP(100) produces evidence for analogous effects in these specimens: a distinct subsurface depletion of P, whereas the Ga/P and In/P elemental ratios in the first atomic layer amount to 0.9 and 0.81, respectively, for steady-state irradiation.

A number of publications have appeared describing the effects of ion irradiation on composition changes in the IV–IV semiconductor SiC [4.196–201]. SiC is largely covalently bonded, but occurs in many polymorphs. The results reported on composition changes are somewhat contradictory. Whereas most often a C enrichment was found, in some studies an indication of an enhanced Si surface concentration was observed [4.198, 201]. In this context, it was proposed [4.201] that the Si–C bonds at the surface undergo a change of the bond type. These authors conclude that the observed changes are due to chemically guided motion of defects rather than ballistic effects (see the discussion in Sect. 3.2.2).

In II–VI compound semiconductors a very different behavior in terms of preferential sputtering and irradiation-induced composition changes has been observed in different specimens [4.202–209]. For example, in CdS a Cd enrichment was anticipated, whereas the experiments [4.207, 208] indicate the opposite, a slightly enhanced ($\sim 10\%$) sulfur surface concentration. In CdTe weakly preferential sputtering of Cd atoms is observed [4.203, 209] which is largely independent of ion mass and ion energy; $(c_{Te}/c_{Cd})_\infty \sim 1.03$–$1.12$. Furthermore, no diffusion or segregation effects were reported, and good LEED patterns were obtained even *after* ion bombardment, indicative of a crystalline surface layer [4.209]. By contrast, in HgCdTe [4.203, 209] loss of Hg atoms upon ion bombardment occurs, with a concomitant damage formation as manifested by increasingly blurred LEED patterns. The Te enrichment in the near-surface region increases with increasing ion energy and ion mass [4.203, 209]. Stahle and Helms [4.209] investigated the occurrence of Gibbsian segregation and/or radiation-induced segregation in HgTe and $Hg_{0.8}Cd_{0.2}Te$ specimens, but concluded that the main mechanism for composition changes in these systems is preferential sputtering. These data showed, furthermore, that the surface of $Hg_{0.8}Cd_{0.2}Te$ is depleted in Hg and enriched in both Cd and Te and that this depletion and enhancement increase as a function of ion energy (0.6–3 keV), being strongest for $Xe^+$ but more moderate (and less energy-dependent) for $Ne^+$ and $Ar^+$ ions. For example, 3 keV $Xe^+$ irradiation results in stationary surface concentrations $c_{Hg} \sim 18\%$, $c_{Cd} \sim 22\%$ and $c_{Te} \sim 60\%$. The authors [4.209] ascribe these findings to the weakness of the Hg–Te bond compared to the Cd–Te bond: this accounts for the higher sputtering yield of HgTe relative to CdTe and, most likely, gives rise to the loss of Hg from the surface. In addition, preferential sputtering is dependent on the chemistry of the $Hg_{1-x}Cd_xTe$, that is, on the amount of HgTe in the specimen.

In summary, the mechanisms leading to bombardment-induced composition changes in compound semiconductors are understood only partially. It still seems to be impossible to make general predictions as to near-surface modifications caused by low-energy ion irradiation for the various classes of semiconductor materials. As demonstrated in the past, studies applying well-defined bombarding conditions and (possibly) combining different surface-sensitive techniques hold great promise to improve insight into the various processes that are operative.

# 5. Ionization Processes
## of Sputtered Atoms and Molecules

The sputtering of atoms and molecules (clusters) from surfaces due to ion bombardment has been described in detail in Sect. 2.3. The examples discussed there referred either to the total yield of sputtered species (i.e. irrespective of their charge state) or, explicitly, to neutral atoms and molecules. On the other hand, the observation of ions in the sputtered flux is well established and has become a point of focus since the early attempts [5.1–6] at employing these ions for materials characterization. Since extensive reviews [5.7–11] of this field exist, this chapter will attempt, after providing some information on the basic ionization mechanisms of sputtered species, to highlight some recent results on sputtered-ion emission. As will be seen, several of these results emerged from the analytical application of these ions in secondary-ion mass spectrometry (SIMS).

When atoms or molecules are sputtered from the surface, the electrons of these species experience strong perturbations. Two major processes are conceivable that may affect the electronic states of the departing atoms or molecules [5.11]: First, the atomic collisions that induce sputtering can create electronic excitations. Second, the valence electrons of the ejected species are subjected to a transition from occupying orbitals in the solid to occupying states of free atoms in a time interval of $10^{-13}$ to $10^{-14}$ s. Electronic excitations can result from such a strong time-dependent perturbation. Hence, a fraction of the sputtered atoms/molecules may be ionized, either positively or negatively, and/or excited above their ground states. The secondary ions produced in this way during sputtering are readily identified by mass spectrometry and are utilized in the important analytical technique of secondary ion mass spectrometry [5.10, 12]. This method is widely used for the determination of the elemental composition of surfaces, interfaces and thin films, providing excellent detection sensitivity (ppb level). (Note that the optical radiation emitted during the deexcitation of atoms sputtered in excited states can be monitored [5.11], but this approach has never led to a widespread analytical application, despite some original hopes [5.13, 14].) Most secondary ions formed during sputtering are singly charged, either positively or negatively. Multiply charged positive ions are formed significantly only for certain samples and bombarding conditions. Doubly charged negative ions have been observed in only one case, namely $C_n$ clusters ($n \geq 7$) sputtered from graphite (see Sect. 5.4).

The emission of secondary ions (and excited species) is very sensitive to the chemical state of the sample surface. After reacting with electronegative elements, most surfaces exhibit drastically enhanced positive secondary-ion yields [5.15–20]; these yields might be higher by two or three orders of magnitude than the yields from the respective clean surfaces. In SIMS analyses, $O_2^+$ bombardment or an increased oxygen ambient at the sputtered surface is widely employed to effect this yield enhancement, thus providing higher detection sensitivities. On the other hand, electropositive elements such as cesium introduced onto or into the surface as bombarding species (e.g. $Cs^+$ ions) enhance negative secondary-ion yields by several orders of magnitude [5.21–25]. The incorporation of these reactive species into the near-surface region of the solid and their influence on the ionization probabilities are discussed later in this chapter (Sect. 5.2).

The extreme sensitivity of the ionization probability to the surface chemical state was realized [5.7–11] to be important for the understanding of the secondary-ion formation processes. It became apparent that the physical mechanisms of secondary-ion emission from chemical compounds and from metals and semiconductors can be distinctly different. This is due to the fact that local bonds are broken when an atom is ejected from systems such as oxides, while delocalized (metallic) bonds break in the sputtering of ions from metals. Yu [5.11] and others proposed to subdivide the experimental data and the theoretical description into three categories: (1) secondary-ion emission from clean metals and semiconductor surfaces where the valence electrons are important; (2) ion emission from chemical systems, where strong local bonding dominates; and (3) secondary-ion formation from systems where inner-shell excitations generated by energetic collisions cause ionization. In the following, the first two of these cases will be discussed and compared to experimental results.

Owing to its fundamental application in SIMS, secondary-ion emission has been studied very extensively in the past. Work done up to the mid-1980s is summarized in the encyclopedic compilation of Benninghoven, Rüdenauer and Werner [5.10]. More recent results can be found in the review by Yu [5.11]. Hence, these data are only briefly recapitulated here to facilitate the comprehension of new developments in secondary-ion emission such as (i) the incorporation of $Cs^+$ ions and the associated changes in surface work function under dynamic irradiation conditions (Sect. 5.2); (ii) the emission of Cs-carrying cations (Sect. 5.4); and (iii) the ionization under multiply-charged-ion bombardment (Sect. 5.5).

The very same charge-exchange processes that are responsible for the formation of sputtered ions play a decisive role both in determining the charge state of species scattered at the surface [5.26–28] (as applied, for example, in ion-scattering spectroscopies) and in the emission of electrons from the surface [5.29, 30] due to ion irradiation. A description of these phenomena goes beyond the scope of this chapter, but comprehensive recent reviews of both topics exist. The proceedings of the biennial Workshops on Inelastic Ion–Surface Collisions [5.31–34] constitute a further reference source to work in these and related fields of ionization processes in ion–surface interactions.

# 5.1 Mechanisms of Secondary-Ion Formation: Ionization Probability of Sputtered Ions

As noted by Yu [5.11], any theoretical approach aiming to describe the ionization of sputtered atoms has to make some assumptions as to the structural, chemical and electronic properties of the emission site on the solid surface. The description of the dynamic electron transfer between the sputtered atom and the surface can then proceed from these assumptions. Major difficulties may arise, however, from the lack of knowledge about the highly perturbed sputtering site for the time ($\sim 10^{-13}$ s) in which the atom leaves the surface. Many models have been proposed which differ in the assumptions they are based on. Generally, they all tend to describe satisfactorily only a rather limited class of materials and experimental conditions, while failing for others. In the following, two theoretical approaches are briefly outlined, as they appear to have a more general applicability; experimental data presented in this chapter support their validity. These are (i) the electron-tunneling model and (ii) the bond-breaking model. They were devised to describe specifically the aforementioned delocalized and localized bonding, respectively. Others models of secondary-ion emission are discussed in [5.11, 35].

## 5.1.1 Electron-Tunneling Model

On metal surfaces, the electronic interaction between a sputtered atom and the substrate is not localized; hence, the ejection of an atom and the breaking of the metallic bond involve a large number of electrons in the valence band. Furthermore, the electron or hole excitation generated in the metal upon the sputtering of charged species would be heavily screened by other electrons [5.11]. The electronic excitation will also not remain localized at the emission site, because of the high electron mobility. The electron-tunneling model envisions the electronic transition as a resonant transfer process (hence "tunneling") between a sputtered atom and the valence band. This many-electron description of the transition is consistent with the delocalized nature of the metallic bond. This approach is thus equivalent to the crossing of the atomic level of the sputtered atom with many electronic levels of the solid. This concept was first proposed by van der Weg and Rol in 1965 [5.36]. More recently, Yu [5.37] and Yu and Lang [5.38] used the electron-tunneling model to interpret the work-function dependence of the ionization probability observed in alkali-metal adsorption experiments [5.24, 37–42].

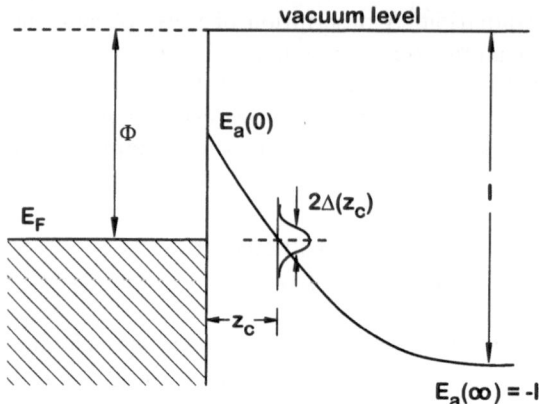

**Fig. 5.1.** Schematic energy diagram of an atom leaving a metal surface. The Fermi level $E_F$ lies below the vacuum level by the work function value $\Phi$. Initially the atomic level $E_a$ is broad and may lie above $E_F$ before the atom is ejected. The variation of the image potential causes $E_a$ to be reduced with separation until $E_a = E_F$ at the crossing point. Electrons in the metal can tunnel out to fill the atomic level once $E_a < E_F$, beyond the crossing point. From [5.35]

In the electron-tunneling model [5.11, 38], the metal is represented by a nearly-free-electron valence band with a constant density of states and a work function $\Phi$. The electron can tunnel between $E_a$, the energy level of the leaving atom, and the electronic levels of the same energy in the valence band; see Fig. 5.1 [5.35]. All excitations are assumed to dissipate rapidly. The tunneling probability is determined by the magnitude of the transition matrix element $V_{ak}$ between the atomic state $|a\rangle$ and the metal state $|k\rangle$. $V_{ak}$ is given by [5.43]

$$V_{ak}(z) = \langle k|V|a\rangle \; , \qquad (5.1)$$

where $V$ is the interaction potential. Since the electron has a finite lifetime in state $|a\rangle$, the atomic level is broadened in energy according to the uncertainty principle. The half-width $\Delta(z)$ is a function of the distance from the surface $z$ and reads, in atomic units [5.11, 43],

$$\Delta(z) = \pi \sum_k |V_{ak}(z)|^2 \, \delta\left[E_k - E_a(z)\right] \; . \qquad (5.2)$$

To calculate the ionization probability $P$ one needs to know how $\Delta$ and $E_a$ vary along the trajectory. Since the wavefunction $|k\rangle$ decays exponentially with $z$, to a good approximation one may assume [5.43]

$$\Delta(z) = \Delta_0 \exp(-\gamma z) \; , \qquad (5.3)$$

where $\gamma^{-1}$ is a characteristic length of about 0.2 nm [5.11]. Owing to the screening by the electrons in the metal, $E_a$ is shifted up (ionization level) or down (electron affinity level) by an image potential $e^2/4(z - z_{im})$, where $z$

is the separation between the atom and the surface and $z_{im}$ is the position of the image plane. For simplicity it is assumed, furthermore, that the metal surface exhibits no lateral inhomogeneity and that the sputtered atom has a constant velocity $v$. Several authors [5.43–46] have derived simple approximate expressions for the ionization probability. In particular, for a crossing of $E_a(z)$ with the Fermi level at $z = z_c$ the following has been found [5.43]:

$$P \cong \exp\left(-\frac{2\Delta(z_c)}{\hbar\gamma v_n(z_c)}\right) , \tag{5.4}$$

where $v_n(= v\cos\theta_0)$ is the normal component of the emission velocity. This is because $P$ is related to the amount of time the sputtered atom spends in the interaction region. The crossing point $z_c$ depends on the position of $E_F$, and hence $P$ is expected to vary with the work function $\Phi$, which defines $E_F$ with respect to the vacuum level. For moderate changes $\Delta\Phi$, an exponential variation of $P$ with $\Phi$ was predicted [5.11, 43, 45],

$$P^+ \propto \exp\left(-\frac{I-\Phi}{\varepsilon_p}\right) , \tag{5.5}$$

and

$$P^- \propto \exp\left(-\frac{\Phi-A}{\varepsilon_n}\right) , \tag{5.6}$$

where $\varepsilon_p$ and $\varepsilon_n$ are proportional to $v\cos\theta_0$ at $z_c$. The essential features of this theoretical prediction, namely the dependence on the ionization potential $I$, the electron affinity $A$ and $\Phi$, have been verified by experimental results [5.24, 38–42, 47, 48] obtained on samples for which the electron-tunneling picture appears applicable: clean surfaces of metals and semiconductors, and surfaces covered with well-defined adsorbates (e.g. alkali metals). With respect to the variation of $P^\pm$ with the normal component of the velocity, the exponential dependence could be verified [5.49–52], but deviations were found at low emission energies [5.40, 52].

One of the most direct verifications of the electron-tunneling model was an experiment by Yu and Lang [5.38]. They monitored the emission of $Cs^+$ from various surfaces and demonstrated the effect of the crossing of $E_a$ with $E_F$. Cs was chosen for its small ionization potential ($I = 3.9\,eV$) and because Cs atoms are known to chemisorb as $Cs^+$ on most metal surfaces at small coverage with the empty 6s level lying above the Fermi level of most metals, and $E_a(z)$ varies with the image potential. If the work function $\Phi$ of the surface is larger than the ionization potential $I$ of Cs (3.9 eV), the Cs level of the sputtered Cs atom will always face empty states of the metal and little neutralization by electron tunneling can occur, i.e. $P^+ = 1$. After lowering of the work function (by adsorbing submonolayers of Li) such that $\Phi < I$, the Cs level would have to cross the Fermi level as the atom escapes, making the neutralization by electron tunneling energetically possible, and $P^+$ should vary according to (5.5). These authors [5.38] tested this concept on different surfaces: $P^+$ for $Cs^+$ ions is constant for $\Phi > I$ but starts to decrease for a shift $\Delta\Phi$ which corresponds roughly to the situation $\Phi < I$. Conversely, for the sputtering of atoms with a higher ionization potential or a smaller change in work function (so that crossing of the Fermi level always occurs), an exponential variation of $P^+$ with $\Phi$ was observed in several experiments. A detailed study into the $\Delta\Phi$ dependence of $P^+$ and $P^-$ has been performed recently by Gnaser [5.53, 54] for dynamic $Cs^+$ incorporation conditions; some of those results are presented in Sect. 5.2.

An important feature of this tunneling model is that the charge state of the sputtered atom is determined in the vicinity of the crossing point $z_c$, which depending on $I - \Phi$ (or $A - \Phi$), can be quite far away from the surface. According to Nørskov and Lundqvist [5.45], any excitation or excess charge on the sputtered atom which was created within or close to the metal surface has very little probability of survival. They argued that with a $\Delta$ of a few eV, the lifetime of these excitations is just too short. For example, with $\Delta(0) = \Delta_0 = 2\,eV$, $\gamma^{-1} = 0.1\,nm$ and $v = 5 \times 10^5\,cm/s$, (5.4) yields $P(\infty) = 10^{-53}$. On the other hand, if the excitation is generated outside the surface, it may have a reasonable chance of survival. When an empty atomic state crosses $E_F$ at $z = z_c$, (5.4) gives the survival probability of this empty state [5.35]. With $z_c = 0.4\,nm$ and the other values used above, an ionization probability $P \approx 0.11$ results.

The electron-tunneling model as described in the foregoing paragraphs does not take into account the perturbations of the substrate due to the sputtering event. Specifically, the following assumptions are made [5.11, 35]: (i) The velocity of the sputtered atom is uniform. Lang [5.43] has corrected for the influence of the surface on the trajectory by introducing a Morse potential for the interaction with the sputtered atom and by considering the dynamics of the collisions. (ii) No surface inhomogeneities are assumed to occur. In reality there is some local variation of the surface potential, especially in the presence of adatoms that induce a shift of the work function due to their electric dipole moments. The inhomogeneous electric field from these individual dipoles was shown [5.35] to decay roughly exponentially with a length $r/2\pi$, where $r$ is the average distance between the dipoles. Hence, the field inhomogeneity would become important only if the crossing distance $z_c$ was comparable to or smaller than the decay distance. Therefore, the local work-function effect would be large for small coverage and/or small $z_c$ (i.e. $I \gg \Phi$). (iii) The third assumption concerns the neglect of any electronic excitation in the specimen by the incoming primary particles.

In a series of papers, Sroubek [5.55–58] has examined the possible influence of such excitations on the ionization probability of sputtered atoms. He postulated that the electrons in the collision cascade are excited to a temperature $T_s$. To compute $T_s$ he estimated the electronic energy density dissipated by the primary ions within the cascade and derives values of a few $10^3$ K. Notably, $T_s$ is smaller for, for example, Cu, with a large carrier concentration, than for Si and for GaAs. Sroubek's model predicts an exponential dependence of the ionization probability on $T_s^{-1}$, $P^+ \propto \exp(-1/T_s)$. The only evidence in favor of the idea that substrate excitations play a role in ionization comes from the primary-ion energy dependence of $P^+$ observed for $Si^+$ ions sputtered from silicon [5.59] and $Ga^+$ and $As^+$ ions emitted from GaAs [5.57, 58]. By contrast, ion emission from metals exhibits little (or no) variation with the impact energy beyond that due to the total sputtering yield.

More recently, Urazgil'din and coworkers [5.60, 61] have tackled the problem of secondary-ion emission from metal surfaces covered with electronegative species (e.g. oxygen). They proposed a mechanism that incorporates, within the framework of electron tunneling, the large ion-yield enhancement effects observed for positive secondary ions from such adsorbate-covered surfaces. In this approach, the presence of an adsorbed species induces an electrostatic potential that causes a shift of the electronic level of the emitted atom; this increases the effective neutralization distance and, concurrently, decreases exponentially the neutralization probability.

Many experimental data have supported the work-function dependence of the ionization probability as predicted by the electron-tunneling model for positive and negative ions sputtered from metals (see (5.5) and (5.6), respectively). While for positive secondary ions the available results are still limited, for negative ions a large set of experiments has been reported [5.24, 38–42, 47, 48], both for static-mode sputtering and for high-fluence bombardment conditions. Generally, $P^-$ may increase by several orders of magnitude upon lowering of the work function (WF) $\Phi$, in agreement with (5.6). The pronounced variation of $P^-$ with $\Delta\Phi$ is exemplified in Fig. 5.2 for $Ni^-$ ions

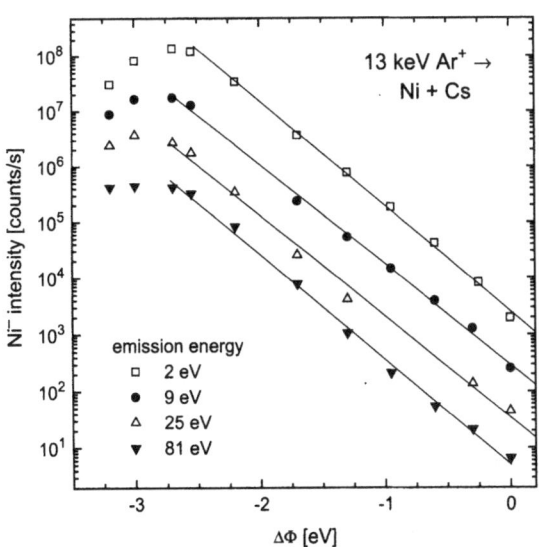

Fig. 5.2. Variation of the $Ni^-$ ion emission from a Ni(100) single crystal bombarded by 13 keV $Ar^+$ ions. Cs adsorption causes a lowering of the surface work function and results in an exponential increase of the ionization probability (indicated by straight-line fits) and saturation at very high Cs coverages. The parameter is the emission energy of the $Ni^-$ ions. Data from [5.48]

sputtered from elemental Ni [5.48]. The change of the WF was effected in this experiment by depositing Cs on the surface. Data for the different emission energies depicted indicate that $\varepsilon_n$ in (5.6) is essentially independent of the emission velocity. Deviations from this simple exponential dependence usually occur at high alkali coverages, where $\Delta\Phi$ approaches its maximum value. A strictly exponential relation was also not observed [5.48] for the sputtering of $Au^-$ and $Cu^-$ from an AuCu alloy, and for $O^-$ emitted from contaminated vanadium. Data for the variation of the ionization probability due to $\Delta\Phi$ shifts induced by Cs incorporation effected by $Cs^+$ ion bombardment are reported in Sect. 5.2.3.

For positive secondary ions, (5.5) predicts also an exponential scaling of $P^+$ with the WF. As mentioned above, the electron-tunneling model predicts [5.38] a constant value of $P^+$ for the case $\Phi > I$; only for $\Phi < I$ should $P^+$ decrease with $\Phi$. These features were nicely demonstrated by Yu in static Cs adsorption experiments [5.47, 62]. He deposited [5.62] submonolayers of Cs on a Si(111) (7×7) surface and monitored the $Cs^+$ ion yield under 500 eV $Ne^+$ sputtering as a function of Cs coverage and the concurrent work-function change. (The ion bombardment in this experiment was performed in a low-fluence mode, i.e. without significantly changing the surface state.) These results [5.62] are shown in Fig. 5.3, which depicts the ionization probability $P^+$ of $Cs^+$ versus $\Delta\Phi$ for two different emission energies of the sputtered $Cs^+$ ions. Two distinct regimes can be discerned: up to $\Delta\Phi = -1\,\text{eV}$, $P^+$ is essentially constant, whereas for a further lowering of the WF, $P^+$ starts

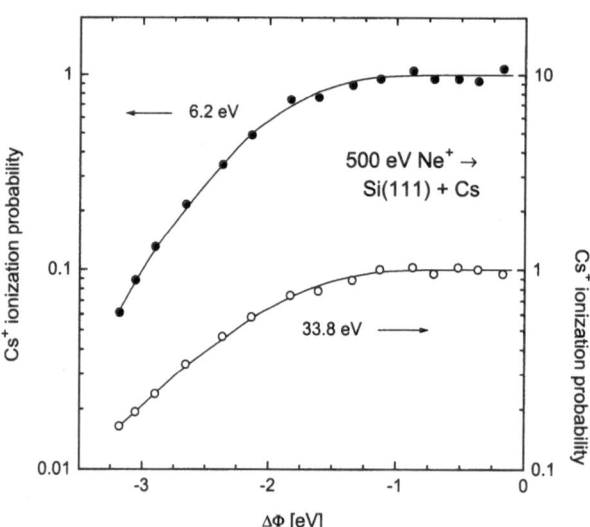

**Fig. 5.3.** The $Cs^+$ ionization probability as a function of the work-function change $\Delta\Phi$ at two emission energies. The $Cs^+$ ions were sputtered from a Si(111) surface covered with a submonolayer of cesium by bombardment with 500 eV $Ne^+$ ions in a low-fluence mode. Data from [5.62]

to decrease and tends to approach an exponential dependence on $\Phi$. The first regime thus corresponds to the $\Phi > I$ range and the second to $\Phi < I$. As the ionization potential of $Cs^+$ is 3.89 eV and the work function of Si is 4.6 eV, the onset of the latter regime is apparently delayed by about 0.4 eV; according to the author, this is due to the fact that $z_c$ in (5.4) has to be small enough (i.e. $I - \Phi$ large enough) to allow effective electron tunneling. Yu noted [5.62] furthermore that the more rapid decrease in $P^+$ with $\Phi$ at the lower emission energy (see Fig. 5.3) is consistent with the dependence on the normal component of the ion's emission velocity, $v_n$. An exponential decrease of $P^+$ with a reduction of the WF was also observed in other adsorption experiments. Data for dynamic Cs incorporation are presented in Sect. 5.2.3.

The situation appears less clear-cut with regard to the emission-velocity dependence of $P$ predicted by the electron-tunneling model. In some experiments [5.63–65] a power-law dependence of $P^+$ on the emission energy was found. On the other hand, Garrett et al. [5.51], Vasile [5.52] and others [5.66–68] reported an exponential scaling of $P^+$ with the normal component of the emission velocity for the high-energy part of the ion energy spectrum. In some experiments [5.52] the deviations at low energy could be corrected by taking account of an appropriate image potential, implying that the atom's velocity at the time of ionization is higher than that measured, because the ion has to overcome the image force on its outbound trajectory. (A similar argument was put forth by Garrison [5.69].) For the sputtering of $O^-$ ions from oxygen-covered vanadium, Yu found [5.40] a linear correlation between $v_n$ and the parameter $\varepsilon_n$ (see (5.6)), but observed deviations at low $v_n$; he ascribed the latter to trajectory modifications by the surface binding energy as discussed above. For this experiment, values of $\varepsilon_n$ were in the range from 0.4 to 1.2 eV. By contrast, experimental data by Bernheim and LeBourse [5.48] showed essentially no velocity dependence for the $P^-$ of negative metallic ions sputtered from the respective metals upon WF changes (cf. Fig. 5.2), while a strong dependence on the emission energy was observed for the emission of negative secondary ions of electronegative elements such as hydrogen, oxygen and phosphorus. These discrepancies have still not been resolved. The energy/velocity dependence of the ionization probability of positive secondary ions sputtered from several metals was investigated by Wucher and Oechsner [5.70]; they derived the emission-energy variation of $P^+$ from measurements of the energy distributions of sputtered atoms and sputtered ions. The results indicate that no single expression can describe the dependence of $P^+$ on either the energy or the velocity of the emitted ions.

## 5.1.2 Bond-Breaking Model

Because the nature of the bonds in samples such as oxides and halides is strongly localized, the electron-tunneling model is not applicable in this case. Rather, the so-called bond-breaking model of secondary-ion formation was proposed. This concept was originally developed by Slodzian [5.71] to explain secondary-ion emission from ionic solids. Later, Williams [5.72] extended this model to compounds, such as oxides, in which the bonds are only partially ionic. Both studies pointed out that the ionization of a sputtered atom via the breaking of the bond with an electronegative atom (e.g. oxygen) at the surface is very similar to the charge-exchange mechanism in the Landau–Zener–Stueckelberg [5.73–75] curve-crossing model for atomic collisions. More recently, Yu and Mann [5.76–78] formulated a bond-breaking scheme in more quantitative terms. The general argument is the following.

Sputtering of a positive $M^+$ ion from the surface creates a cation vacancy X; it may trap the excess electron for at least the sputtering time (about $10^{-13}$ s), with an electron affinity $A$. In this case $M^+$ is emitted. Charge exchange then happens at the crossing of the *diabatic covalent* potential curve $M^0 + X^0$ and the *diabatic ionic* potential curve $M^+ + X^-$ at a distance $R_c$ from the surface [5.78]. This is illustrated in Fig. 5.4. According to Landau and Zener [5.73, 74, 79], the transition probability is determined by the wave functions and the shape of the diabatic curves at the crossing point, which is some distance away from the equilibrium position. As discussed by Williams [5.72], the transition probability is not sensitive to the actual equilibrium bonding. Since the covalent force is short-ranged, the assumption is made that in the region of crossing the covalent curve is independent of $R$, i.e. the distance between the sputtered atom and the surface vacancy. Furthermore, the ionic curve is determined by the Coulombic attraction between $M^+$ and the negatively charged vacancy. At infinity the ionic curve lies above the covalent curve by $I - A$, where $I$ is the ionization potential of the sputtered M atom. At the crossing point, the Coulomb potential exactly balances this energy difference [5.76, 78]. Hence the crossing distance is simply equal to

$$R_c = \frac{e^2}{I - A} \ . \tag{5.7}$$

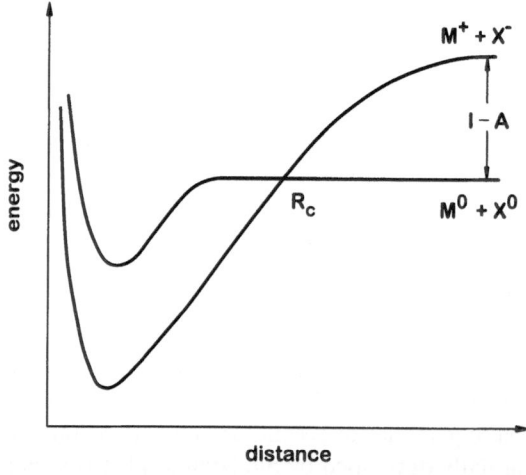

Fig. 5.4. Schematic energy diagram showing the crossing of the covalent energy curve $M^0 + X^0$ and the ionic potential curve $M^+ + X^-$ at the crossing point $R_c$. In the bond-breaking model, charge exchange can occur at the crossing of these diabatic curves at the distance $R_c$. From [5.78]

To compute the ionization probability $P^+$, the Landau–Zener formula is used [5.76]:

$$P^+ \cong G \exp\left(-\frac{2\pi H_{12}^2}{v\,|a|}\right)_{R=R_c}, \tag{5.8}$$

where $H_{12}$ is the transition matrix element, $v$ is the emission velocity at the crossing point and $|a|$ is the difference in the first derivatives of the potential curves [5.79],

$$|a| = \frac{\mathrm{d}}{\mathrm{d}R}\left(V_{\mathrm{cov}} - V_{\mathrm{ion}}\right) = \frac{e^2}{R_c^2}. \tag{5.9}$$

To a good approximation, $G$ is given by the ratio of the degeneracies of $M^+$ and $M^0$, i.e. $G \approx g_+/g_0$. Yu has pointed out [5.78] that the wavefunction for the atomic level of the sputtered atom can be simulated by the decaying region of a hydrogen 1s wavefunction at the crossing distance; conversely, the wavefunction of the electron trapped at the cation vacancy site is assumed to be identical to that of a negative ion in the presence of $M^+$, with an electron affinity $A$ and an amplitude parameter $C$. Both quantities depend on the chemical bonding.

In (5.8), $P^+$ depends directly on the velocity at the crossing point $R_c$, which is related to the emission energy $E_0$ measured in an experiment by [5.76]

$$v\left(R_c\right) = \left[\frac{2\left(E_0 + I - A\right)}{M}\right]^{1/2}, \tag{5.10}$$

where $M$ is the mass of the sputtered atom. Obviously, at very low $E_0$, $P^+$ converges to a constant (and finite) value corresponding to $v\left(R_c\right) \cong \left[2\left(I - A\right)/M\right]^{1/2}$. At very high $E_0$, $P^+$ approaches an exponential dependence

$$P^+ \propto \exp\left(-v_0/v\right), \tag{5.11}$$

with $v \cong \left(2E_0/M\right)^{1/2}$. For an intermediate energy range, a power-law dependence on $v$ might follow; such a relation has been observed in some experiments [5.78]. On the other hand, the data on isotopic mass effects on the ionization probability presented in Sect. 5.3 corroborate the predictions of a constant $P^+$ at low energies and the exponential velocity dependence at high emission energies.

In general, the bond-breaking model [5.78, 80] predicts that $P^+$ decreases rapidly (i.e. exponentially) with increasing $I$. Such a relation is reminiscent of one of the first SIMS quantification schemes reported [5.81–83], which utilized a Boltzmann-like relation with the ionization potential, assuming a local thermal equilibrium for sputter emission. Apart from this similarity, the bond-breaking scheme of ion formation features a dependence on the sputtered ion's mass via (5.10). Yu reports [5.78, 84] a reasonable agreement with existing experimental data for the dependence of the ionization probability $P^+$ on the atom's ionization potential $I$ for 11 elements of the fourth period, but deviation for elements from other periods. The comparison, however, was restricted to the emission of positively charged ions, which this model specifically addresses. Its applicability to negative secondary ions, molecular ions and multiply charged ions has not been tested. Furthermore, isotopic mass effects were investigated for only a single set of data.

The parameters $A$ and $C$ mentioned above are properties of the cation vacancy whose chemical and structural characteristics would depend on the original chemical bonding with the metal atom. Phenomenologically, the secondary-ion yield $I_M^+$ of the atom M is the sum of the contributions from different bonding configurations $i$ [5.76, 78]:

$$I_M^+ \propto \sum_i f_i P_i^+ Y_i \,, \tag{5.12}$$

where $f_i$, $P_i^+$ and $Y_i$ are the fractional concentration, ionization probability and partial sputtering yield of the M atoms bonded in the $i$th configuration. Such a site-dependent ion emission has been invoked by several authors [5.35]. Mann and Yu [5.77] measured ion yields from silicon oxides and suboxides and reported a consistency with the predictions of the bond-breaking model. The linearity between $I_M^+$ and $f_i$ expressed in (5.12) is the unique feature of this localized interaction described by the bond-breaking picture. Further verification was provided by low-fluence sputter experiments on various chemical systems. The influence of differences in chemical bonding was observed also in studies of isotope sputtering from oxygen-rich specimens (cf. Sect. 5.3); in these experiments the differences became manifest in varying values of the parameter $v_0$, which absorbs all the variables in the exponent of (5.8) except $v(R_c)$.

Secondary-ion yields have most often been investigated under high-fluence conditions, either at an elevated oxygen partial pressure at the surface or by incorporating oxygen atoms into the near-surface region as the bombarding species. While in this situation different bonding configurations are difficult to identify, all of those data report a strong enhancement of the ionization probability with oxygen concentration $c_O$, although considerable divergence in terms of the actual correlation between $P^+$ and $c_O$ is evident. Both exponential [5.85–87] and power-law dependences have been reported [5.18, 88, 89]. Oechsner and Sroubek [5.90] have proposed a model which may apply to the dynamic sputtering conditions typically encountered in SIMS. Assuming a localized electron transfer, the average ionization probability is the sum of the contributions from sites with different numbers of oxygen neighbors. With a random distribution of the oxygen atoms, every oxygen neighbor induces the same ion-yield enhancement factor $K$ and the ionization probability has a power-like dependence on $c_O$ [5.90],

$$P^+ \propto [c_O (K - 1) + 1]^N \,, \tag{5.13}$$

where $N$ is the maximum number of oxygen neighbors. These authors studied Ta$^+$ emission from Ta under 4 keV Ar$^+$ bombardment and simultaneous oxygen exposure and report a good agreement of the experimental data with (5.13) for a wide range of $c_O \leq 0.6$. Generally, $P^+$ will depend critically on the distribution of the $f_i$ in (5.12); for dynamic sputtering, the latter, in turn, is determined by the experimental bombardment conditions. Diverging data on the oxygen-concentration dependence of Si$^+$ yields sputtered from oxygenated silicon can be due to a distribution of sites with different oxygen coordination numbers.

For the specific, albeit important case of Si$^+$ emission from oxygen-ion-bombarded silicon, Alay and Vandervorst [5.91] proposed an ion formation mechanism that combines features from the electron-tunneling and the bond-breaking model. They investigated the ion-beam-induced oxidation of Si using X-ray photoelectron spectroscopy. From a detailed analysis of the Si 2p core level the authors [5.91] inferred the presence of suboxide chemical states for ion incidence angles larger than 30°, whereas for smaller angles (i.e. near-normal incidence) the Si$^{4+}$ chemical state corresponding to silicon oxide dominates. Furthermore, a comparison of the Si and SiO$_2$ valence bands indicated that the valence bands for the altered (oxygen-containing) layers are the result of the combination of these end states. Since Si–Si bonds are present in the suboxide molecules, the tops of the new valence bands are formed by the corresponding 3p–3p Si-like subbands, which extend up to the Si Fermi level. The authors [5.91] concluded also that small variations in energy position for this subband have a dramatic influence on the intensity of the Si$^+$ emission during oxygen-ion sputtering of silicon.

Note that in addition to oxygen, the influence of other electronegative elements (e.g. fluorine [5.20]) on the ionization probability of positive secondary ions has been investigated.

## 5.2 Sputtering with Cs$^+$ Ions

It was realized a long time ago [5.21] that sputtered negative ion emission is strongly enhanced by the presence of alkali metals at the ions' emission site on the surface. This finding is widely utilized [5.12] in secondary-ion mass spectrometry (SIMS) for the sensitive detection of electronegative elements by monitoring their respective negatively charged secondary ions. (Detection limits as low as $10^{12}$ atoms/cm$^3$ have been realized [5.92] in this way.) Most often, this enhancement is accomplished by employing a Cs$^+$ primary ion beam for sputtering, thus loading the near-surface region of the bombarded specimen with cesium. In general, the resulting amount of Cs at the surface is not known but can be expected to depend on a variety of parameters such as the projectile's energy and incidence angle, the specimen's sputtering yield and, possibly, others [5.43, 45]. (Under steady-state conditions, the balance of Cs incorporation and concurrent removal by sample erosion results in a

stationary Cs surface concentration.) The increase of the ionization probability of sputtered negative secondary ions in the presence of Cs has been
ascribed to a lowering of the sample's work function (WF) (in fact, its surface contribution is lowered because of the development of a surface dipole
layer, with the positive charge of $Cs^+$ farther away and an electron donated
to the substrate). The minimum amount of energy required to transfer an
electron from the solid to an atom at infinity to form a negative ion is $\Phi - A$
[5.11], where $A$ is the atom's electron affinity. Hence the formation probability
$P^-$ of negative secondary ions should depend on this quantity. Theoretical
treatments of sputtered-ion emission (see Sect. 5.1.1) predict an exponential
scaling $P^- \propto \exp\left[(\Phi - A)/\varepsilon_n\right]$. The parameter $\varepsilon_n$ was anticipated [5.43] to
vary with the normal component of the ion's emission velocity but to be
roughly constant for (moderate) WF changes; its magnitude should fall in
the range 0.5–1 eV [5.43]. As discussed in Sect. 5.1.1, experimental data produce strongly divergent results in terms of the velocity dependence; in fact,
some of them [5.48] indicate $\varepsilon_n$ to be largely independent of the emission
velocity, with values of 0.2–0.4 eV.

An exponential dependence on the work function $\Phi$ was verified by static
alkali-metal adsorption experiments (see Fig. 5.2). Generally, in these investigations the amount of alkali atoms is well controlled and can be derived from
Auger electron spectroscopy data, for example. By contrast, for the *dynamic*
SIMS conditions described above ($Cs^+$ is used as the bombarding species),
the situation is less obvious [5.25] and no results on the Cs surface concentration have been available. More recently, Gnaser [5.53, 54] has investigated
the WF changes associated with this Cs incorporation (see Sect. 5.2.2).

The presence of alkali atoms (in particular Cs) on the surface strongly in-
fluences also the emission of positive ions [5.38, 62]. A dependence on the WF
is again observed (cf. Fig. 5.3); however, the direction of variation is reversed
for positive ions: a *lowering* of the work function *reduces* the probability of
forming a *positive* secondary ion. While for usual SIMS analyses this effect is
of little concern (oxygen primary beams are commonly used for positive-ion
detection [5.12]), it does apparently influence the emission of so-called $MCs^+$
molecular ions (M stands for an atom of the sample material). These species
are (abundantly) formed under $Cs^+$ bombardment (i.e. from Cs-loaded surfaces) from essentially all elements [5.93] and their formation mechanisms
are discussed in Sect. 5.4.1. Their analytical usefulness lies in the observation
that matrix effects (that is, drastic variations of ions yields with sample composition), which are common for atomic ions [5.10, 11], appear to be largely
absent for $MCs^+$ [5.94, 95]. This has been rationalized by their possible formation mechanism: the association of a *neutral* M atom with a $Cs^+$ ion in
the sputtering event. Clearly, for such a process the amount of sputtered $Cs^+$
is of the utmost importance as it should determine the number of $MCs^+$ ions
actually formed. Since little information is available as to the equilibrium Cs
concentration building up at the surface upon dynamic $Cs^+$ bombardment

(and the WF changes induced thereby), predictions of the Cs$^+$ yield and its variations in different substrates (e.g. in a depth profile analysis) are virtually impossible.

Theoretical descriptions of secondary-ion emission such as the electron-tunneling model [5.38] outlined in Sect. 5.1.1 predict an exponential dependence of the ionization probability of positive ions, $P^+$, on the WF and the atom's ionization potential $I$: $P^+ \propto \exp\left[-\left(I - \Phi\right)/\varepsilon_p\right]$. The parameter $\varepsilon_p$ is considered to scale with the normal component of the ion's emission velocity. Again considerable uncertainties prevail as to the velocity dependence of $\varepsilon_p$. While some experiments [5.40, 52] appear to confirm this prediction, others [5.48] find $\varepsilon_p$ to be (almost) independent of velocity. Also, the former data tend to exhibit a leveling off (towards a constant value) for low velocities. Furthermore, a velocity dependence in (5.5) should become manifest in *pronounced* differences of the secondary ions' energy distributions with small changes of $(I - \Phi)$ [5.96]. The experiments by Gnaser [5.54] indicate, however, that the shape of the energy spectra does not change (at least in the low-energy portion) with moderate $\Phi$ variations (see Sect. 5.2.2).

Very important in this context is the aforementioned experimental finding (and the associated theoretical description) of Yu and Lang that the ionization probability is unity for $\Phi > I$ [5.38]. Since $I_{Cs} = 3.89$ eV, one may expect $P_{Cs}^+ = 1$ for most clean surfaces, with *no* dependence on emission velocity. Only a lowering of the WF due to Cs incorporation might reduce $P_{Cs}^+$. Conversely, if this reduction is not sufficient to reach $I > \Phi$ (because of, for example, too low a stationary Cs concentration), $P_{Cs}^+$ should remain unity. The data presented in the following sections will illustrate both of these possibilities.

### 5.2.1 Cesium Implantation Distribution

Ion implantation in the presence of concurrent sample erosion results in a stationary near-surface concentration of the implanted species [5.97–99]. The actual surface concentration depends on a variety of parameters such as the ion's range (and range straggling), the sputtering yield of the substrate and the implant atoms and, possibly, factors that influence the mobility of the implanted species in the sample. The importance of the latter, in particular, appears difficult to assess. In the case of Cs$^+$ ion bombardment, segregation of Cs atoms at (towards) the surface might be induced by the presence of impurity species (e.g. oxygen) at the surface; such processes might be held responsible for the considerable discrepancies reported for the Cs surface concentration in Cs$^+$-bombarded specimens [5.100–104].

To obtain information on the Cs surface concentration during its gradual incorporation in the initial stages of irradiation, computer simulations were performed [5.54] using the T-DYN code [5.105]. This program determines dynamically the implantation and buildup of the bombarding species in the target. Figure 5.5a exemplifies this process by depicting, for different fluences,

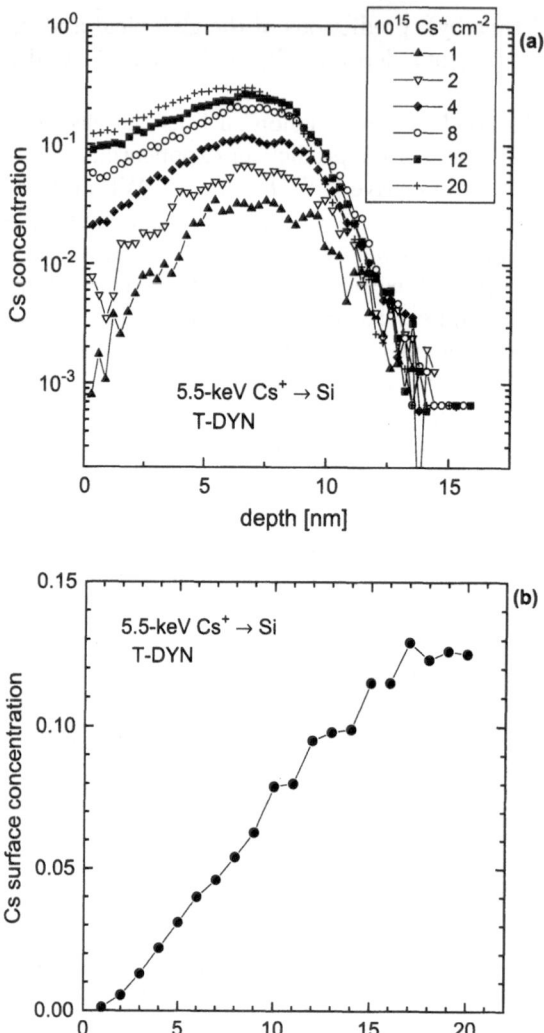

**Fig. 5.5a,b.** Computer simulation data for 5.5 keV Cs impact on Si. (a) Cs concentration versus depth. The parameter is the Cs fluence. (b) Cs surface concentration as a function of Cs fluence. From [5.54]

the Cs concentration as a function of depth in Si bombarded by 5.5 keV Cs. At low fluences the distribution is close to a standard implantation profile (roughly Gaussian, with a mean range of about 7 nm); with increasing fluence the distributions are found to approach a stationary state. Figure 5.5b shows the Cs surface concentration versus Cs fluence derived from the implant distributions (Fig. 5.5a and others). It is seen that the Cs surface concentration increases roughly in proportion to the implantation fluence up to $\sim 1.5 \times 10^{16}$ Cs atoms/cm$^2$, but saturates for a fluence of about $2 \times 10^{16}$ Cs atoms/cm$^2$. Then, the Cs concentration amounts to $\sim 12$ at%. Such a linear increase appears to agree with experiments [5.101, 102] studying, via ion backscattering spectroscopy, the incorporation of Cs in Si, albeit at somewhat different bombarding energies and angles.

It should be noted, however, that these simulations [5.105] account only for ballistic processes such as preferential sputtering and collisional mixing (see Sect. 3.2), while the possible occurrence of Gibbsian or radiation-induced segregation cannot be modeled. Furthermore, the surface binding energies that are required as input parameters are not well known for Cs-loaded surfaces. Such an influence became apparent in recent computations by Eckstein et al. [5.106]. They studied the incorporation of Cs into silicon for a range of impact energies from 0.1 to 100 keV. Depending on the surface binding energy values and the interaction potentials chosen, the *steady-state* Cs surface concentration $c_{Cs}^{(1)}$ for 4 keV Cs bombardment exhibited a large variation, between $\sim 7\%$ and $\sim 47\%$. The Cs distribution within the near-surface region is, however, more uniform ($c_{Cs} = 0.37$–$0.45$) and resembles the profile depicted in Fig. 5.5a. These simulations again monitored solely ballistic relocation processes.

### 5.2.2 Work-Function Changes Due to Cs Incorporation

Cs$^+$ ion irradiation of several metal surfaces at low ion energies (30–900 eV) indicated [5.107, 108] that the induced work-function change is impact-energy-dependent: the lower the Cs$^+$ energy the larger the lowering of the WF. In qualitative terms, this finding relates to the observation that a low sputtering yield (at low energies) results in a high stationary Cs concentration and hence a large $\Delta\Phi$. For the lowest energies, the observed $\Delta\Phi$ values [5.107, 108] come close ($\sim 2.5$ eV) to those reported for typical Cs adsorption on the surface [5.109]. At higher impact energies (some keV) the situation appears to be somewhat ambiguous [5.100–104], as mentioned in the previous section.

Recent work by Gnaser [5.53, 54] was aimed at determining in situ, under dynamic Cs$^+$ irradiation conditions, transient work-function changes as well as the yield variations of positive (Cs$^+$ and MCs$^+$) and negative secondary ions. To this end, pristine surfaces of elemental samples were exposed to the Cs$^+$ primary ion beam (impact energy 5.5 keV or 14.5 keV) in an incremental fashion, and energy distributions of various secondary ions were recorded after each fluence increment. Their energy shift is indicative of a variation of the contact potential between the sample (and thus of the variation of the WF) and the energy analyzer. WF changes of about 0.1 eV were detectable by these means (see below). Note that this onset method of a (relative) WF determination is often employed using secondary electrons [5.110, 111], but has also been utilized in static Cs adsorption experiments [5.48, 112]. The work of Gnaser [5.53, 54] however, appears to constitute the first application to dynamic Cs implantation conditions which are relevant to common SIMS

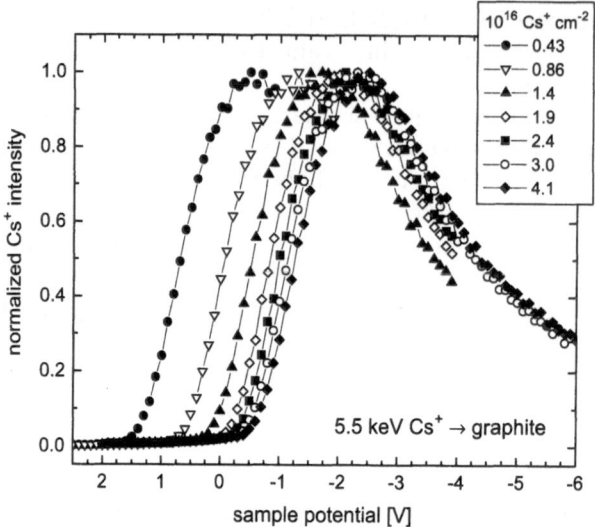

**Fig. 5.6.** Normalized energy spectra of $Cs^+$ sputtered from graphite under 5.5 keV $Cs^+$ impact. The sample potential is given relative to the value of 4500 V and the ions' emission energy increases from left to right in the plot. The parameter is the $Cs^+$ fluence

analyses. Furthermore, it attempted to correlate $Cs^+$ and $MCs^+$ ion yields during the gradual cesium buildup at the surface.

A distinct shift of the secondary-ion energy spectra with increasing $Cs^+$ fluence was observed [5.54]. This is exemplified in Fig. 5.6 for $Cs^+$ ions sputtered from graphite under 5.5 keV $Cs^+$ bombardment. The observation that the *complete* distributions are shifted, whereas their shape stays essentially unchanged, is important. These shifts can be ascribed, therefore, to WF changes induced by the transient Cs incorporation and were used to determine the relative changes of the work function induced by the gradual Cs buildup. Typically, this was accomplished by fitting tangents to the (steeply rising) low-energy portions of the energy spectra; their intercepts with the energy scale yielded the relative WF change. The energy distributions of $Cs^+$ showed that the measured $Cs^+$ intensity increases initially with fluence, passes through a maximum and decreases before reaching a stationary value. At a fluence of about $4 \times 10^{16}$ $Cs^+/cm^2$ a saturation of the Cs concentration is reached and the value of $\Phi$ levels off. The maximum shift then amounts to $\Delta\Phi \approx 2.1$ eV. Similar results were obtained for a variety of elemental specimens (C, Al, Si, Ge, Nb, Au). (In this and the following section, data for graphite and silicon specimens are mostly shown as in these materials, because of their rather low sputtering yields, the $\Delta\Phi$ variations are very pronounced and the associated effects are most clearly evident.)

The evolution of the ion intensity and the work function with increasing $Cs^+$ fluence is seen more clearly in Fig. 5.7 which depicts, as a function of $Cs^+$

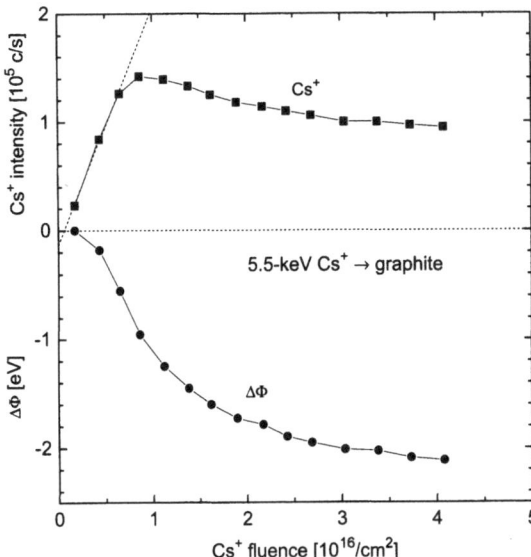

**Fig. 5.7.** The intensities of Cs$^+$ ions and the WF change $\Delta\Phi$ as a function of fluence for a graphite sample bombarded by 5.5 keV Cs$^+$. The *dashed line* in the *upper panel* indicates an expected linear increase of the Cs$^+$ intensity with concentration

fluence, the WF shift $\Delta\Phi$ and the intensities of Cs$^+$ ions for the sputtering of graphite; the latter values were extracted from the maxima of the respective energy distributions. The values of $\Delta\Phi$ have been evaluated from the Cs$^+$ data (both normalized and absolute energy spectra yield the same values of $\Delta\Phi$ within the accuracy of the results) and are given relative to the largest value of $\Phi$. Figure 5.7 also displays the general behavior observed for most of the other elements for dynamic Cs$^+$ incorporation: with increasing fluence, the work function is lowered and a stationary value is reached at fluences of some $10^{16}$ Cs$^+$ cm$^{-2}$. This fluence value is in agreement with the above-mentioned computer simulations. (Note that in graphite a Cs$^+$ fluence of $1 \times 10^{16}$ cm$^{-2}$ corresponds to the removal of a layer of about 2 nm if the bulk sputtering yield is assumed to be applicable.)

The observed reductions of the WF ($\sim 2.1$ eV for graphite, $\sim 1.4$ eV for both Al and Si, and 0.4 eV for Au) appear to be smaller than those typically found in Cs adsorption experiments (about 3 eV [5.113]). The difference, very probably, is due to the lower Cs concentration in the present case and, perhaps, a different site configuration of the Cs atoms (ions). While in Cs adsorption the Cs$^+$ ions sit on top of the substrate [5.114], this is not necessarily the case in the present implantation situation, where Cs is replenished by implantation concurrently with sputter removal, which exposes, at the surface, previously injected Cs atoms. Notwithstanding these differences, the WF changes drastically influence ion yields even under dynamic conditions. The data show that, upon lowering of the WF, the ion intensities start to deviate from the initially linear increase with fluence (see Fig. 5.7) and, with a further reduction of $\Phi$ due to the still increasing Cs surface concentration, pass through a maximum and then saturate. Such a behavior (see also [5.25]) can then be ascribed to a *reduction* of the ionization probability for positive secondary ions (here $Cs^+$) with *decreasing* WF. A similar yield evolution with $\Delta\Phi$ has been reported for a static adsorption/desorption experiment [5.62]. In terms of the formation of the Cs-carrying molecular ions discussed in Sect. 5.4.1, the finding that the intensity of these ions (e.g. $SiCs^+$) closely follows that of $Cs^+$ which supports the proposed formation process of these $MCs^+$ species (an association of a M atom with a $Cs^+$ ion), appears also to be rather important.

The influence of $Cs^+$ incorporation on the ionization probability of negative secondary ions is exemplified in Fig. 5.8, which depicts the gradual Cs buildup for graphite bombarded with 14.5 keV $Cs^+$. The intensity of $C^-$ is plotted (open circles, left-hand scale) as a function of $Cs^+$ fluence; for this measurement the energy slit was completely open ($\Delta E \sim 120\,eV$) in order to detect (almost) all ions irrespective of their emission energy. After passing through a regime of almost constant intensity (Cs atoms are implanted into the bulk but there is still little Cs at the surface), the ionization probability $P^-$ and therefore the yield rise with increasing Cs content at the surface and saturate at a fluence of $\sim 2.5 \times 10^{16}$ $Cs^+/cm^2$, indicative of equilibrium conditions (the yield enhancement at the surface might be due to the Cs incorporation beneath the surface in the initial stage of bombardment). In order to determine WF changes, the energy slit was closed to obtain a narrow passband ($\sim 2$–$3\,eV$) and partial energy spectra were recorded at different bombardment fluences as described above. A distinct shift of the onset of the spectra towards *lower* energies (higher sample potentials) with increasing fluence is observed and determines the relative changes of the work function $\Delta\Phi$ induced by the Cs buildup. The variation of $\Delta\Phi$ with fluence is shown

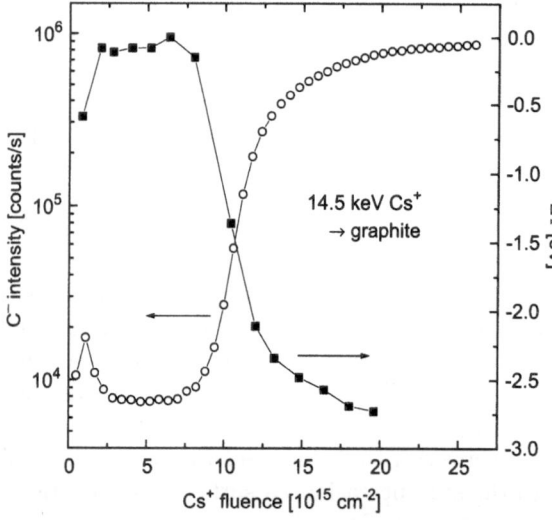

**Fig. 5.8.** $C^-$ intensity (*open circles, left-hand scale*) and WF variations $\Delta\Phi$ (*closed squares, right-hand scale*) as a function of fluence for 14.5 keV $Cs^+$ bombardment of graphite. Data from [5.53]

in Fig. 5.8 (closed squares, right-hand scale). It can be seen that the work function of Cs-loaded graphite is *lowered* by 2.75 eV upon reaching steady-state sputtering and effects a considerable yield enhancement. In addition to graphite, such experiments were also performed for Si, Nb and Au [5.54].

### 5.2.3 Correlation Between Work Function and Ionization

The linear increase of the Cs surface concentration obtained in the simulations (Fig. 5.5) has to be compared with the deviation from linearity of the Cs$^+$ intensity (see Fig. 5.7) at fluences of $> 2 \times 10^{15}$ cm$^{-2}$, i.e. at a value where the concentration is still increasing. Ascribing this deviation to a reduction of the Cs$^+$ ionization probability [5.54], it is possible to determine $P^+_{\mathrm{Cs}}$ and to correlate it with the associated changes $\Delta\Phi$. Such data are shown in Fig. 5.9; they indicate that $P^+_{\mathrm{Cs}}$ is indeed constant (it is assumed here that $P^+_{\mathrm{Cs}} = 1$) for small $\Delta\Phi$, but decreases drastically for $\Delta\Phi > 0.4$ eV, being, for Si, for example, a factor of seven lower for the maximum WF change observed. From the measured Cs$^+$ intensity at equilibrium (when the partial Cs sputtering yield is unity), the primary ion current and the instrument transmission ($\sim 15\%$), an experimental value of $P^+_{\mathrm{Cs}} \sim 0.17$ is found, which is in good agreement with the result depicted in Fig. 5.9 for Si. These data are in general agreement with results of Yu and Lang [5.38], who find a constant value of $P^+_{\mathrm{Cs}}$ for $\Phi > I$ and a pronounced reduction for $\Phi$ slightly below $I_{\mathrm{Cs}}$ (see also Fig. 5.3). Unlike these experiments, which determined an absolute scale for $\Phi$, the former study [5.54] could only derive relative changes of the WF. Nevertheless, a value $\Delta\Phi \approx 1$ eV, which corresponds to the factor-of-seven reduction of $P^+_{\mathrm{Cs}}$, is close to what is reported by Yu and Lang [5.38].

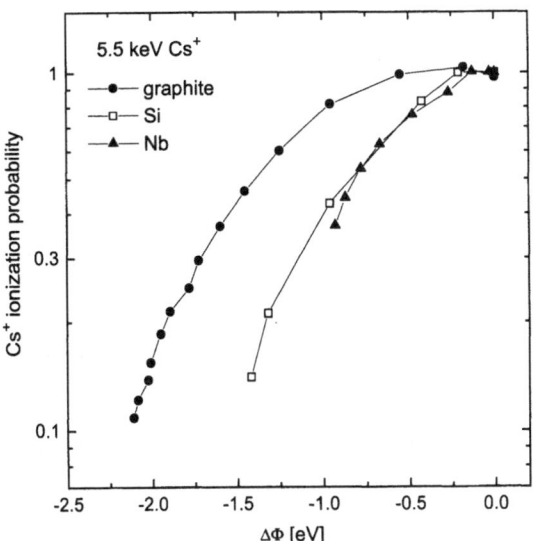

**Fig. 5.9.** The Cs$^+$ ionization probability $P^+$ versus work-function changes $\Delta\Phi$ for graphite, Si and Nb specimens. The values of $P^+$ are derived from the deviations of Cs$^+$ intensities from a linear increase with concentration (see Fig. 5.7 and text)

According to the simulations, a stationary Cs surface concentration is reached at a fluence of some $10^{16}$ $Cs^+/cm^2$; in the experiment, both $\Delta\Phi$ and the ion yields approach constant values at this fluence. The maximum WF shifts found [5.54] are typically less than that observed in adsorption studies [5.113–115], which produce shifts of $\sim 3\,eV$, albeit at considerably higher cesium coverages. Taking these investigations as a guideline, a value of $\Delta\Phi \approx 1.4\,eV$ would correspond to a Cs surface coverage of about 0.15 of a monolayer; the latter figure is close to the steady-state Cs concentration obtained from the computer simulations. This agreement should not be stressed too much, however, because of the possible difference between static and dynamic conditions with regard to Cs occupation sites discussed above.

The positive-ion emission under $Cs^+$ irradiation obtained for Au was distinctly different from those of the other elemental specimens (C, Al, Si, Ge, Nb) studied [5.54]. Because of its higher sputtering yield ($Y_{Au} \sim 12$ atoms/ion, as compared, for example, to $Y_{Si} \sim 2.3$ atoms/ion [5.116]), the equilibrium Cs concentration in Au should be correspondingly lower (under steady-state conditions, on average one Cs species is reemitted together with $Y$ sample atoms; so the concentrations might scale like $1/(1 + Y)$ to first order). This is reflected in the experiments, which find a much smaller WF change ($\Delta\Phi = -0.4\,eV$), reached already at a fluence of $4 \times 10^{14}$ $Cs^+/cm^2$, and a $Cs^+$ intensity which increases *without* passing through a maximum. Apparently, owing to the small value of $\Delta\Phi$, the work function is not reduced below $I_{Cs}$ (for the pristine Au surface, $\Phi_{Au} = 5.1\,eV$); therefore, $P_{Cs}^+$ is unity in the regime accessible under the present bombardment conditions and the $Cs^+$ intensity reflects directly the buildup of the surface Cs concentration.

The correlation between the measured ion yields of $C^-$ and $Si^-$ (which, apart from a transmission factor, represent the ionization probability $P^-$) and the $\Delta\Phi$ values is depicted in Fig. 5.10 [5.53]. A scaling in agreement with (5.6), i.e. an exponential dependence of $P^-$ on the work function, is observed. From the slopes values of $\varepsilon_n = 0.59$ and $\varepsilon_n = 0.49$ were derived for $C^-$ and $Si^-$, respectively, which are compatible with the theoretical predictions [5.43]. Employing the electron affinities of C and Si $A_C = 1.27\,eV$ and $A_{Si} = 1.39\,eV$ [5.117], the work functions of the pristine surfaces $\Phi_C = 5.0\,eV$ and $\Phi_{Si} = 4.85\,eV$ [5.118] and the stationary WF change due to Cs loading (see Fig. 5.10), the ionization probabilities for *equilibrium* conditions derived from (5.6) amount to $P_{th}^- = 0.19$ for $C^-$ and to $P_{th}^- = 0.093$ for $Si^-$, assuming the constant of proportionality to be unity (this value will be compared below with estimates obtained from experimental yield data).

Measurements essentially identical to those for C and Si have also been carried out for Ge and Au [5.53]. Although the general features are comparable, there exist clear element-specific differences. In all cases the ionization probability is enhanced during Cs buildup and saturates for steady-state sputtering; these equilibrium fluences decrease from that for silicon ($2 \times 10^{16}$ $Cs^+/cm^2$) through that for Ge ($1.5 \times 10^{16}$ $Cs^+/cm^2$) to that for Au ($3 \times 10^{15}$

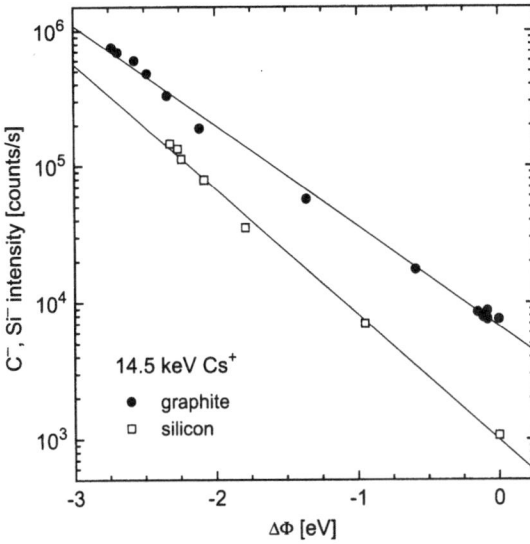

**Fig. 5.10.** Intensity versus WF change $\Delta\Phi$ for $C^-$ ions sputtered from graphite and for $Si^-$ ions sputtered from silicon. Data from [5.53]

$Cs^+/cm^2$). Also, the magnitude of the enhancement of $P^-$ is distinctly different: a factor of $\sim 150$ for $Si^-$ but only about 9 for $Ge^-$ and $\sim 5.8$ for $Au^-$. These differences scale with the maximum WF shifts observed between the virgin and the steady-state surfaces: $\Delta\Phi = 2.3\,eV$ for Si, $\Delta\Phi = 0.84\,eV$ for Ge and $\Delta\Phi = 0.62\,eV$ for Au. Using the corresponding $\Delta\Phi$ values and yield data obtained during the gradual incorporation of Cs, correlations between $P^-$ and $\Delta\Phi$ could be established for all elements investigated in this work [5.53] and a scaling in accordance with the theoretical predictions (cf. (5.6)) was found. From these data, values of $\varepsilon_n$ were derived and are compiled in Table 5.1. Using the electron affinities ($A_{Si} = 1.39\,eV$, $A_{Ge} = 1.2\,eV$ and $A_{Au} = 2.31\,eV$ [5.117]) and the work functions ($\Phi_{Si} = 4.85\,eV$, $\Phi_{Ge} = 5.0$ and $\Phi_{Au} = 5.1\,eV$ [5.118]) together with the maximum shifts $\Delta\Phi$ derived experimentally (listed in Table 5.1), the ionization probabilities $P_{th}^-$ for steady-state Cs implantation conditions were evaluated from (5.6) and are also listed in Table 5.1. It is seen that $P_{th}^-$ is rather high for $C^-$ (19%) and $Si^-$ (9.3%) but distinctly lower for $Au^-$ and $Ge^-$. Obviously, this finding can be ascribed to the much smaller $\Delta\Phi$ values for the two latter elements, since the values of $(\Phi - A)$ for the pristine surfaces investigated here are almost identical. By means of the measured ion intensities, the primary ion current and the sputtering yields given above, the number of *detected* secondary ions per sputtered atom can be determined. This quantity constitutes the product of $P^-$ and the instrumental transmission factor $\eta$. Employing $\eta \sim 0.20$, the experimental data $P_{exp}^-$ for the *steady state* have been determined; they are listed in Table 5.1 and are in excellent agreement with the theoretical values derived via (5.6).

**Table 5.1.** The maximum WF change $\Delta\Phi$; the value of $\varepsilon_n$ derived from plots of the ion yield versus $\Delta\Phi$; and the steady-state ionization probability evaluated from (5.6), $P_{th}^-$, and from the experimental ion yield data, $P_{exp}^-$. From [5.53]

|     | $-\Delta\Phi/\mathrm{eV}$ | $\varepsilon_n/\mathrm{eV}$ | $P_{th}^-$ | $P_{exp}^-$ |
|-----|------|------|------|------|
| C   | 2.75 | 0.59 | 0.19 | 0.17 |
| Si  | 2.3  | 0.49 | 0.093 | 0.088 |
| Ge  | 0.84 | 0.46 | $1.6 \times 10^{-3}$ | $2.2 \times 10^{-3}$ |
| Au  | 0.62 | 0.42 | $5.7 \times 10^{-3}$ | $5.3 \times 10^{-3}$ |

## 5.3 Isotopic Mass Effects in Sputtered-Ion Formation

The influence of mass differences between two isotopes of a given element on preferential sputtering was outlined in Sect. 3.4; the effects reported in that context were traced to mass-dependent differences in the stopping cross sections. An isotopic mass effect might be expected also for the *ionization* of sputtered species. In Sect. 5.1.1 it was noted that the ionization probability may depend in some way on the ion's emission velocity, although the experimentally derived dependences are not fully consistent. In the following, investigations by Gnaser and Hutcheon [5.119–122] of the isotopic mass effect in secondary-ion formation are reported. These authors studied a variety of specimens that are also of geological relevance, such as oxides, silicates and related minerals, and compared the determined isotope fractionations in the ionization process with the bond-breaking model [5.76, 77]. (In fact, high-precision isotope measurements [5.123] in geological and cosmochemical studies must take account of such fractionation effects related to ion formation.)

The bond-breaking model of ion emission predicts that the ionization probability $P$ for an atom to be sputtered as an ion depends exponentially on the atom's emission velocity $v$ [5.76–78] (5.11): $P \propto \exp(-v_0/v)$, where $v_0$ is a constant for a given ion–substrate system and is related to the transition matrix element. A dependence of $P$ on $v$ formally equivalent to (5.11) has also been proposed in other models of secondary-ion emission (see [5.11]). For two isotopes $i$ and $j$ of an element having masses $M_i$ and $M_j$ (e.g. $M_i < M_j$), one can define the isotopic fractionation $F_{ji}$ as the relative difference of the ionization probabilities of isotopes $i$ and $j$ [5.119, 120],

$$F_{ji} \equiv (P_j - P_i)/P_i \tag{5.14}$$

or, by applying (5.11),

$$F_{ji} = \frac{\exp(-v_0/v_j)}{\exp(-v_0/v_i)} - 1. \tag{5.15}$$

Replacing $v_j$ by $v_i (M_i/M_j)^{1/2}$ and expanding (5.15) in a series for the exponential function yields

$$F_{ji} = 1 + \left(-\frac{v_0 M_0}{v_i}\right) + \cdots - 1 , \tag{5.16}$$

where $M_0 = (M_j/M_i)^{1/2} - 1$. Since $M_0 \sim 10^{-2}$ and $v_i$ is approximately equal to $v_0$ for the range of velocities investigated here [5.119–121], second- and higher-order terms in (5.16) can be neglected and

$$F_{ji} \cong -\frac{v_0 M_0}{v_i} . \tag{5.17}$$

In the bond-breaking model the velocity $v$ is evaluated at the crossing point of the covalent and ionic potential-energy curves. This velocity usually will be different from the velocity measured experimentally, i.e. far away from the surface. According to this concept [5.76],

$$v = \left[2 \left(E_0 + I - A\right)/M\right]^{1/2} , \tag{5.18}$$

where $E_0$ is the ion energy *measured*, $I$ is the ionization potential of the sputtered atom and $A$ is the electron affinity of the cation vacancy created in the solid. For the latter, Yu [5.78] proposed to use the electron affinity of the electronegative species (e.g. oxygen) in the sample. While this choice is perhaps somewhat arbitrary, it appears that the term $(I - A)$ is typically a few eV and is relevant only for low energies $E_0$. Then, according to (5.11) and (5.17), $P$ and consequently $F$ will converge to constant values. Because of the uncertainties associated with the proper choice of $A$, $v$ in the following was computed directly from the measured values of $E_0$, i.e. $v = (2E_0/M)^{1/2}$.

Following the concept outlined in the previous paragraph, the isotopic fractionation of secondary ions is expected to show three characteristic features:

(i) for different isotopes of an element, $F$ should depend linearly on $M_0$, i.e. to *first* order $F$ is proportional to the relative mass difference of the isotopes;

(ii) the measured fractionation $F_{ji}$ of two isotopes of an element emitted from a given sample should be inversely proportional to the emission velocity $v$ and, according to (5.18), should approach a constant value for low velocities;

(iii) for two isotopes of an element sputtered from *different* substrates with the *same* energy, $F_{ji}$ should scale linearly with the $v_0$ of the respective substrate.

The results obtained in the experiments mentioned above [5.119–121] demonstrate that all three dependences are valid for a great variety of ion species and samples. Furthermore, the measurements allow a precise determination of the parameter $v_0$ entering the bond-breaking model (see (5.11) and (5.16)). Values of $F_{ji}$ for a given ion velocity were calculated from the measured secondary-ion intensity ratios (usually the mean of a set of 50 to 100 individual ratios) compared with the "true" abundance ratios [5.119]:

$$F_{ji} = \left( R_{ji}^{m}/R_{ji}^{0} - 1 \right) \times 1000 \,, \tag{5.19}$$

where $R_{ji}^{m}$ is the measured intensity ratio and $R_{ji}^{0}$ is the abundance ratio for isotopes $i$ and $j$. The latter was taken from the literature since isotopic variations are usually small (less than one part per thousand) for most elements in terrestrial materials; even if they were to occur, they would not change the essential features of the experiment as outlined above. In the notation of (5.19) the isotopic fractionation $F$ is reported in parts per thousand (permil or $\%_0$). The experiments of Gnaser and Hutcheon [5.119–122] were carried out on a secondary-ion mass spectrometer modified for high-precision isotope measurements [5.123]; the data presented in the following refer to steady-state conditions, when the composition of the sputtered flux is identical to the bulk abundances of the substrate. The kinetic energy (and therefore the velocity) of the secondary ions was selected by offsetting the sample potential and keeping the other parameters of the instrument constant. Since the instrument has a rather large angular acceptance, the small differences in emission angle for two isotopes (see Chap. 3) are not expected to play a significant role. Thus, the pronounced isotopic fractionation observed for secondary ions is due to the *ionization processes*.

To investigate the predictions of isotopic fractionation in secondary-ion formation, different sets of experiments were performed. The dependence of $F$ on $M_0$ was examined [5.120] by measuring the magnitude of Ca isotope fractionation as a function of mass in calcite ($CaCO_3$). Calcium was chosen because of its high ion yield and the presence of many (6) isotopes covering a wide range in relative mass difference (20%). The measured values of $F$ (in parts per thousand) versus mass number, with $^{40}Ca$ as the reference isotope, exhibit an accurate dependence on $M_0 = (M_j/M_i)^{1/2} - 1$, where $M_i = 40$ and $M_j$ is the mass of one of the other Ca isotopes. For small mass differences the dependence of $F$ is essentially linear in the mass, while deviations from linearity become important for larger mass differences. The measurements demonstrated that a precision of a few permil is achievable at (moderately) high mass resolution ($M/\Delta M = 4500$). While Ca isotopes are exceptionally well suited to establish the mass dependence of $F$, the validity of this relationship was also verified for other elements which have more than two isotopes, e.g. Mg, Si and Ti.

The dependence of $F$ on the emission velocity $v$ was studied [5.119–121] by measuring the steady-state isotope fractionation for a variety of ion species ($Li^+$, $Li_2^+$, $B^+$, $Mg^+$, $Si^+$, $Cl^-$, $Ca^+$, $Ca^{2+}$, $Ti^+$) as a function of ion energy (velocity). Typical results of these experiments are shown in Fig. 5.11, which depicts the measured fractionation $F_{ji}$ versus $1/v_i$ for two $Si^+$ isotopes ($^{28}Si$ and $^{29}Si$) sputtered from silicon. $F_{ji}$ is plotted in parts per thousand and represents the mean of a set of individual ratios; the errors are $\pm 2\sigma_{mean}$. The errors associated with the velocity result from the slight shifts in energy due to sample charging and the uncertainties related to the origin of the energy scale. The straight line is a linear least-squares fit to the data. For

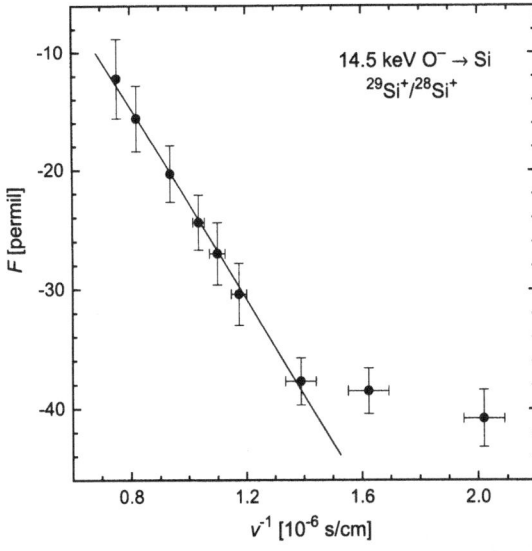

**Fig. 5.11.** Fractionation versus the inverse emission velocity $v^{-1}$ for $Si^+$ isotopes emitted from silicon. The isotope fractionation increases with decreasing ion velocity. Data from [5.119]

all ion species a negative fractionation is found, which, in the present notation, indicates a *depletion* of the *heavier* isotopes in the secondary-ion flux relative to the normal composition. This depletion *increases* with *decreasing* ion velocity. Apart from the deviations at low energies (see the discussion below), the correlation between $F$ and $v^{-1}$ is in excellent agreement with (5.17) and therefore seems to support the exponential dependence of $P$ on $v$ in (5.11). Similarly good agreement was observed for a great number of ions emitted from various substrates. The species investigated are listed in Table 5.2 together with the samples used and the values of $v_0$ derived from the straight-line fits in the respective $F$ versus $v^{-1}$ plots. The errors in $v_0$ represent the quality of the fits and, as can be seen, $v_0$ can be determined with a precision of better than 10% in most cases. As was anticipated from (5.11), the values of $v_0$ depend on the ion species and are clearly *higher* for ions for which the measured ion intensity is *lower* (e.g. $Li_2^+$ and $Ca^{2+}$).

Using the experimentally determined values of $v_0$ and assuming that the preexponential factor in (5.11) is of order unity, the ionization probability $P$ can be derived; for an emission energy of 7 eV, values of $P$ range from $\sim 10^{-4}$ for $Ca^{2+}$ to a few tens of percent for $Li^+$ and $Cl^-$. While a direct comparison of these numbers with the actual measured ion intensities is difficult since the instrumental transmission and the sputtering yields of these samples are not known, they generally fall in the range to be expected for oxygen-rich (and strongly ionic) specimens. For $Be^+$ sputtered from BeO, for example, a value $P \sim 0.03$ (for $E_0 = 7$ eV) was reported [5.64]. Also, for $B^+$ emitted from oxygen-saturated silicon, a useful yield (the product of $P$ and the instrument transmission) of $2 \times 10^{-2}$ was determined [5.124]. A transmission of the order of 0.1 yields $P \sim 0.2$, a value essentially identical to the present result ($P =$

**Table 5.2.** The ion species studied, together with the respective samples and the values of $v_0$ derived from the respective plots of isotopic fractionation $F$ versus inverse velocity $v^{-1}$ [5.120]

| Ion | Sample | $v_0/\mathrm{cm\,s^{-1}}$ |
|-----|--------|---------------------------|
| $Li^+$ | LiF | $(1.86 \pm 0.08) \times 10^6$ |
| $Li_2^+$ | LiF | $(4.03 \pm 0.10) \times 10^6$ |
| $B^+$ | Tourmaline* | $(1.97 \pm 0.08) \times 10^6$ |
| $Mg^+$ | $MgAl_2O_4$ | $(1.23 \pm 0.10) \times 10^6$ |
| $Si^+$ | Si | $(2.27 \pm 0.09) \times 10^6$ |
| $Cl^-$ | NaCl | $(5.94 \pm 0.11) \times 10^5$ |
| $Ca^+$ | $CaF_2$ | $(1.94 \pm 0.08) \times 10^6$ |
| $Ca^{2+}$ | $CaF_2$ | $(5.65 \pm 0.15) \times 10^6$ |
| $Ti^+$ | $TiO_2$ | $(1.65 \pm 0.07) \times 10^6$ |

* Composition approximately $Na(Li,Al)_3Al_6B_3Si_6O_{27}(OH,F)_4$

0.18 for $E_0 = 7\,\mathrm{eV}$). Ionization probabilities of some tens of percent have also been reported for various secondary ions sputtered from oxidized metal surfaces [5.10]. The overall range of $v_0$ determined in [5.120, 121] is in good agreement with previously reported determinations of $v_0$ [5.52] which utilized (5.11) directly, i.e. by measuring energy spectra of ions and neutrals species (or making assumptions about the latter).

The variation in the velocity dependence of isotope fractionation for a given ion ejected from different matrices was investigated [5.120] for $Ca^+$ ions sputtered from a variety of Ca-bearing minerals. In addition to the fluorite ($CaF_2$) listed in Table 5.2, five other samples were studied. The specimens, together with the values of $v_0$ derived again from the respective plots of $F$ versus $v^{-1}$, are listed in Table 5.3. For three of them this functional dependence is shown in Fig. 5.12. Again, $F$ is linearly correlated with $1/v$ in all cases but the slope of the correlation line (i.e. $v_0$) varies by about a factor of 50 between calcite and anorthite. A common feature of these measurements is evident from Fig. 5.12; at low ion velocities (corresponding to energies of less than $\sim 5\,\mathrm{eV}$), $F$ approaches a constant value. This behavior is expected

**Table 5.3.** The samples for which isotope effects for $Ca^+$ ions were determined. The values of $v_0$ are derived from the plots of $F$ versus $v^{-1}$ [5.120]

| Sample | Composition | $v_0/\mathrm{cm\,s^{-1}}$ |
|--------|-------------|---------------------------|
| Anorthite | $CaAl_2Si_2O_8$ | $(4.77 \pm 1.90) \times 10^4$ |
| Hibonite | $CaAl_{12}O_{19}$ | $(2.80 \pm 0.48) \times 10^5$ |
| Sphene | $CaTiSiO_5$ | $(6.95 \pm 0.25) \times 10^5$ |
| Perovskite | $CaTiO_3$ | $(1.00 \pm 0.05) \times 10^6$ |
| Fluorite | $CaF_2$ | $(1.94 \pm 0.08) \times 10^6$ |
| Calcite | $CaCO_3$ | $(1.96 \pm 0.15) \times 10^6$ |

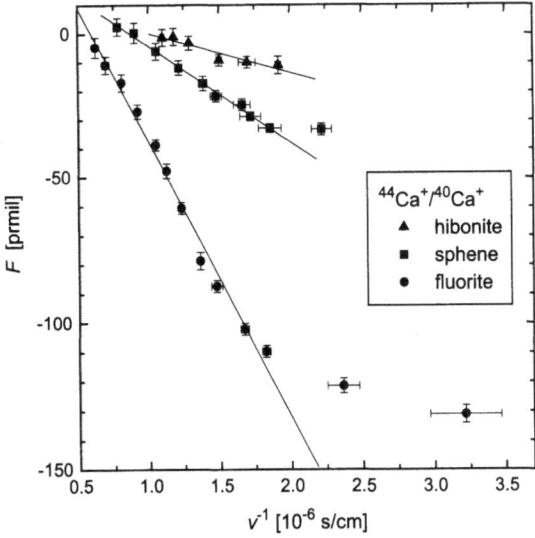

**Fig. 5.12.** Fractionation $F$ of $Ca^+$ ions sputtered from hibonite, fluorite and sphene versus $1/v$. Data from [5.120]

to the extent that the bond-breaking model is valid. According to (5.18), for small $E_0$, $v$ is proportional to $I - A$, i.e. approaches a constant value for a given ion. Thus, at low energy, $P$ and therefore $F$ will converge to a constant value. For these conditions, a rough estimate of the term $(I - A)$ in (5.18) may be obtained from the velocity at which $P$ and $F$ level off to a constant value. In a few cases this could be identified unambiguously. The values of $(I - A)$ (in eV) derived are: 5.5 (for $Li^+$ and $Li_2^+$), 6.4 (for $Si^+$), 3.6 ($Cl^-$), 7.3 ($Ti^+$ in $TiO_2$), 8.2 ($Ti^+$ in Ti) and 4.7–5.2 ($Ca^+$ in the various Ca-containing samples).

It follows from (5.17) that, for a given velocity $v$, values of $F$ for the same isotopes in different substrates should depend linearly on the respective $v_0$. This correlation was verified [5.120] by plotting the measured values of $F$ (at a given ion velocity) in the various Ca-bearing samples (cf. Table 5.3) versus the respective $v_0$ for two emission energies $E_0$. Over the wide range associated with $v_0$, an excellent correlation was found for different emission velocities. This finding thus constitutes additional (independent) support for (5.17) and the bond-breaking model, since the values of $F$ and $v_0$ were derived from six different samples.

The variations of $v_0$ among the different Ca-bearing samples are (roughly) correlated with a change in the measured ion intensity and of the energy distribution, such that with increasing $v_0$ the ion intensity (corrected for the respective Ca concentration) decreases and the energy spectrum becomes broader. Both findings are in qualitative agreement with (5.11). The aforementioned discrepancies at low energies (cf. Fig. 5.12) might, to some extent, be caused by the (incorrect) evaluation of $v$ in that regime (see (5.18)). By evaluating $v$ from (5.18) instead of from $(2E_0/M)^{1/2}$, it is found that for

$(I - A) \sim 6 \pm 1\,\mathrm{eV}$ the correlation between $F$ and $1/v$ (see (5.17)) is much improved for the low-energy data. Unfortunately, for most ion species such an evaluation was not comparably successful.

The samples investigated by Gnaser and Hutcheon [5.119–121] were, at least to some extent, ionic in nature, and therefore the bond-breaking model appears well suited to describe secondary-ion emission in such cases. Only a few of the previous studies [5.125–129] on isotope effects in secondary-ion sputtering were carried out on similar materials. The general trends observed in those studies appear to agree with the present findings, namely a decrease in the isotopic fractionation with increasing ion velocity. For other specimens, such as metals and some semiconductors, the isotopic-fractionation dependence on the ion's emission velocity (energy) often displays more complex patterns [5.122, 130–132] and an evaluation in terms of the bond-breaking model is apparently not possible.

## 5.4 Formation of Molecular Ions in Sputtering

As discussed in Sect. 2.3.4, the flux of particles sputtered from a surface may contain a large fraction of molecules and clusters. While monitoring neutral clusters became possible only fairly recently with the advent of efficient methods of post-ionization, the detection of charged clusters has a long history (see Sect. 2.3.4). Honig [5.3] first observed dimers in the mass spectrum of positive ions. Krohn [5.21] and later Hortig and Müller [5.133] reported the emission of negatively charged $Ag_n$ clusters containing a considerable number of atoms. Katakuse and coworkers [5.134, 135] found $Ag_n^+$ cluster ions up to $n = 240$ sputtered from polycrystalline silver bombarded by $10\,\mathrm{keV}$ $Xe^+$ ions. Many studies revealed, furthermore, strong oscillations in the abundance distributions of cluster ions for certain elements (see Fig. 2.20). Quite commonly, these abundance distributions are determined by two effects: (i) the stability of the cluster against fragmentation and (ii) the different ionization potentials (or electron affinities). Many charged clusters exhibit a rather high probability of spontaneous fragmentation by evaporation of one (or sometimes two) atoms. This may happen on timescales which are short compared to the time needed to detect these species (e.g. the flight time from the sample through the mass spectrometer). Hence, the detected mass distribution may reflect the clusters' stability rather then their genuine emission distribution. This problem was recognized [5.136, 137] quite frequently and many investigations were performed to address these fragmentation processes (see Sect. 2.3.4).

Apart from spontaneous fragmentation effects, the measured abundance distributions of positive cluster *ions* are determined by variations of the ionization potential with cluster size. According to the spherical-drop model [5.138], the ionization potential of a positively charged cluster $I_{clust}$ exhibits a monotonic fall-off inversely proportional to the cluster radius $R_{clust}$ and, with increasing cluster size, approaches asymptotically the bulk work function $\Phi$ of that element:

$$I_{\text{clust}} = \Phi + \kappa \left( \frac{e^2}{R_{\text{clust}}} \right) , \tag{5.20}$$

where $\kappa$ is a constant, usually found to be in the range between 0.3 and 0.5. Experiments and more elaborate computations have demonstrated, however, that $I_{\text{clust}}$ may show significant alternations around the values given by (5.20). These are well documented for alkali metals but have also been reported for some transition metals and carbon [5.137]. It is thought that these irregularities in the ionization potentials of clusters are caused by the electronic shell structure and may be prominent at the shell closings (e.g. $n = 8, 18, \ldots$ for positive clusters of the alkali metals).

Wahl and Wucher [5.139, 140] have performed detailed investigations of the ionization and fragmentation cross sections of sputtered $Ag_n$ (see Fig. 2.22) and other metal clusters. They employed single-photon ionization (at a wavelength of $\lambda = 157\,\text{nm}$, corresponding to an energy of 7.9 eV) to post-ionize sputtered atoms and clusters, monitoring them in a time-of-flight mass spectrometer. For $Ag_n$ clusters containing more than two atoms the total variation of the ionization cross section $\sigma_I$ is less than a factor of two around an average value of about $1.5 \times 10^{-16}\,\text{cm}^2$, whereas the values for $Ag_2$ and Ag atoms are smaller by factors of three and ten, respectively. The authors [5.139, 140] noted also that the ionization cross sections of clusters do not follow the simple additivity rule $\sigma_I \propto n$ or a geometrical rule $\sigma_I \propto n^{2/3}$ [5.141] which was derived under the assumption of a spherical shape of the clusters (see (5.20)). Wahl and Wucher stressed, furthermore, the importance of fragmentation processes: the measured fragmentation cross sections $\sigma_F$ of $Ag_n$ clusters were found to exhibit distinct variations as a function of cluster size and they identified a slight odd–even alternation, $\sigma_F$ being larger for clusters containing an even number of atoms, with the exception of $Ag_2$, $Ag_6$ and $Ag_{14}$. The authors conclude that the magnitude of the fragmentation cross section is determined by the relative stability of the $Ag_n^+$ cluster *ions* [5.142].

By comparing the fragmentation-corrected yields of sputtered neutral clusters with the yields of singly positive-charged clusters, Wahl and Wucher [5.139] derived the ionization probability $P^+$ of $Ag_n$ clusters sputtered from polycrystalline silver by 5 keV $Ar^+$ ions at an incidence angle of 45°. Figure 5.13 depicts these data; two features can be recognized. First, $P^+$ increases rapidly with increasing cluster size $n$ for $n < 10$ and becomes essentially independent of cluster size for $n > 10$. Second, the large majority of all clusters investigated are emitted in the neutral state. Furthermore, pronounced even–odd alternations are observed for $n < 10$: $Ag_n$ clusters with an odd number of atoms have a much higher ionization probability than the even-numbered ones. This finding matches the strong oscillations reported for the ionization potential of silver clusters in this size range [5.143]: the

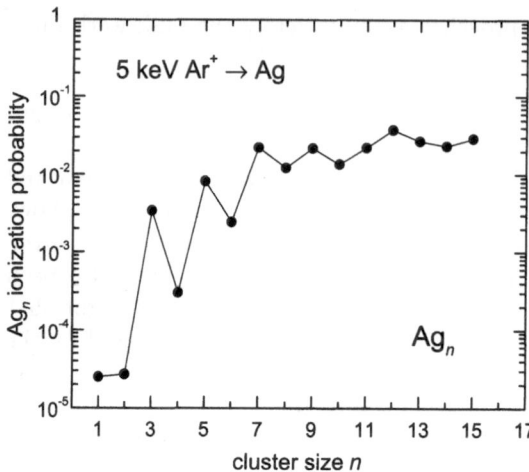

**Fig. 5.13.** Ionization probability of positive $Ag_n^+$ clusters sputtered from silver by 5 keV $Ar^+$ ions (incidence angle 45°) as a function of cluster size $n$. Data from [5.139]

ionization potential is much smaller for $Ag_n$ clusters with an odd number of constituents. For example, $I(Ag_5) = 6.35\,eV$, whereas $I(Ag_6) = 7.15\,eV$. For cluster sizes $n > 10$ these oscillations of the ionization potential become less pronounced, in agreement with the leveling off seen for $P^+$ in Fig. 5.13.

By choosing a laser wavelength in the VUV spectral region, Wahl and Wucher [5.139, 140] were able to ionize all sputtered clusters and the atomic species by a single-photon absorption process and, hence, with comparable ionization efficiency. This provided the possibility to determine the ionization and fragmentation cross sections; using the latter to correct the measured intensities for fragmentation, the authors presented yields of sputtered neutral metal clusters and the absolute fraction of sputtered silver atoms that are ejected in a bound state.

The question of ionization and fragmentation of large cluster ions will be discussed in Sect. 5.4.2 for singly and doubly charged negative $C_n$ clusters. The pertinent mechanisms of cluster formation in sputtering have been outlined in Sect. 2.3.4; these would apply also to charged clusters, with the added complexity of the size-dependent ionization potential. The present section on cluster ions will also present results on the emission of diatomic cations carrying a Cs atom (Sect. 5.4.1); the formation of these species is closely related to the Cs incorporation processes discussed in Sect. 5.2.

### 5.4.1 Cs-Carrying Diatomic Cations

It has been known for some time [5.93] that $Cs^+$ ion bombardment of surfaces produces a fair number of molecular species $MCs^+$, where M stands for a sample constituent. Since the yield of these ions was found [5.94, 95, 144, 145] to exhibit little (or even no) dependence on sample composition (for atomic ions, these variations, called the matrix effect [5.10], have severely limited the

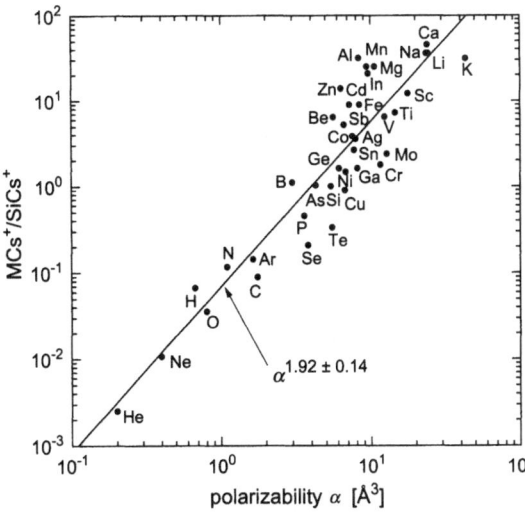

**Fig. 5.14.** Yields of sputtered $MCs^+$ ions from various low-concentration elements in silicon as a function of the polarizability of M, $\alpha$. The *solid line* is a linear least-squares fit with the slope indicated. Data from [5.146]

quantification of SIMS data), these ions are frequently utilized for analytical purposes [5.12, 94, 95, 144, 145]. The reduction (or absence) of matrix effects for these species was rationalized by their possible formation/ionization mechanism, namely the combination of a *neutral* M atom with a $Cs^+$ *ion* in the sputtering event. The implanted Cs atoms generally have very high ion yields (close to unity) and, in many cases, will be present as ions at the surface [5.114] (see Sect. 5.2.2). Thus, under steady-state conditions the flux of $MCs^+$ ions will reflect (via that of M atoms) the atomic concentration of the element M. Obviously, some dependence on the actual $Cs^+$ yield is expected and, in fact, found experimentally. The proposed formation process for $MCs^+$ ions was corroborated by recent experiments [5.146] that produced evidence for a correlation of the yield of $MCs^+$ secondary ions with the atomic polarizability $\alpha$ of the species M (see Fig. 5.14). These data were obtained on a secondary ion mass spectrometer and utilized 5.5 keV $Cs^+$ primary ions (impact angle $42°$ off the surface normal), monitoring $MCs^+$ molecular ions for low-concentration dopants in silicon [5.146]. From the linear least-squares fit (the solid line in Fig. 5.13) a power-law dependence $Y_{MCs^+} \propto \alpha^n$, with $n = 1.92 \pm 0.14$, was derived. A very similar exponent ($n = 2.07 \pm 0.20$) has been determined from a second data set, measuring $MCs^+$ yields sputtered from a variety of elemental and compound semiconductors [5.146].

The experimentally established correlation between the yields and the polarizability can be explained in a quantitative fashion by employing an appropriate interaction potential between $Cs^+$ and M; the argument is based on a distribution function $f^{(2)}$ developed for the emission of small molecules in sputtering according to the double-collision scenario (see Sect. 2.3.4). For a dimer this function is, in the limit of large $E_0$ [5.147],

$$f^{(2)}(E_0, \theta_0) \propto D^{3/2} E_0^{-5} \cos^2 \theta_0 , \tag{5.21}$$

where $E_0$ and $\theta_0$ are the dimer's emission energy and angle, respectively, and $D$ is its dissociation energy. While the experimental evidence for the energy and angle dependences in (5.21) was discussed in Sect. 2.3.4, apparently no such data exist for the scaling with $D^{3/2}$. It is this dependence on the dissociation energy which will be utilized to demonstrate the validity of the experimentally observed correlation between MCs$^+$ yields and polarizability (see Fig. 5.14).

The interaction between an alkali ion and a polarizable neutral atom (or molecule) is a focus of long-standing research in gas-phase reactions [5.148]. The mobility of an alkali ion in a gas and the attachment of neutral gas species onto it are but two questions addressed frequently. These experiments were rather successfully modeled by applying for the alkali-ion–neutral-atom interaction a potential of the form [5.149]

$$V(r) = Ar^{-12} - 0.5\alpha e^2 r^{-4} , \qquad (5.22)$$

where $r$ is the distance between the neutral atom and the ion, $\alpha$ is the polarizability of the former and $A$ is a constant which characterizes the repulsive term. The attraction is of the Langevin type [5.148] and is that arising between the ion and the ion-induced dipole moment of the neutral species. In the following the potential given by (5.22) will be used to model the interaction relevant to the binding of MCs$^+$ ions. Of particular concern is the depth of the potential well (i.e. the value of $V(r)$ at the equilibrium distance $r_m$), which is the dissociation energy for the MCs$^+$ molecule. From (5.22) it is seen that the potential minimum is given by

$$D \equiv V(r_m) \propto \alpha^{3/2} A^{-1/2} . \qquad (5.23)$$

Assuming MCs$^+$ ions to form according to the double-collision mechanism outlined above, then (5.21) should be applicable. Together with (5.23), the dimer distribution function $f^{(2)}$ will exhibit a dependence on the polarizability $\alpha$ given by

$$f^{(2)} \propto \alpha^{9/4} A^{-3/4} , \qquad (5.24)$$

with an exponent for $\alpha$ ($n = 2.25$) which is very close to what has been observed experimentally (see Fig. 5.14). This rather good agreement seems to support the two main premises the above treatment is based on [5.146]: first, the formation of MCs$^+$ ions proceeds via a combinative association of a Cs$^+$ ion and a neutral M atom; second, the binding strength of these species can be modeled by a potential of the type given in (5.22), with the polarizability of M playing the key role. Thus, the data also appear to constitute the first direct verification of the $D^{3/2}$ scaling of the dimer formation probability as predicted in (5.21).

Clearly, this theoretical approach includes some assumptions: the interaction potential employed has been used successfully to describe gas-phase reactions, mostly at very high gas densities; its transferability to interactions occurring at the surface of a solid is not obvious. On the other hand, the pronounced dependence on the polarizability found in the experiments indicates that the latter is a key parameter in any kind of formation mechanism of MCs$^+$. Also, (5.24) shows the dimer distribution function to depend (albeit weakly) on the parameter $A$ of the repulsive part of the interaction. Conceivably, $A$ may not be identical for the different MCs$^+$ ions investigated here. In fact, the experiments show, for some elements, distinct deviations from the power-law correlation which, in part, may reflect a variation of the value of $A$. Using a value $A = 1.1 \times 10^5$ eV Å$^{12}$ which was derived from the aforementioned gas-phase clustering experiments and $\alpha = 10$ Å$^3$, the interaction potential yields an equilibrium distance $r_m = 2.8$ Å and a well depth $V(r_m) = 0.71$ eV. For a comparison it is noted that the sum of the ionic radius of Cs$^+$ and the atom radius of Si amounts to 2.84 Å.

## 5.4.2 Doubly Charged Negative $C_n^{2-}$ Clusters

The existence of positively and negatively charged clusters of carbon in the sputtered flux is well documented. For positive secondary ions, doubly (and multiply) charged atoms and molecules are a common observation, in particular for light elements (e.g. Al, Si, Mg); their abundant emission from these elements at sufficiently high bombarding energies ($\sim 10$ keV) was explained by an ion formation mechanism involving inner-shell excitations [5.150, 151]. By contrast, doubly charged negative ions have been observed only very recently [5.152, 153]. Sputtering of graphite produced small doubly negative carbon clusters, with $C_7^{2-}$ being the smallest one detected. The abundance distribution of these doubly charged $C_n^{2-}$ cluster ions is shown in Fig. 5.15 as a function of cluster size. Also depicted is the abundance distribution of the singly charged $C_n^-$ clusters. Both types of ion species exhibit pronounced odd–even alternations with cluster size. For singly charged $C_n^-$ clusters this finding agrees with previous results; of note is the change in periodicity observed around $n = 10$: even-numbered and odd-numbered $C_n^-$ clusters dominate below and above that value, respectively. This observation has been ascribed [5.154] to a change in cluster geometry: according to ultraviolet photoelectron spectra, clusters with $n < 10$ have a linear chain-like structure, while larger ones appear to form monocyclic rings. This change in configuration is reflected also in a variation of the electron affinities which favors the formation of even- or odd-numbered clusters, respectively.

With increasing cluster size $n$, the abundance of doubly charged clusters is seen to approach that of singly charged ones. (The maximum size detectable is determined for both species by the mass range of the instrument, which amounts to about 280 amu × charge.) Obviously, with increasing cluster size the accommodation of two negative charges becomes more probable. In fact, stable doubly charged $C_{60}^{2-}$ ions have been produced by laser desorption [5.155]. Theoretical computations by Cederbaum and coworkers [5.156–159] have shown that small doubly charged negative clusters might indeed be stable in the gas phase, albeit with rather small binding energies. Possible

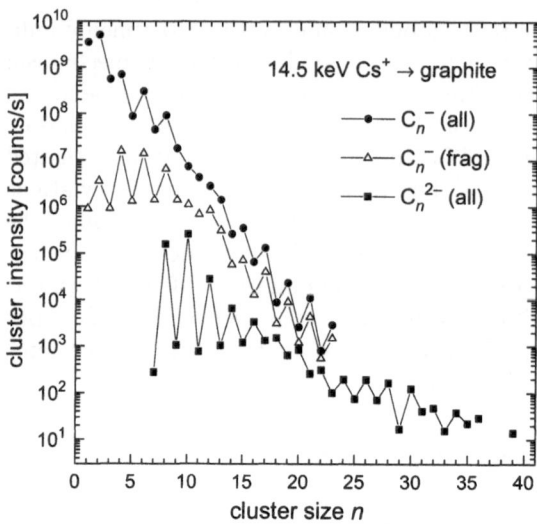

**Fig. 5.15.** Ion intensities of singly and doubly charged negative carbon clusters sputtered from graphite by 14.5 keV $Cs^+$ ion irradiation as a function of cluster size $n$

candidate molecules identified in these calculations were of the general type $XY_3^{2-}$, where X and Y are alkali and halogen atoms, respectively; the most stable ones reported were $LiF_3^{2-}$ and $KF_3^{2-}$. An even higher electron affinity (and hence formation probability) was found for doubly charged negative ions of the type $X_2Y_4^{2-}$. These and other authors [5.159, 160] also computed electron affinities $A$ for doubly charged negative $C_n$ clusters. The corresponding $A$ values show pronounced alternations, with the even-numbered ones being considerably higher, but with a general tendency to increase with increasing $n$. Interestingly, these calculations predicted electronic instability for $C_n^{2-}$ clusters with $n < 10$, which apparently contradicts experimental findings. This discrepancy was clarified in a more recent study [5.159, 161]: the structurally and electronically stable ground states of $C_7^{2-}$ and $C_9^{2-}$ are represented by the "center-ligand sphere" structures $[C(C_2)_3]^{2-}$ and $[(C_4)C(C_2)_2]^{2-}$, respectively. The electron affinities of these $C_n^-$ cluster ions, i.e. the energy required to form the corresponding $C_n^{2-}$ ions, are shown in Fig. 5.16. (For further details on the computations see [5.161].)

Owing to the very energetic nature of the collision cascade which effects the sputtering of atoms and clusters, the latter are generally ejected with a high degree of internal (vibrational/rotational) excitation (see the results at the beginning of Sect. 5.4). This internal energy plays a dominant role in cluster fragmentation processes, both for neutral and for charged clusters; a considerable number of cluster decomposition experiments have been performed and have elucidated pertinent features [5.137]. Decay times of metastable clusters and the kinetic energy release were determined in various experiments. Theoretical modeling of such data attempted to shed light on the fragmentation kinetics and to correlate them with thermodynamic cluster properties. While these calculations put the emphasis on considera-

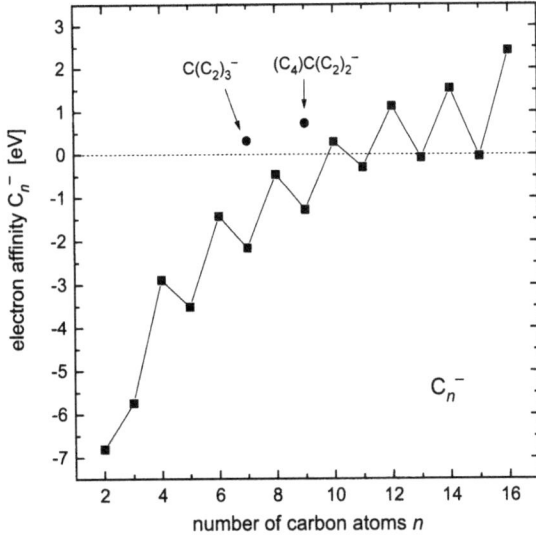

**Fig. 5.16.** Computed electron affinities of carbon clusters $C_n^-$ (i.e. the energy to create a doubly charged cluster) versus cluster size. A special geometrical arrangement for $C_7$ and $C_9$ clusters yields higher electron affinities (*closed circles*). Data from [5.161]

tion of cluster stability derived from binding energies, this approach could be extended to include the structure visible in cluster ionization potentials. The experimental results are supported by MD simulations of keV Ar sputtering of Ag (see Sect. 2.3.4). In these computations, the majority of the emitted trimers and virtually all larger clusters were found to fragment spontaneously within the first nanosecond after emission. This behavior could be clearly traced back to the high internal energies with which these clusters are "born".

Fragmentation of singly negative $C_n^-$ clusters was observed recently by Bekkerman et al. [5.162]. These authors report the occurrence of different fragmentation channels. Apart from the most common one

$$C_n^- \to C_{n-m}^- + C_m \;, \tag{5.25}$$

with $m = 1, 2, 3$, the clusters may decompose into fragments of approximately equal size:

$$C_n^- \to C_5^- + C_{n-5} \;, \qquad n = 8\text{--}12 \;, \tag{5.26a}$$
$$C_n^- \to C_{n-5}^- + C_5 \;, \qquad n = 9\text{--}14 \;. \tag{5.26b}$$

Computed fragmentation energies [5.162] indicate that it requires slightly more energy to remove a single carbon atom from the even-numbered anions than from the odd-numbered ones. In this sense, the former may be considered more stable, which is in accord with their higher electron detachment energies.

The experimental detection of the fragmentation of sputtered cluster ions can be performed in a suitable arrangement of energy- and mass-selective devices (e.g. a combination of electric and magnetic sector fields). Usually, the ions sputtered from the surface are accelerated before entering the instrument. Decay of cluster ions during their passage through this accelerating

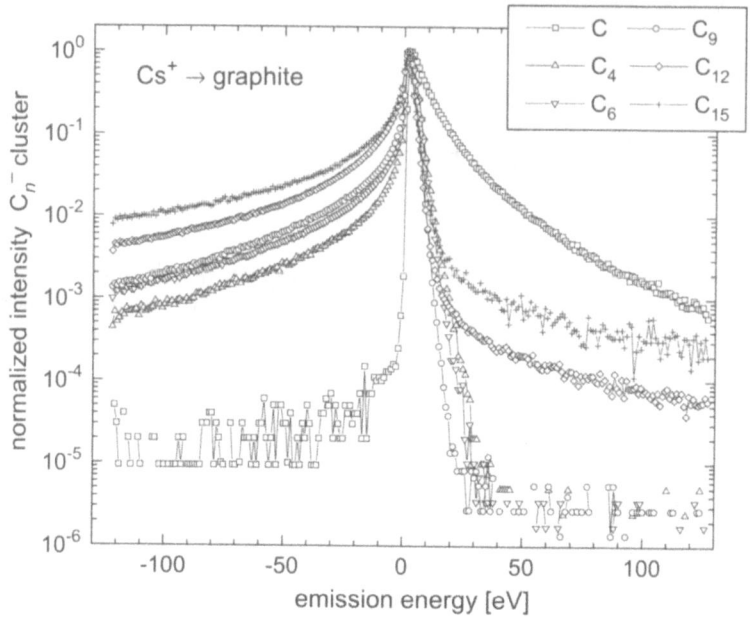

**Fig. 5.17.** Normalized energy distributions of selected $C_n^-$ clusters emitted from graphite under 14.5 keV $Cs^+$ irradiation. Intensities at "negative" energy values are due to daughter cluster ions whose parents have decomposed during the acceleration stage in the mass spectrometer (see text)

field results in daughter ions which carry only a fraction of the full accelerating energy. This energy deficit becomes manifest in the energy distributions because these fragmentation products appear at "negative" energy values as compared to those intact species that experience the full acceleration potential. Figure 5.17 depicts such energy spectra for atomic $C^-$ and several $C_n^-$ clusters. While the $C^-$ ions exhibit the typical distribution of sputtered ions, the "negative-energy" contributions of $C_n^-$ clusters are seen to increase with increasing cluster size. Apparently, for larger clusters the fraction of fragmented species is dominant in the flux of clusters of any given size. Comparing the total yields (i.e. integrated over all emission energies) of $C_n^-$ cluster ions depicted in Fig. 5.15 (labeled "all") with the yield in the "negative" portions of the spectra (labeled "frag" in Fig. 5.15), it is noted that for $n \geq 12$ the latter make up a large fraction ($\sim 50\%$) of the total yield. In other words, the major part of the large clusters has experienced decomposition into smaller daughter fragments, possibly via the processes given in (5.25) or (5.26).

Not surprisingly, fragmentation processes have also been observed for doubly negative carbon clusters. From their measured energy distributions it is obvious that a decay into doubly negative daughters does not occur (i.e. no negative tails are observed). On the other hand, doubly charged ions decomposing during the acceleration may produce singly charged fragments that have a *higher* energy than the corresponding singly charged ions emitted directly from the surface. These decay products thus appear at higher energies and cause a rather gradual decrease of the measured signal in the energy spectra. These features are evident in Fig. 5.17 for $C_{12}^-$ and $C_{15}^-$, but were observed also for other, larger $C_n^-$ clusters.

The aforementioned calculations show that the larger doubly negative clusters are substantially more stable than two smaller singly negative ions. Furthermore, a clear distinction between the odd- and even-numbered dianions is found: $C_8^{2-}$ and $C_{10}^{2-}$ are more stable to decomposition to monoanions than $C_7^{2-}$ and $C_9^{2-}$. Unfortunately, no comparable experimental data are available.

### 5.4.3 Sputtering of Metastable $N_2^-$ and $CO^-$ Anions

Investigations into the formation of atomic and molecular negative ions have been a field of active research for many decades [5.163]. For most atomic ions the electron affinities $A$ are known rather precisely (e.g. with relative uncertainties as low as $5 \times 10^{-7}$ for O and S) and are summarized in various compilations [5.117, 164, 165]. The situation is more complex for molecular ions: for many not even their existence as negative ions has, as yet, been established. Among these, two have been investigated in recent work by Gnaser [5.166]: $N_2^-$ and $CO^-$. Massey [5.163] has previously put forward theoretical arguments against the stability of $N_2^-$ and of the isoelectronic $CO^-$. The reasoning was that on going from the $N_2$ closed-shell electron configuration to $N_2^-$, adding an additional electron in an antibonding $\pi_g 2p$ orbital would reduce the binding energy by about 1.25 eV. Thus, the lowest level of $N_2^-$ should be above that of $N_2$ by this amount and the vibrational levels of the $N_2^-$ ground state would be unstable towards autodetachment. In a recent review [5.161] a value of $A = -2.4$ eV for both $N_2^-$ and $CO^-$ was quoted. The formation of (short-lived) intermediate $N_2^-$ states was inferred from resonances observed in electron-scattering data (cf. [5.163]), but the level widths derived were indicative of a lifetime of the order of a vibration period. Experiments employing a two-step electron capture in alkali-vapor targets [5.167] also appeared to show that $N_2^-$ is not stable, although the data were considered to be not completely conclusive.

The recent work by Gnaser [5.166] established sound evidence for the existence of $N_2^-$ and $CO^-$ anions. In that study sputtering of a solid by an energetic 14.5 keV $Cs^+$ ion beam was utilized and the ions formed were analyzed by high-resolution mass spectrometry. The emitted negative ions were accelerated to an energy of 4.5 keV. The field strength in the acceleration region is 10 kV/cm. After passing through the extraction electrode, the negative-ion beam is focused by electrostatic lenses onto the entrance slit of the mass spectrometer proper. This consists of an electrostatic (spherical condenser) and a magnetic sector field, both with a 90° deflection of the beam. Ions of the correct energy and mass-to-charge ratio pass through the

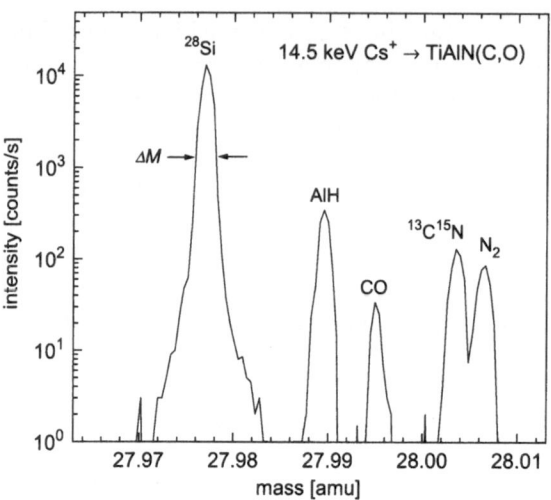

**Fig. 5.18.** High-resolution mass spectrum around 28 amu obtained by 14.5 keV $Cs^+$ sputtering of a TiAlN(C,O) specimen. The arrows around the $^{28}$Si peak mark the width of the 10% intensity level; the resulting value $\Delta M_{0.1} = 0.00217$ amu yields a mass resolution $M/\Delta M_{0.1} = 13\,000$. From [5.166]

mass spectrometer exit slit. All results were obtained at a fairly high mass resolution of $M/\Delta M_{0.1} \sim 13\,000$ ($\Delta M_{0.1}$ corresponds to the width of a mass peak at 10% of its maximum intensity).

Figure 5.18 displays a high-resolution mass spectrum obtained in the 28 amu range. Five distinct peaks are observed in this 0.05 amu-wide mass region. An unambiguous peak assignment is possible by determining from the spectrum the differences between individual mass peaks. These mass differences (relative to the peak with the lowest mass in Fig. 5.18) were evaluated from several separate spectra recorded in this mass range and were compared with the nominal values of the following ion species (in order of ascending mass): $^{28}$Si$^-$, AlH$^-$, CO$^-$, $^{13}$C$^{15}$N$^-$ and N$_2^-$ (see Table 5.4). Both the excellent agreement of the numerical values and the fact that no other series of ion species which might occur in this mass range (e.g. $^{26}$MgH$_2$, $^{12}$C$^{15}$NH, BOH, $^{13}$CNH) fit the pattern observed in Fig. 5.18 (and in the other spectra recorded) indicate the correctness of the peak identification. It was argued in [5.166] that the combination of energy and mass/charge selection carried out in this experiment virtually excludes interferences from fragmenting ion species. These observations provide convincing evidence that metastable N$_2^-$ and CO$^-$ ions are present in the sputtered flux. While the experimental setup is not suited to derive the lifetime of these ions, the flight time from the sample through the spectrometer ($\sim 10\mu$s) can be regarded as a lower limit as to their stability.

The author proposed [5.166] that the metastable state of N$_2^-$ (and possibly, CO$^-$) is derived from an electronically excited neutral parent state relative to which it is bound (i.e. $A > 0$). In fact, available data on the emission of sputtered atoms and molecules show that they very often are emitted in electronically excited states [5.11, 168]. Also, rotational and vibrational excitation have been reported frequently [5.11]. Snowdon and Heiland [5.169]

**Table 5.4.** Negative-ion species detected in the mass range around 28 amu (see Fig. 5.18) and their masses. The nominal mass differences $\Delta$ of the peaks relative to that of $^{28}$Si and the averaged values $\Delta_{exp}$ derived experimentally from several individual spectra are given. The numbers in parentheses give the statistical errors of the last two digits. All values are in amu. From [5.166]

| Ion species | Mass | $\Delta$ | $\Delta_{exp}$ |
|---|---|---|---|
| $^{28}$Si | 27.976929 | – | – |
| AlH | 27.989364 | 0.01244 | 0.01238(13) |
| CO | 27.994915 | 0.01799 | 0.01789(10) |
| $^{13}$C$^{15}$N | 28.003462 | 0.02653 | 0.02658(21) |
| N$_2$ | 28.006148 | 0.02922 | 0.02924(16) |

investigated optical emission spectra of excited species in the sputtering of nitrogen- and carbon–oxygen-implanted silicon. Specifically, they report the identification of either or both of the $N_2$ $B^3 \Pi_g$ and $N_2^+$ $A^2 \Pi_u$ states. The first gives rise via the $N_2$ $B^3 \Pi_g$–$A^3 \Sigma_u^+$ transition (first positive system) to emission features which imply the population of high vibrational states ($v' = 6$–12). They ascribe [5.169] these observations to the Lewis–Rayleigh afterglow proceeding via the recombination of two $N(^4S^0)$ atoms: $N(^4S^0) + N(^4S^0) \leftrightarrow N_2{}^5\Sigma_g^+$, followed by the stabilization $N_2{}^5\Sigma_g^+$ + surface $\rightarrow N_2$ $B^3 \Pi_g$ + surface, where the surface acts as the third body. Such a process would be indicative of an association mechanism in the sputtering event which can produce the observed vibrational- and broad rotational-level populations. Other authors [5.170] observed a strong emission only from the $N_2$ $C^3 \Pi_u$–$B^3 \Pi_g$ bands. The sputtering conditions employed would favor an ejection in a single-collision event, that is, by a recoil from an individual atom at the surface, which is especially applicable to molecules with strong bonding and equal (or similar) masses [5.146, 147]. Both of these conditions fit $N_2^-$ and $CO^-$. The situation was less clear-cut in the case of CO. Here the CO $C^1\Sigma$–$A^1\Pi$ and/or $CO^+$ $B^2\Sigma$–$A^2\Pi$ transitions were tentatively identified [5.169].

In view of these results there is good reason for the assumption that a considerable fraction of $N_2$ and CO is in an excited state upon leaving the solid, which might constitute the precursor for the negative ions. A possible process might be as follows. At the rather low work function of the surface (due to cesium incorporation), these excited molecules may capture an electron, thus forming a temporary negative-ion state (Feshbach resonance). A similar mechanism was proposed [5.171] to explain electron emission observed in slow collisions of $N_2^+$ with cesiated low-work-function surfaces as resulting from autodetachment of the resonance state $N_2^{-*}$ $E^2\Sigma_g^+$, which was created by electron transfer to the projectile previously neutralized into a Rydberg state. As the lifetime of such states is, very probably, not compatible with the flight time of $N_2^-$ ions observed in the present experiment (about 10 μs), metastable quartet states of $N_2^-$ and $CO^-$ might be invoked to explain their rather long lifetimes. In fact, very recent computations by Sommerfeld and Cederbaum [5.172] indicated that a $^4\Pi_u$ state is by far the most likely candidate for the $N_2^-$ ions observed in the experiment.

## 5.5 Ion Emission
## Under Multiply-Charged-Ion Bombardment

The charge state of the primary ion has apparently no influence on sputtering and ion emission from *metals* [5.173, 174]. By contrast, in sputtering of insulators such an influence can definitely be expected [5.174, 175]. A widely quoted explanation invokes the so called "Coulomb explosion" model introduced by Bitensky et al. [5.176], who observed secondary-ion yields from silicon to increase with the primary-ion charge. These authors postulated also a similar behavior for the total sputtering yield of Si, but for impact of 20 keV $Ar^{q+}$ ions ($q \leq 9$) on Si the sputter yield was found [5.174] to be independent of $q$. In this study, the $Si^+$ *ion yield* was reported to increase for $q \geq 6$. For the irradiation of LiF and NaCl with slow multicharged ions ($Ar^{q+}$, with $q \leq 14$) [5.177–179], the *total sputtering yield* strongly increases with the projectile charge (by up to a factor of ten at low kinetic energies); this "potential sputtering" suggested as a major sputtering mechanism the creation of electronic defects such as "self-trapped excitons" and color centers. A similar defect-mediated sputtering process was claimed to be responsible in the case of $SiO_2$ irradiated with multiply charged ions, associated again with a pronounced enhancement of the sputter yield with increasing ion charge [5.180]. On the other hand, Varga and coworkers [5.179] do not find any dependence of the sputtering yield on the charge state ($Ar^{q+}$ with $q \leq 9$) for either MgO or the semiconductors Si and GaAs. On the basis of these data, the authors argued against the occurrence of a Coulomb explosion in the sputtering of insulators with highly charged ions but, rather, favor a "defect-mediated" sputtering model [5.177], see below.

Highly charged ions [5.181–183] carry a rather large amount of potential energy. (One of the ions with the highest charge used in surface interaction studies, $Th^{75+}$, has a potential energy of about 200 keV [5.183].) In the case of slow collisions with the solid surface, this potential energy can greatly exceed the ion's kinetic energy and may dominate, therefore, the interaction process. Transfer of such a large energy onto a comparatively small surface area (typically 1 $nm^2$) within the rather short interaction time ($\leq 100$ fs) corresponds to a huge power flux of $\geq 10^{14}$ $W/cm^2$; this can give rise to various phenomena (e.g. the formation of "hollow atoms") which release, ultimately, the potential energy via emission of electrons and photons (X-rays), [5.184, 185] and possibly lead to surface modifications and sputtering. As stressed recently, these processes may also have considerable practical relevance. For example, high sputtering yields at very low impact energies would allow material removal from surfaces without producing radiation defects deeper in the solid ("soft sputtering").

For a multiply charged ion approaching a surface (see [5.184–186] for further details), inside a critical distance classically allowed resonance neutralization (RN) to highly excited projectile states can take place and will rapidly convert the multicharged ion into a so-called "hollow atom". Auger deexcitation towards inner shells is too slow, generally, for the time the ion spends in front of the surface. RN and "peeling off" of already-captured electrons may partially reionize the hollow atom closer to the surface. Upon penetration into the first layer of the specimen, the ion quickly becomes neutralized by resonance and Auger transitions from the target into the (by now much excited) ion states. This neutralization is held responsible for the production of electron holes and electron–hole pairs in alkali halide surfaces and causes the large total sputtering yields.

While the emission of electrons from the surface in response to the impact of highly charged ions has been investigated intensively [5.184, 185], information on sputtered secondary ions is still rather scarce [5.183, 187]. Neidhart et al. [5.177, 178] studied ion emission from LiF under $Ar^{q+}$ ($q \leq 9$) ion irradiation for kinetic energies $E_k$ between $\sim 25\,\mathrm{eV}$ and $500\,\mathrm{eV}$. Generally, the observed ion yields are at least two orders of magnitude below the yields of neutral species. Very different ion-yield dependences on the charge state were reported for the various secondary ions. The $F^-$ yield is essentially independent of the $Ar^{q+}$ charge and increases by about two orders of magnitude with increasing impact kinetic energy. Conversely, $F^+$ ions are strongly dependent on the charge state and their yield, at a given kinetic energy, is enhanced by a factor of roughly 100 between $q = 2$ and $q = 9$. These data [5.178] are depicted in Fig. 5.19. For $Li^+$, a strong charge-state dependence is found only for kinetic energies below $\sim 200\,\mathrm{eV}$.

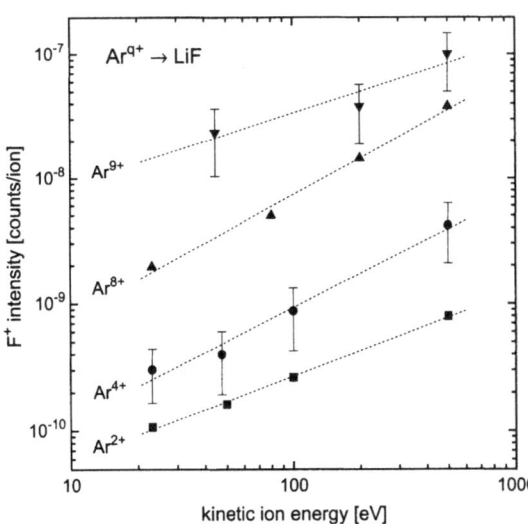

**Fig. 5.19.** The yield of $F^+$ ions sputtered from LiF by $Ar^{q+}$ ions with different charge states ($q = 2$, 4, 8 and 9) as a function of the impact kinetic energy. Data from [5.178]

As pointed out in [5.178], Auger transitions between the projectile and the $F^-$ 2p valence band are primarily responsible for $F^+$ production at low impact energies ($E_k \leq 100 \, eV$). At higher $E$, $F^+$ can also be produced by electron promotion in close encounters of the projectile with target atoms. Emission of $F^+$ ions is probably dominated by Coulomb repulsion from the surface and appears possible upon neutralization of only two or three next-neighbor F atoms. This neutralization is effected by resonant and Auger neutralization processes into highly excited projectile states and becomes more efficient for higher charge states of the primary ion. For impact energies below 1 keV, the probability of production of neutrals at the surface increases roughly with the ion's potential energy. An estimate indicates that for the neutralization of $Ar^{9+}$ about 40 electron holes in the LiF valence band need to be created. Since the ion penetration at these low impact energies is very shallow, all these holes are produced at the surface, creating thereby a strongly electron-depleted region near the ion's impact site. This charge depletion can directly induce $F^+$ emission.

The emission of positive and negative secondary ions from a $SiO_2$ layer on Si and from a thin hydrocarbon film on Au was investigated by Schenkel et al. [5.182, 183, 188–190], using multiply charged ions of very high charge states (up to $Xe^{52+}$, $Au^{69+}$ and $Th^{70+}$). Generally, the total production of positive or negative ions was found to increase drastically with increasing charge state: by about two orders of magnitude between $q \sim 1$ and $q \sim 70$. For selected ion species the dependence could be fitted by power laws (e.g. for $SiO_3^-$, the intensity is proportional to $q^4$). Secondary-ion yields induced by bombardment with multiply charged ions of very high charge states ($q \leq 69$) are shown in Fig. 5.20. Schenkel et al. [5.183] irradiated a thin $SiO_2$ layer (50 nm) on Si with $O^{7+}$, $Xe^{q+}$ ($q = 15$–$52$) and $Au^{q+}$ ($q = 60, 69$) ions and recorded positive and negative secondary-ion spectra with a time-of-flight mass spectrometer. Figure 5.20 depicts the integrated positive and negative ion yields per incoming highly charged ion. As the instrumental transmission ($\sim 10\%$) is not included in these data, the actual number of *emitted* ions would be correspondingly higher. (The most abundant positive ion species ejected from $SiO_2$ under highly-charged-ion impact were $Si^+$, $O^+$ and $SiO^+$, whereas in the spectrum of negative secondary ions $Si_2^-$, $SiO_3^-$ and $O^-$ dominate, with a sizable emission of $(SiO_2)_n^-$ clusters.) For reference purposes these authors [5.183] also used 2.75 keV/amu ($\sim 300$ keV) Xe ions, which, after passing through a 10 nm thick carbon foil, have an equilibrium charge state of $q \cong 1.5$. These ions thus provide an assessment of the collisional contribution to the ion yield (the data corresponding to these bombarding conditions are shown in Fig. 5.20 at $q = 1.5$). For the multiply charged ions shown in Fig. 5.20 the impact *kinetic* energy increases (3.5 keV $\times$ $q$ and 10 keV $\times$ $q$ for positive and negative ions, respectively) with increasing charge state $q$. Hence, from $q = 7$ to $q = 69$ the kinetic energy of the impinging primary ions

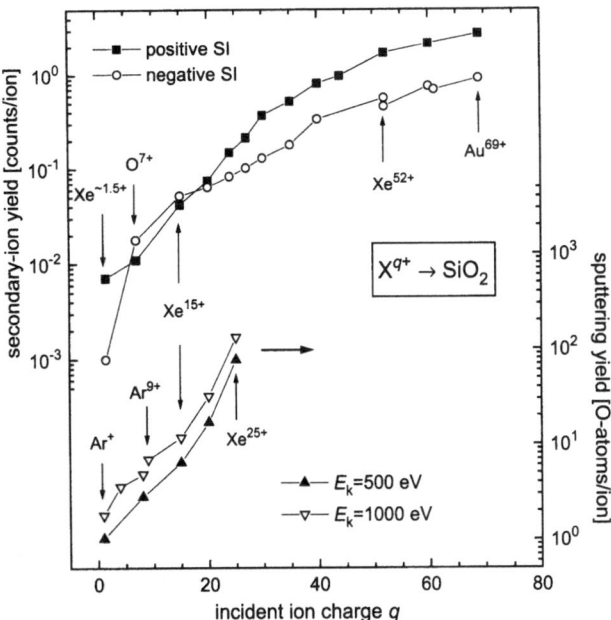

**Fig. 5.20.** Integrated positive and negative secondary ion intensity/ion [5.183] and partial sputtering yield (O atoms/ion) [5.180] as a function of incident ion charge $q$ for bombardment of $SiO_2$ with highly charged ions. The ion yields were measured with a time-of-flight SIMS instrument with the kinetic energy of the impinging ions varying as $3.5\,keV \times q$ and $10\,keV \times q$ for positive and negative secondary-ion detection, respectively, and the sputtering yields were derived from mass-loss measurements with the kinetic energy fixed at 0.5 and 1 keV

changes from 24.5 keV and 70 keV to 241.5 keV and 690 keV for positive and negative secondary-ion detection, respectively.

This energy variation may, of course, result in a corresponding variation of the sputtering yield. Since for heavy ions (such as Xe and Au) at an energy around 100 keV the stopping cross section and the sputtering yield are in a regime (close to their maximum value, cf. Fig. 2.1) where the energy-dependent variations are small, the contributions of collision-cascade sputtering probably contribute little to the ion yield variations seen in Fig. 5.20. On the other hand, Sporn et al. [5.180] observed a drastic increase of the sputtering yield of $SiO_2$ with increasing charge state of the bombarding ion: the yield increases by a factor of $\sim 100$ between $Ar^+$ and $Xe^{25+}$ ions with the *same* kinetic energy (1 keV). These data are included in Fig. 5.20. Apparently, the magnitude of that increase of the sputtering yield is very similar to that of the ion-yield data of Schenkel et al. [5.183]; a direct comparison is, however, difficult because of very different kinetic energies used in these experiments.

Although Schenkel et al. [5.183] argued that signatures in their experimental data were consistent with the Coulomb-explosion model of electronic sputtering as advanced by Bitensky et al. [5.176], they stress also that the secondary-ion enhancement with ion charge might constitute a convolution of an increase in the total material removal rate and a higher ionization probability. It was not possible in that experiment to single out the individual contributions of these effects. Similar arguments would apply to the observation of an enhanced production of cluster ions with increasing charge state. This effect was demonstrated [5.183, 190] for $(SiO_2)_n O^-$ (with $n \leq 14$) and $C_n^-$ ($n \leq 16$) ions and appears plausible in view of the common notion that the relative yields of large clusters scale with the total sputtering yield. The variation of the cluster yields with the cluster size $n$ can be described by a power law for the $(SiO_2)_n O^-$ species ($Y \propto n^{-2.5}$), but by an exponential dependence for $C_n^-$ ions ($Y \propto \exp(-0.47 \times n)$). While the former would be in agreement with the Coulomb-explosion (shock wave) model, which predicts a scaling with $n^{-2}$, the latter relation is indicative of a statistical cluster formation process.

In a related study [5.187] the yield of several positive secondary ions sputtered from $SiO_2$, CsI and some organic layers under irradiation with multiply charged $Ar^{q+}$ ions ($q \leq 11$) were reported. Different charge-state dependences were found: whereas the $H^+$ yield (probably originating from surface impurities) generally increases with $q$ (e.g. $I(H^+) \propto q^3$ for the CsI specimen), the yields of matrix-related secondary ions ($Si^+$ or $Cs^+$) were independent of the charge state.

In view of the rather limited and (sometimes conflicting) data sets available for ion emission from surfaces under multiply-charged-ion impact, definite conclusions as to the ion formation mechanisms operative would appear premature.

# Appendix

This appendix gives a brief outline of the experimental techniques and the methods of computer simulation that are commonly used to study ion irradiation effects at surfaces and in solids, and that are also mentioned repeatedly in this book. This overview is not meant to give a comprehensive description of these approaches; rather, it is intended to summarize some of their basic features and to provide the reader with suitable references containing more detailed information on the pertinent techniques and methods.

## A.1 Experimental Techniques

Many different experimental approaches are used to investigate the effects low-energy ion irradiation causes in the near-surface region of the solid. As most of the phenomena which are described in this volume are limited to the topmost atomic layers, *surface sensitivity* is a distinct requirement. Many textbooks [A.1–10] have surveyed these techniques and described their application in the fields of surface and thin-film science/analysis. The present section is intended to summarize briefly the pertinent features of those methods that have been mentioned frequently in the preceding chapters.

These techniques can be naturally separated into those that characterize the structural and compositional state of the surface after ion bombardment, and those that monitor the flux of atoms (or molecules) emitted from the surface. Obviously, the latter approach is well suited to study sputtering phenomena but can also yield, under certain conditions, information on composition changes at the surface. Probably the most versatile techniques in this second group are *mass-spectrometric* methods. Apart from their inherent species selectivity, they quite often provide means for an energy analysis and (less often) for angular analysis. Since the conversion of the measured (ion) signal into an absolute flux of sputtered atoms or molecules is not straightforward and is frequently very difficult, they typically produce *relative* yields of sputtered species. The principal types of mass spectrometer employed are quadrupole, magnetic sector field and time-of-flight instruments [A.11]. Because of the rather low energies of sputtered species, these mass spectrometers require charged species for a sensitive detection. *Secondary-ion mass spectrometry* (SIMS) utilizes the charged fraction of the sputtered flux [A.11–13].

By combining electric and magnetic sector fields in a double-focusing mass spectrometer arrangement, an intrinsically broad energy passband and very high mass resolution ($M/\Delta M \sim 10^4$) are achievable; the latter may become important in high-precision isotope measurements (Sect. 5.3). To enhance the ionization probability of the sputtered species (Sect. 5.1), the commonly used primary ions are $O_2^+$ and $Cs^+$ for positive and negative secondary-ion detection, respectively. While SIMS is routinely applied for depth profiling of multilayer systems and ion implanted distributions [A.10, 12], its use for studying composition changes is limited by the unknown (and often strongly changing) ionization probabilities of sputtered ions. As was discussed in Chap. 5, these difficulties result from the pronounced dependence of the ionization probability on the (local) sample composition at the ion's emission site; hence, minute variations may influence the yield of detected ions. (Typical SIMS instrumentation and applications are described in comprehensive textbooks [A.11, 12].)

The aforementioned problems are usually avoided when detecting the sputtered atoms (and molecules) in *secondary-neutral mass spectrometry* (SNMS) applying a separate post-ionization step via the interaction of energetic electrons or photons with the emitted species. The first attempts to use an energetic electron beam for this post-ionization were made in the 1950s, but were largely abandoned until much later [A.14]. The use of the electron component of a low-pressure plasma as an ionizing agent for the detection of sputtered neutral species was introduced by Oechsner [A.15, 16] and applied successfully both in basic studies of emission processes in sputtering and in surface and thin-film analysis. A typical SNMS instrument of this kind [A.17] utilizes a low-pressure ($\sim 2 \times 10^{-3}$ mbar of Ar or Xe) rf plasma (electron temperatures corresponding to $\sim 7$–$10\,\mathrm{eV}$ and electron density of some $10^{10}\,\mathrm{cm}^{-3}$) to post-ionize the sputtered particle flux. By biasing the specimen negatively with respect to the plasma, plasma ions ($Ar^+$ or $Xe^+$) can be utilized for sputtering, making very low impact energies ($\sim 30\,\mathrm{eV}$) accessible.

Photoionization of sputtered neutral atoms is a more recent, but promising development in this field of secondary-neutral mass spectrometry [A.18–21]. Owing to the very high laser powers that can be achieved, the ionization process can be close to saturation, at least for those species that are intercepted by the laser in a small spatial volume. Depending on the photon energy, the laser power available and the specific problem addressed, various ionization schemes are employed in this variant of SNMS. Because of the pulsed nature of the ionizing laser, most often time-of-flight (TOF) mass spectrometry is chosen for ion detection.

All of these mass spectrometric techniques remove material from the surface because of sputtering. Monitoring the intensity of ion species as function of the erosion time (depth) provides thus the depth-dependent composition of those species [A.2, 10, 22]. Note that this procedure, commonly called depth profiling, may induce a (further) modification of the state of the surface and the near-surface region due to any of the processes which are outlined in Chap. 3. Therefore, depth profiling might not be an adequate method for investigating the surface changes induced by a prior ion irradiation step. Transient changes in the sputtered flux can be employed to monitor surface composition variations.

Another approach frequently used to detect the sputtered flux is by means of a suitably arranged collector and the subsequent analysis of the deposited material. Although fluence-resolved measurements are tedious (albeit feasible), the technique lends itself naturally to angular-distribution measurements. The properties of the collector and the amount of collected material have to be matched carefully to the characteristics of the probing method, such as sensitivity and accessible depth range.

Composition changes at the surface can be monitored by, for example, various techniques of electron spectroscopy [A.1, 3, 4, 8, 9, 23, 24] and of ion scattering [A.24–28]. In the former category, *Auger electron spectroscopy* (AES) is probably used most often [A.1, 3, 9]. AES is an electron core-level spectroscopy, with the excitation effected by a primary electron beam. The Auger process results in secondary electrons with relatively well-defined energies that relate to differences in core-level energies. In the solid, electronic bands may be involved additionally in the electronic transitions. Measurement of the electron energies by means of an energy analyzer provides the possibility to identify the presence of a particular type of atom at or near the surface. The characteristic energy of the Auger electrons that are excited by an energetic primary electron beam and manage to escape from the surface provides information on the local near-surface composition. Because the inelastic mean free path length $\lambda$ of the electrons is energy-dependent [A.1, 3, 8, 9], different Auger transitions probe different depth ranges in the target and, furthermore, the signal is attenuated exponentially with depth, $I(x) \propto \exp(-x/\lambda)$; $\lambda$ has a broad minimum around 50–100 eV with $\lambda \sim 0.5$ nm, rising to about 1.5–2 nm around 1 keV electron energy [A.1, 3, 6, 8, 9]. Low-energy Auger electrons thus probe the uppermost surface layers, while high-energy transitions are more characteristic of an average composition in some 5 to 10 topmost layers.

*X-ray photoelectron spectroscopy* (XPS) [A.1, 3, 4, 8, 9, 23, 24] is also used to study surface composition changes. XPS is based on the well-known photoelectric effect. The solid is irradiated by monochromatic photons (X-rays in XPS or UV photons in UPS), which excite electrons from occupied states into empty states within the solid from where they can leave into the vacuum and be detected by energy analysis. Apart from providing chemical information, photoemission spectroscopies are suited to map bulk and surface band structures of occupied electronic states [A.1, 3, 8, 9, 23, 24]. XPS has an information depth comparable to high-energy AES, i.e. it may range from about 1 nm to some tens of nanometers. In both techniques, additional depth resolution can be obtained from angle-resolved measurements. Similarly to the mass-spectrometric techniques, both AES and XPS can be combined with sample erosion, using an independent low-energy ion beam; composition depth profiles can be recorded in this way.

Ion-scattering spectroscopy utilizes the energy loss the impinging ion experiences in a (binary) collision with a target atom. Because of the large cross section (and the rapid neutralization upon penetration into the solid), low-energy *ion-scattering spectroscopy* (ISS) [A.27, 28], using ions of a few keV and less, is essentially sensitive to the outermost surface layer only, but virtually no information is obtainable from the underlying region. In several studies the complementary depth ranges of AES and ISS have been utilized to establish concentration gradients within the near-surface range of the specimen. While these techniques can monitor composition changes over the depth scales accessible, information about deeper-lying modifications can only be obtained by depth profiling of the target; as mentioned above, such additional irradiation can alter the original element distributions.

*Rutherford backscattering spectroscopy* (RBS) [A.25] typically employs light ions ($H^+$, $He^+$) at incident energies in the MeV range (heavier projectiles and/or higher energies are in occasional use); it yields, therefore, information over a considerably larger depth range (some 100 nm) in a largely nondestructive and quantitative manner. As ion scattering under these conditions proceeds largely in the Rutherford regime, cross sections and energy losses are rather well established. RBS has been employed to determine composition changes induced by high-energy bombardment. Because of the moderate depth resolution it is not suited to investigate changes in a shallow region near the surface. The depth resolution is much improved in *medium-energy ion scattering* [A.26], to the extent that it can resolve surface reconstructions and the location of adatoms, but the data evaluation appears prohibitively complex in this energy regime. Irrespective of ion energy, the mass resolution is rather moderate for both ion-scattering methods, hence only systems with appreciable mass differences are accessible.

*Low-energy electron diffraction* (LEED) [A.3, 29] employs the elastic scattering of electrons from a surface to derive information about the structural state of this surface. Since the resulting diffraction pattern corresponds essentially to the surface reciprocal lattice, the reverse transformation yields the periodicity (i.e. the surface unit mesh) in real space. The determination of atomic locations within this unit requires, however, a detailed measurement of diffracted intensities and the evaluation is rather complex because of the strong interaction of slow electrons with a solid. Surface sensitivity is a consequence of the low energy ($\sim 20$ to $300\,eV$) of the primary electrons. The elastically backscattered electrons give rise to diffraction (Bragg) spots that can be imaged on a suitable device (e.g. a phosphor screen). The diffraction pattern derives from the well-known Ewald construction but, in contrast to the 3D scattering case, for the scattering from the (ideal) 2D network of surface atoms a LEED pattern is obtained for every electron energy and scattering geometry. Since the primary electron beam deviates from an ideal plane wave (owing to the finite energy width and the angular spread), only atoms within a so-called coherence length (or radius) contribute to the formation of the diffraction pattern. For standard LEED experiments this coherence length is of the order of some 10 nm. Examples of the application of LEED to study the modification of crystalline semiconductor surfaces due to low-energy ion bombardment are given in Chap. 4.

*Electron energy loss spectroscopy* (EELS) refers [A.1, 3, 8, 30, 31], in a general sense, to techniques that utilize low-energy inelastic electron scattering to monitor the excitation of surfaces and of thin solid films. These excitations may cover a wide range of energy from some meV to more than 1000 eV; therefore, different requirements in terms of monochromaticity of the primary beam and of the energy resolution of the analyzer occur. At very low primary electron energies (below a few eV), probably the widest application of this high-resolution EELS is concerned with the study of vibrations of adsorbed atoms or molecules on surfaces. The major loss mechanisms in this regime are excitations of surface phonons, while on semiconductor surfaces typical free-carrier concentrations ($10^{17}$ electrons/cm$^3$ in the conduction band) give rise to bulk and surface plasmons with energies from 20 to 100 meV. At higher primary energies (100–500 eV) EELS reveals loss features due to both bulk and surface scattering; they lie in the energy range from 1 to 100 eV and are due to valence-electron plasma excitations (surface and bulk plasmons) and electronic interband transitions. Core-level transitions may give rise to energy losses of even higher energy (100 to 1000 eV). Owing to the less stringent requirements with regard to energy resolution, this variant of low-resolution EELS is often performed in standard equipment employed, for example, for AES. Chapter 4 presents examples of the use of EELS to monitor ion-bombardment-induced modifications of semiconductor surfaces.

The application of *scanning tunneling microscopy* (STM) [A.32, 33] to study the state of solid surfaces has experienced a very rapid increase in recent years. Many STM data and technical details of STM instruments have been compiled in several textbooks and review articles. In this technique, real-space images of a surface with atomic resolution can be obtained by moving a tiny metal tip across the sample surface and recording the electron tunneling current between tip and specimen as a function of position. Apart from information concerning the surface structure (i.e. the corrugation), STM may also provide details about the electronic structure; this is accomplished by studying the dependence of the STM signal on the sign and magnitude of the tip–sample voltage: for a (sufficiently) positive sample bias (relative to the tip), electrons can tunnel only from occupied metal states into empty surface states (or conduction-band states in semiconductors). Depending on the bias direction, therefore, occupied or empty states of the surface can be probed. Furthermore, by measuring the dependence of the tunnel current on the applied voltage, the state distribution can be mapped. STM investigations of surfaces modified by ion irradiation are presented in this book both for metals (Sects. 2.2.2 and 3.3.1) and for semiconductors (Sects. 4.1.1 and 4.1.2).

Other well-established techniques such as electron microscopy [A.34–36] and X-ray diffraction [A.37] are often employed to characterize the structure and morphology of surfaces and thin films [A.5] and the modifications induced by (low-energy) ion irradiation.

## A.2 Computer Simulations

Although a major fraction of the results presented and discussed in this book was obtained by experiments (in the classical meaning of the word), considerable space is devoted to the outcome of computer simulations. Similarly to other areas of science, such "computer experiments" are now finding a rather widespread use in the field of ion–surface interactions, as discussed in the present book. This is largely due to major advances in computer hardware, increasing both the cost-effectiveness and the complexity of the problems that can be reasonably tackled [A.38]. While a detailed account of this progress is clearly beyond the scope of this work, it might be worthwhile to outline briefly the different computer simulation models to which reference is made at many instances throughout this book. Comprehensive reviews highlighting various viewpoints of ion–solid (or ion–surface) interaction can be found in [A.38–42].

A large number of programs have been developed to simulate the effects of ion irradiation of solids. According to Robinson [A.38], four main categories of models can be identified. The first are codes which integrate the classical equations of motion of a large number of particles simultaneously, commonly called molecular-dynamics (MD) models. In this type of simulation, atomic collisions initiated by some primary event (in the present context, the impact of an ion species) are followed until some terminating condition is met (e.g. the initially supplied energy is dissipated in the solid). A general account of MD models including the basic physics and some practical details is given in [A.42–45]. Concurrently with the advances in computer power, the interaction potentials employed were refined, progressing from a purely two-body interaction to many-body potentials. The various interaction potentials in current use for ion–surface irradiation studies are thoroughly discussed in [A.39, 41, 42].

Two types of code [A.38] employ the binary-collision approximation (BCA) to solve the equations of motion of the projectile and of the target atoms. At low kinetic energies this approximation of purely binary collisions loses its validity. One group of these BCA programs uses a definite target structure and also conserves, as do MD codes, the number of particles (although not the energy and momentum). The most widely used program in this category is MARLOWE [A.46–48].

Stochastic (aleatory) methods are utilized in the other group of BCA simulations to determine the locations of target atoms and to select the impact parameters or scattering angles. These so-called Monte Carlo (MC) codes conserve energy and momentum in individual collisions, but do not conserve particle number. Furthermore, the targets are structureless. The TRIM code and its various derivatives are major representatives of this group [A.41, 49–52].

The fourth category of simulation programs comprises intermediate codes that combine different aspects of MD and BCA models. An example of this approach is the dynamic code DYACAT [A.53], which includes collisions between moving particles and keeps track of the time in the collision cascade. Furthermore, several hybrid codes have been proposed. A very detailed account of the BCA codes has been given by Eckstein [A.41].

Apart from this coarse categorization, the programs used in computer simulations of ion irradiation effects in solids may differ in many other quite important aspects, such as the interaction potential applied, the handling of inelastic energy losses or the modeling of surface binding in sputtering; these features were recently described in great detail by Robinson [A.38], Eckstein [A.41] and others [A.40, 54–58].

# References

**Chapter 1**

1.1 J.W. Rabalais (Ed.): *Low Energy Ion–Surface Interactions* (Wiley, Chichester, 1994).

1.2 M. Kaminsky: *Atomic and Ionic Impact Phenomena on Metal Surfaces* (Springer, Berlin, Heidelberg 1965).

1.3 G. Leibfried: *Bestrahlungseffekte in Festkörpern* (Teubner, Stuttgart, 1965).

1.4 G. Carter, J.S. Colligon: *Ion Bombardment of Solids* (Heinemann, London, 1968).

1.5 Ch. Lehmann: *Interaction of Radiation with Solids and Elementary Defect Production* (North-Holland, Amsterdam, 1977).

1.6 R. Behrisch (Ed.): *Sputtering by Particle Bombardment I* (Springer, Berlin, Heidelberg, 1981).

1.7 R. Behrisch (Ed.): *Sputtering by Particle Bombardment II* (Springer, Berlin, Heidelberg, 1983).

1.8 O. Auciello, R. Kelly (Eds.): *Ion Beam Modifications of Surfaces* (Elsevier, Amsterdam, 1984).

1.9 R.A. Johnson, A.N. Orlov (Eds.): *Physics of Radiation Effects in Crystals* (North-Holland, Amsterdam, 1986).

1.10 A. Gras-Marti, H.M. Urbassek, N.R. Arista, F. Flores (Eds.): *Interaction of Charged Particles with Solids and Surfaces* (Plenum, New York, 1991).

1.11 R. Behrisch, K. Wittmaack (Eds.): *Sputtering by Particle Bombardment III* (Springer, Berlin, Heidelberg, 1991).

1.12 P. Sigmund (Ed.): *Fundamental Processes in Sputtering of Atoms and Molecules*, K. Dan. Vidensk. Selsk. Mat. Fys. Medd. **43** (1993).

1.13 M. Nastasi, J.W. Mayer, J.K. Hirvonen: *Ion Solid Interactions: Fundamentals and Applications* (Cambridge University Press, Cambridge, 1996).

1.14 K.K. Schuegraf (Ed.): *Handbook of Thin Film Deposition Processes and Techniques* (Noyes, Park Ridge, 1988).

1.15 T. Itoh (Ed.): *Ion Beam Assisted Film Growth* (Elsevier, Amsterdam, 1989).

1.16 J.J. Cuomo, S.M. Rossnagel, H.R. Kaufman: *Handbook of Ion Beam Processing Technology* (Noyes, Park Ridge, 1989).

1.17 R.F. Bunshah (Ed.): *Handbook of Deposition Technologies for Films and Coatings* (Noyes, Westwood, 1994).

1.18 H. Oechsner (Ed.): *Thin Film and Depth Profile Analysis* (Springer, Berlin, Heidelberg, 1984).

1.19 D.P. Woodruff, T.A. Delchar: *Modern Techniques of Surface Science* (Cambridge University Press, Cambridge, 1986).

1.20 L.C. Feldman, J.W. Mayer: *Fundamentals of Surface and Thin Film Analysis* (North-Holland, New York, 1986).

1.21 D.J. O'Connor, B.A. Sexton, R.S.C. Smart (Eds.): *Surface Analysis Methods in Materials Science* (Springer, Berlin, Heidelberg, 1992).

1.22    E. Taglauer: Appl. Phys. A **51**, 283 (1990).

1.23    M. Henzler, W. Göpel: *Oberflächenphysik des Festkörpers* (Teubner, Stuttgart, 1991).

1.24    H. Lüth: *Surfaces and Interfaces of Solid Materials* (Springer, Berlin, Heidelberg, 1995).

1.25    P. Ehrhart, K.H. Robrock, H.R. Schober: in *Physics of Radiation Effects in Crystals,* ed. by R.A. Johnson, A.N. Orlov (North-Holland, Amsterdam, 1986), p. 3.

1.26    J.Y. Tsao, E. Chason, K.M. Horn, D.K. Brice, S.T. Picraux: Nucl. Instrum. Methods B **39**, 72 (1989).

1.27    P. Mazzoldi, G.W. Arnold (Eds.): *Ion Beam Modification of Insulators* (Elsevier, Amsterdam, 1987).

1.28    R.E. Johnson: *Energetic Charged Particle Interactions with Atmospheres and Surfaces* (Springer, Berlin, Heidelberg, 1990).

1.29    C.T. Reimann: K. Dan. Vidensk. Selsk. Mat. Fys. Medd. **43**, 351 (1993).

1.30    R.E. Johnson, J. Schou: K. Dan. Vidensk. Selsk. Mat. Fys. Medd. **43**, 403 (1993).

1.31    G. Dearnaley, J.H. Freeman, R.S. Nelson, J. Stephan: *Ion Implantation* (North-Holland, Amsterdam, 1973).

1.32    R. Behrisch: Ergebn. Exakt. Naturw. **35**, 295 (1964).

1.33    A. Benninghoven, F.G. Rüdenauer, H.W. Werner: Secondary Ion Mass Spectrometry (Wiley, New York, 1987).

1.34    M.L. Yu: in *Sputtering by Particle Bombardment III,* ed. by R. Behrisch, K. Wittmaack (Springer, Berlin, Heidelberg, 1991), p. 91.

1.35    P. Sigmund: Phys. Rev. **184**, 383 (1969); Phys. Rev. **187**, 768 (1969).

1.36    F. Besenbacher, J.U. Andersen, A.H. Sørensen (Eds.): Proc. 13th Int. Conference on Atomic Collisions in Solids [Nucl. Instrum. Methods B **48**, 1–642 (1990)].

1.37    J.A. van den Berg, P.C. Zalm, G.A. Stephens (Eds.): Proc. 14th Int. Conference on Atomic Collisions in Solids [Nucl. Instrum. Methods B **67**, 1–664 (1992)].

1.38    M. Zinke-Allmang, W.N. Lennard, G.R. Massoumi (Eds.): Proc. 15th Int. Conference on Atomic Collisions in Solids [Nucl. Instrum. Methods B **90**, 1–610 (1994)].

1.39    D. Semrad, P. Bauer, O. Benka (Eds.): Proc. 16th Int. Conference on Atomic Collisions in Solids [Nucl. Instrum. Methods B **115**, 1–598 (1996)].

1.40    Z.L. Wang, B.R. Shi, K.M. Wang (Eds.): Proc. 17th Int. Conference on Atomic Collisions in Solids [Nucl. Instrum. Methods B **135**, 1–573 (1998)].

1.41    S. Namba, N. Itoh, M. Iwaki (Eds.): Proc. 6th Int. Conference on Ion Beam Modifications of Materials [Nucl. Instrum. Methods B **39**, 1–816 (1989)].

1.42    S.P. Withrow, D.B. Poker (Eds.): Proc. 7th Int. Conference on Ion Beam Modifications of Materials [Nucl. Instrum. Methods B **59/60**, 1–1476 (1991)].

1.43    S. Kalbitzer, O. Meyer, G.K. Wolf (Eds.): Proc. 8th Int. Conference on Ion Beam Modifications of Materials [Nucl. Instrum. Methods B **80/81**, 3–1510 (1993)].

1.44    J.S. Williams, R.G. Elliman, M.C. Ridgeway (Eds.): Proc. 9th Int. Conference on Ion Beam Modifications of Materials [Nucl. Instrum. Methods B **106**, 1–670 (1995)].

1.45    J.C. Barbour, M. Nastasi (Eds.): Proc. 10th Int. Conference on Ion Beam Modifications of Materials [Nucl. Instrum. Methods B **127/128**, 1–1026 (1997)].

## Chapter 2

2.1    N. Bohr: K. Dan. Vidensk. Selsk. Mat. Fys. Medd. **18** (8) (1948).

2.2    J. Lindhard, V. Nielsen, M. Scharff: K. Dan. Vidensk. Selsk. Mat. Fys. Medd. **36** (10) (1968).

2.3    J. Lindhard, M. Scharff, H.E. Schiøtt: K. Dan. Vidensk. Selsk. Mat. Fys. Medd. **33** (14) (1963).

2.4    J. Lindhard, M. Scharff: Phys. Rev. **124**, 128 (1961).

2.5    K.B. Winterbon, P. Sigmund, J.B. Sanders: K. Dan. Vidensk. Selsk. Mat. Fys. Medd. **37** (14) (1970).

2.6    J.W. Corbett: Surf. Sci. **90**, 205 (1979).

2.7    P. Sigmund: Rev. Roum. Phys. **17**, 823, 969, 1079 (1972).

2.8    P. Sigmund: Phys. Rev. **184**, 383 (1969); Phys. Rev. **187**, 768 (1969).

2.9    R. Behrisch (Ed.): *Sputtering by Particle Bombardment I, II* (Springer, Berlin, Heidelberg, 1981, 1983).

2.10   R. Behrisch, K. Wittmaack (Eds.): *Sputtering by Particle Bombardment III* (Springer, Berlin, Heidelberg, 1991).

2.11   P. Sigmund: in *Sputtering by Particle Bombardment I*, ed. by R. Behrisch (Springer, Berlin, Heidelberg, 1981), p. 9.

2.12   J.F. Ziegler, J.P. Biersack, U. Littmark: *The Stopping and Range of Ions in Matter*, Vol. 1 (Pergamon, New York, 1985).

2.13   W. Eckstein: *Computer Simulation of Ion Solid Interactions* (Springer, Berlin, Heidelberg, 1991).

2.14   H.H. Andersen, P. Sigmund: Nucl. Instrum. Methods **38**, 238 (1965); Risø Rep. no. 103 (1965).

2.15   W.D. Wilson, L.G. Haggmark, J.P. Biersack: Phys. Rev. B **15**, 2458 (1977).

2.16   M. Vicanek, J.J. Jimenéz-Rodríguez, P. Sigmund: Nucl. Instrum. Methods B **36**, 124 (1989).

2.17   H.A. Bethe: Ann. Phys. (Leipzig) **5**, 325 (1930).

2.18   O.S.Oen, M.T. Robinson: Nucl. Instrum. Methods **132**, 647 (1976).

2.19   H.E. Schiøtt: K. Dan. Vidensk. Selsk. Mat. Fys. Medd. **35** (9) (1966).

2.20   H.E. Schiøtt: Radiat. Eff. **6**, 107 (1970).

2.21   G. Dearnaley, J.H. Freeman, R.S. Nelson, J. Stephan: *Ion Implantation* (North-Holland, Amsterdam, 1973).

2.22   D.K. Brice: *Ion Implantation Range and Energy Deposition Distributions*, Vol. 1 (IFI/Plenum, New York, 1975).

2.23   K.B. Winterbon: *Ion Implantation Range and Energy Deposition Distributions*, Vol. 2 (IFI/Plenum, New York, 1975).

2.24   S. Kalbitzer, H. Oetzmann: Radiat. Eff. **47**, 57 (1980).

2.25   U. Littmark, J.F. Ziegler: *Handbook of Range Distributions for Energetic Ions in all Elements* (Pergamon, New York, 1980).

2.26   J.F. Ziegler: *Handbook of Stopping Cross Sections for Eneregtic Ions in all Elements* (Pergamon, New York, 1980).

2.27   H. Gnaser, H.L. Bay, W.O. Hofer: Nucl. Instrum. Methods B **15**, 49 (1986).

2.28   H. Gnaser, H.L. Bay, W.O. Hofer: J. Vac. Sci. Technol. A **5**, 1194 (1987).

2.29   M.T. Robinson, O.S. Oen: Phys. Rev. **132**, 2385 (1963); Appl. Phys. Lett. **2**, 30 (1963).

2.30   J. Lindhard: K. Dan. Vidensk. Selsk. Mat. Fys. Medd. **34** (14) (1965); Phys. Lett. **12**, 126 (1964).

2.31   H. Ryssel, I. Ruge: *Ionenimplantation* (Teubner, Stuttgart, 1978).

2.32   J.F. Ziegler (Ed.): *Handbook of Ion Implantation* (North-Holland, Amsterdam, 1992).

2.33   P. Sigmund: Can. J. Phys. **46**, 731 (1968).

2.34 W. Eckstein, J.P. Biersack: Z. Phys. A **310**, 1 (1986); Z. Phys. B **63**, 471 (1986).

2.35 J. Bøttiger, J.A. Davies, P. Sigmund, K.B. Winterbon: Radiat. Eff. **11**, 69 (1971).

2.36 H.H. Andersen: Radiat. Eff. **3**, 51 (1970); Radiat. Eff. **7**, 179 (1971).

2.37 D. Hildebrandt, R. Manns: Phys. Stat. Sol. (a) **38**, K155 (1976); Radiat. Eff. **31**, 153 (1977).

2.38 H.H. Andersen, T. Lenskjær, G. Sidenius, H. Sørensen: J. Appl. Phys. **47**, 13 (1976).

2.39 J. Schou, H. Sørensen, U. Littmark: J. Nucl. Mat. **76**, 359 (1978).

2.40 W.R. Gesang, H. Oechsner, H. Schoof: Nucl. Instrum. Methods **132**, 687 (1976).

2.41 H.F. Winters, H. Coufal, C.T. Rettner, D.S. Bethune: Phys. Rev. **41**, 6240 (1990).

2.42 H. Coufal, H.F. Winters, H.L. Bay, W. Eckstein: Phys. Rev. B **44**, 4747 (1991).

2.43 H.L. Bay, H.F. Winters, H.J. Coufal, W. Eckstein: Appl. Phys. A **55**, 274 (1992).

2.44 H. Gades, H.M. Urbassek: Appl. Phys. A **61**, 39 (1995).

2.45 M.W. Thompson: *Defects and Radiation Damage in Metals* (Cambridge University Press, Cambridge, 1969).

2.46 P. Jung: in *Atomic Defects in Metals*, Landolt-Börnstein III/25, ed. by H. Ullmaier (Springer, Berlin, Heidelberg, 1991), p. 1.

2.47 R.S. Averback, T. Diaz de la Rubia, R.S. Benedek: Nucl. Instrum. Methods B **33**, 693 (1988).

2.48 K.L. Merkle, L.R. Singer, R.K. Hart: J. Appl. Phys. **34**, 2800 (1963).

2.49 J.R. Beeler: J. Appl. Phys. **35**, 2226 (1964); J. Appl. Phys. **37**, 3000 (1966); Phys. Rev. **150**, 470 (1966).

2.50 M.T. Robinson: Philos. Mag. **12**, 741 (1965); Philos. Mag. **17**, 639 (1968).

2.51 G.H. Kinchin, R.S. Pease: Rep. Prog. Phys. **18**, 1 (1955).

2.52 P. Sigmund: Appl. Phys. Lett. **14**, 114 (1969).

2.53 R.S. Averback, R. Benedek, K.L. Merkle: Phys. Rev. B **18**, 4156 (1978).

2.54 M.W. Guinan, J.H. Kinney: J. Nucl. Mat. **103&104**, 1319 (1981).

2.55 K.L. Merkle, W.E. King, A.C. Baily, K. Haga, M. Meshi: J. Nucl. Mat. **117**, 4 (1983).

2.56 W.E. King, K.L. Merkle, M. Meshi: J. Nucl. Mat. **117**, 12 (1983).

2.57 W.E. King, R. Benedek: J. Nucl. Mat. **117**, 26 (1983).

2.58 J.A. Brinkman: J. Appl. Phys. **25**, 961 (1954).

2.59 J.B. Gibson, A.N. Goland, M. Milgram, G.H. Vineyard: Phys. Rev. **120**, 1229 (1960).

2.60 C. Erginsoy, G.H. Vineyard, A. Englert: Phys. Rev. **133**, A595 (1964).

2.61 C. Erginsoy, G.H. Vineyard, A. Shimizu: Phys. Rev. **139**, A118 (1965).

2.62 F. Seitz, J.S. Koehler: in *Solid State Physics*, Vol. 2, ed. by F. Seitz, D. Turnbull (Academic, New York, 1956), p. 305.

2.63 N.Q. Lam: Scanning Microscopy, Suppl. **4**, 311 (1990).

2.64 H.L. Heinisch: J. Nucl. Mat. **103&104**, 1325 (1981); J. Nucl. Mat. **117**, 46 (1983).

2.65 T. Diaz de la Rubia, R.S. Averback, R. Benedek, W.E. King: Phys. Rev. Lett. **59**, 1930 (1987).

2.66 T. Diaz de la Rubia, R.S. Averback, H. Hsieh, R.S. Benedek: J. Mater. Res. **4**, 579 (1989).

2.67 H. Hsieh, T. Diaz de la Rubia, R.S. Averback, R.S. Benedek: Phys. Rev. B **40**, 9986 (1989).

2.68   R.P. Webb, D.E. Harrison: Phys. Rev. Lett. **50**, 1478 (1983).
2.69   D.E. Harrison, R.P. Webb: Nucl. Instrum. Methods **218**, 727 (1983).
2.70   D.E. Harrison: Radiat. Eff. **70**, 1 (1983); Crit. Rev. Solid State Mater. Sci. **14**, S1 (1988).
2.71   F. Karetta, H.M. Urbassek: J. Appl. Phys. **71**, 5410 (1992).
2.72   H. Gades, H.M. Urbassek: Phys. Rev. B **50**, 11 167 (1994).
2.73   T. Michely, G. Comsa: Phys. Rev. B **44**, 8411 (1991).
2.74   T. Michely, C. Teichert: Phys. Rev. B **50**, 11 156 (1994).
2.75   T. Michely, T. Land, U. Littmark, G. Comsa: Surf. Sci. **272**, 204 (1992).
2.76   N.J. Zheng, I.H. Wilson, U. Knipping, D.M. Burt, D.H. Krinsley, I.S.T. Tsong: Phys. Rev. B **38**, 12 780 (1988).
2.77   I.H. Wilson, N.J. Zheng, U. Knipping, I.S.T. Tsong: Phys. Rev. B **38**, 8444 (1988); Appl. Phys. Lett. **53**, 2039 (1988).
2.78   C.A. Lang, C.F. Quate, J. Nogami: Appl. Phys. Lett. **59**, 1696 (1991).
2.79   T. Michely, K.H. Besocke, G. Comsa: Surf. Sci. **230**, L135 (1990).
2.80   T. Michely, G. Comsa: Surf. Sci. **256**, 217 (1991); J. Vac. Sci. Technol. B **9**, 862 (1991).
2.81   J.C. Girard, Y. Samson, S. Gauthier, S. Rousset, J. Klein: Surf. Sci. **302**, 73 (1994).
2.82   J.R. Hahn, H. Kang: Surf. Sci. **357/358**, 165 (1996).
2.83   J. Naumann, J. Osing, A.J. Quinn, I.V. Shvets: Surf. Sci. **388**, 212 (1997).
2.84   C. Teichert, M. Hohage, T. Michely, G. Comsa: Phys. Rev. Lett. **72**, 1682 (1994).
2.85   M. Ghaly, R.S. Averback: Phys. Rev. Lett. **72**, 364 (1994).
2.86   M. Ghaly, R.S. Averback, T. Diaz de la Rubia: Nucl. Instrum. Methods B **102**, 51 (1995).
2.87   T.J. Colla, H.M. Urbassek: Radiat. Eff. Def. Solids **142**, 439 (1997); private communication (1998).
2.88   H.M. Urbassek: Nucl. Instrum. Methods B **122**, 427 (1997).
2.89   T. Michely, G. Comsa: Nucl. Instrum. Methods B **82**, 207 (1993).
2.90   B. Poelsema, L.K. Verheij, G. Comsa: Phys. Rev. Lett. **53**, 2500 (1984).
2.91   B. Poelsema, R. Kunkel, L.K. Verheij, G. Comsa: Phys. Rev. B **41**, 11609 (1990).
2.92   M. Müller: Doctoral Thesis, Universität Kaiserslautern (1997).
2.93   W. Jäger, K.L. Merkle: 9th Int. Cong. Electr. Micros., Vol. I, (Toronto, 1978), p. 378.
2.94   W. Jäger: J. Microsc. Spectrosc. Electron. **6**, 437 (1981).
2.95   K.L. Merkle, W. Jäger: Philos. Mag. A **44**, 741 (1981).
2.96   W. Jäger, K.L. Merkle: Philos. Mag. A **57**, 479 (1988).
2.97   R.S. Averback, M. Ghaly: Nucl. Instrum. Methods B **90**, 191 (1994).
2.98   R.S. Averback, M. Ghaly: Nucl. Instrum. Methods B **127/128**, 1 (1997).
2.99   M. Morgenstern, T. Michely, G. Comsa: Philos. Mag. A (1998), in press.
2.100  P. Sigmund (Ed.): *Fundamental Processes in Sputtering of Atoms and Molecules*: K. Dan. Vidensk. Selsk. Mat. Fys. Medd. **43** (1993).
2.101  P. Sigmund: in *Inelastic Ion–Surface Collisions*, ed. by N.H. Tolk, J.C. Tully, W. Heiland (Academic, New York, 1977), p. 121.
2.102  P. Sigmund: Nucl. Instrum. Methods B **27**, 1 (1987).
2.103  M.T. Robinson: Philos. Mag. **12**, 145 (1965).
2.104  W. Brandt, R. Laubert: Nucl. Instrum. Methods **47**, 201 (1967).
2.105  M.W. Thompson: Philos. Mag. **18**, 377 (1968); Phys. Rep. **69**, 335 (1981).
2.106  M.W. Thompson: Nucl. Instrum. Methods B **18**, 411 (1987).
2.107  R. Kelly: Radiat. Eff. **80**, 273 (1984).

2.108  K.B. Winterbon, H.M. Urbassek, P. Sigmund, A. Gras-Marti: Phys. Scr. **36**, 689 (1987).
2.109  Y.T. Chang, M.A. Nicolet, W.L. Johnson: Phys. Rev. Lett. **58**, 2083 (1987).
2.110  F. Rossi, D.M. Parkin, M. Nastasi: J. Mater. Res. **4**, 137 (1989).
2.111  H.H. Andersen: Nucl. Instrum. Methods B **18**, 321 (1987).
2.112  J.P. Biersack: Nucl. Instrum. Methods B **27**, 21 (1987).
2.113  M.T. Robinson: K. Dan. Vidensk. Selsk. Mat. Fys. Medd. **43**, 27 (1993).
2.114  H.H. Andersen, H.L. Bay: in *Sputtering by Particle Bombardment I*, ed. by R. Behrisch (Springer, Berlin, Heidelberg, 1981), p. 145.
2.115  H.H. Andersen: Nucl. Instrum. Methods B **33**, 466 (1988).
2.116  W.O. Hofer: in *Sputtering by Particle Bombardment III*, ed. by R. Behrisch, K. Wittmaack (Springer, Berlin, Heidelberg, 1991), p. 15.
2.117  G. Betz, K. Wien: Int. J. Mass Spectrom. Ion Proc. **140**, 1 (1994).
2.118  R.V. Stuart, G.K. Wehner: J. Appl. Phys. **35**, 1819 (1964).
2.119  H. Oechsner, L. Reichert: Phys. Lett. **23**, 90 (1966).
2.120  H. Oechsner: Phys. Rev. Lett. **24**, 583 (1970); Z. Phys. **238**, 433 (1970).
2.121  J. Dembowski, H. Oechsner, Y. Yamamura, M. Urbassek: Nucl. Instrum. Methods B **18**, 464 (1987).
2.122  R.A. Brizzolara, C.B. Cooper, T.K. Olson: Nucl. Instrum. Methods B **35**, 36 (1988).
2.123  R.A. Brizzolara, C.B. Cooper: Nucl. Instrum. Methods B **43**, 136 (1989).
2.124  M. Urbassek: Nucl. Instrum. Methods B **4**, 356 (1984).
2.125  W. Eckstein: Nucl. Instrum. Methods B **18**, 344 (1987).
2.126  D.P. Jackson: Radiat. Eff. **18**, 185 (1973); Can. J. Phys. **53**, 1513 (1975).
2.127  B.J. Garrison, N. Winograd, D. Lo, T.A. Tombrello, M.H. Shapiro, D.E. Harrison: Surf. Sci. **180**, L129 (1987).
2.128  R. Kelly: Nucl. Instrum. Methods B **18**, 388 (1987).
2.129  H. Gades, H.M. Urbassek: Nucl. Instrum. Methods B **69**, 232 (1992).
2.130  J.P. Biersack, W. Eckstein: Appl. Phys. A **34**, 73 (1984).
2.131  K.A. Gschneidner: in *Solid State Physics*, Vol. 16 ed. by F. Seitz, D. Turnbull (Academic, New York, 1964), p. 275.
2.132  G.K. Wehner, D. Rosenberg: J. Appl. Phys. **31**, 177 (1960).
2.133  K. Rödelsberger, A. Scharmann: Nucl. Instrum. Methods **132**, 355 (1976).
2.134  T. Okutani, M. Shikata, S. Ichimura, R. Shimizu: J. Appl. Phys. **51**, 2884 (1980).
2.135  H. Tsuge, S. Esho: J. Appl. Phys. **52**, 4391 (1981).
2.136  H.H. Andersen, J. Chevallier, V. Chernish: Nucl. Instrum. Methods **191**, 241 (1981).
2.137  H.H. Andersen, B. Stenum, T. Sørensen, H.J. Whitlow: Nucl. Instrum. Methods B **6**, 459 (1985).
2.138  R.G. Allas, A.R. Knudson, J.M. Lambert, P.A. Treado, G.W. Reynolds: Nucl. Instrum. Methods **194**, 615 (1982).
2.139  F.R. Vozzo, G.W. Reynolds: Nucl. Instrum. Methods **209/210**, 555 (1983).
2.140  T.K. Cini, M. Tanemura, F. Okuyama: Nucl. Instrum. Methods B **119**, 387 (1996).
2.141  M.T. Robinson: J. Appl. Phys. **40**, 4982 (1969).
2.142  C. Schwebel, C. Pellet, G. Gautherin: Nucl. Instrum. Methods B **18**, 525 (1987).
2.143  T.J. Whitaker, P.L. Jones, A. Li, R.O. Watts: Rev. Sci. Instrum. **64**, 452 (1993).
2.144  M. Szymonski, W. Huang, J. Onsgaard: Nucl. Instrum. Methods B **14**, 263 (1986).
2.145  W. Huang: Surf. Sci. **202**, L603 (1988); Phys. Lett. A **134**, 269 (1989).

2.146  Y. Yamamura, C. Mößner, H. Oechsner: Radiat. Eff. **103**, 25 (1987); Radiat. Eff. **105**, 31 (1987).
2.147  G. Betz, R. Dobrozemsky, F.P. Viehböck: Ned. Tijdschr. Vacuumtech. **8**, 203 (1970).
2.148  H. Oechsner: Z. Phys. **261**, 37 (1973); Appl. Phys. **8**, 185 (1975).
2.149  H.L. Bay, J. Bohdansky: Appl. Phys. **19**, 421 (1979).
2.150  J.E. Westmoreland, P. Sigmund: Radiat. Eff. **6**, 187 (1970).
2.151  W. Eckstein: Nucl. Instrum. Methods B **33**, 489 (1988).
2.152  U. Conrad, H.M. Urbassek: Nucl. Instrum. Methods B **48**, 399 (1990).
2.153  M. Hou, W. Eckstein: J. Appl. Phys. **71**, 3975 (1992).
2.154  M.T. Robinson: Nucl. Instrum. Methods B **90**, 509 (1994).
2.155  G. Betz, R. Kirchner, W. Husinsky, F. Rüdenauer, H.M. Urbassek: Radiat. Eff. Def. Solids **130/131**, 251 (1994).
2.156  G. Falcone, P. Sigmund: Appl. Phys. **25**, 307 (1981).
2.157  P. Sigmund, M.T. Robinson, M.I. Baskes, M. Hautala, F.Z. Cui, W. Eckstein, Y. Yamamura, S. Hosaka, T. Ishitani, V.I. Shulga, D.E. Harrison, I.R. Chakarov, D.S. Karpuzov, E. Kawatoh, R. Shimizu, S. Valkealahti, R.M. Nieminen, G. Betz, W. Husinsky, M.H. Shapiro, M. Vicanek, H.M. Urbassek: Nucl. Instrum. Methods B **36**, 110 (1989).
2.158  A.R. Krauss, D.M. Gruen, A.B. DeWald: J. Nucl. Mater. **121**, 398 (1984).
2.159  M.F. Dumke, T.A. Tombrello, R.A. Weller, R.M. Housley, E.H. Cirlin: Surf. Sci. **124**, 407 (1983).
2.160  K.M. Hubbard, R.A. Weller, D.L. Weathers, T.A. Tombrello: Nucl. Instrum. Methods B **36**, 395 (1989); Nucl. Instrum. Methods B **40/41**, 278 (1989).
2.161  J.W. Burnett, J.P. Biersack, D.M. Gruen, B. Jørgensen, A.R. Krauss, M.J. Pellin, E.L. Schweitzer, J.T. Yates, C.E. Young: J. Vac. Sci. Technol. A **6**, 2064 (1988).
2.162  M.J. Pellin, J.W. Burnett: Pure Appl. Chem. **65**, 2361 (1993).
2.163  S. Tang, N.Q. Lam: Surf. Sci. **223**, 179 (1989).
2.164  N.Q. Lam, H.A. Hoff, H. Wiedersich, L.E. Rehn: Surf. Sci. **149**, 517 (1985).
2.165  N.Q. Lam, H.A. Hoff: Surf. Sci. **193**, 353 (1988).
2.166  K. Wittmaack: Phys. Rev. B **56**, R5701 (1997).
2.167  N.J. Sack, M. Akbulut, T.E. Madey: Phys. Rev. B **51**, 4585 (1995); Phys. Rev. Lett. **73**, 794 (1994); Nucl. Instrum. Methods B **90**, 451 (1994).
2.168  P. Klein, H.M. Urbassek, M. Vicanek: Phys. Rev. B **51**, 4597 (1995).
2.169  G. Carter, B. Navinsek, J.L. Whitton: in *Sputtering by Particle Bombardment II*, ed. by R. Behrisch, (Springer, Berlin, Heidelberg, 1983), p. 231.
2.170  B.M.U. Scherzer: in *Sputtering by Particle Bombardment II*, ed. by R. Behrisch, (Springer, Berlin, Heidelberg, 1983), p. 271.
2.171  J.W. Burnett, M.J. Pellin, W.F. Calaway, D.M. Gruen, J.T. Yates: Phys. Rev. Lett. **63**, 562 (1989).
2.172  G. Betz, M.J. Pellin, J.W. Burnett, D.M. Gruen: Nucl. Instrum. Methods B **58**, 429 (1991).
2.173  T.J. Colla, H.M. Urbassek: Nucl. Instrum. Methods B **117**, 361 (1996).
2.174  R. Smith, D.E. Harrison, B.J. Garrison: Phys. Rev. B **40**, 93 (1989).
2.175  R. Weissmann, P. Sigmund: Radiat. Eff. **19**, 7 (1973).
2.176  R.V. Stuart, G.K. Wehner: Phys. Rev. Lett. **4**, 409 (1960); Phys. Rev. **33**, 2345 (1962).
2.177  D.E. Harrison, G.D. Magnuson: Phys. Rev. **122**, 1421 (1961).
2.178  W.O. Hofer, J. Roth (Eds.): *Physical Processes of the Interaction of Fusion Plasmas with Solids* (Academic, New York, 1996).
2.179  W. Eckstein: Surf. Interf. Anal. **14**, 799 (1989).
2.180  M. Hou, M.T. Robinson: Appl. Phys. **18**, 381 (1979).

2.181  R. Behrisch, G. Maderlechner, B.M.U. Scherzer, M.T. Robinson: Appl. Phys. **18**, 391 (1979).
2.182  Y. Yamamura, J. Bohdansky: Vacuum **35**, 561 (1985).
2.183  Y. Yamamura, T. Takiguchi, H. Kimura: Nucl. Instrum. Methods B **78**, 337 (1993).
2.184  H.F. Winters, P. Sigmund: J. Appl. Phys. **45**, 4760 (1974).
2.185  W. Eckstein, C. Garcia-Rosales, J. Roth, J. Laszlo: Nucl. Instrum. Methods B **83**, 95 (1993).
2.186  W. Eckstein, C. Garcia-Rosales, J. Roth, W. Ottenberger: Report IPP 9/82 (MPI für Plasmaphysik, Garching, 1993).
2.187  C. Garcia-Rosales, W. Eckstein, J. Roth: J. Nucl. Mat. **218**, 8 (1994).
2.188  J. Bohdansky, J. Roth, H.L. Bay: J. Appl. Phys. **51**, 2861 (1980); J. Appl. Phys. **52**, 1610 (1981).
2.189  J. Bohdansky: Nucl. Instrum. Methods B **2**, 587 (1984).
2.190  N. Matsunami, Y. Yamamura, Y. Itikawa, N. Itoh, Y. Kazumata, S. Miyagawa, K. Morita, R. Shimizu: Radiat. Eff. Lett. **57**, 15 (1980).
2.191  Y. Yamamura: Nucl. Instrum. Methods **194**, 515 (1982).
2.192  Y. Yamamura, N. Matsunami, N. Itoh: Radiat. Eff. **68**, 83 (1982); Radiat. Eff. **71**, 65 (1983).
2.193  N. Matsunami, Y. Yamamura, Y. Itikawa, N. Itoh, Y. Kazumata, S. Miyagawa, K. Morita, R. Shimizu, H. Tawara: At. Data Nucl. Data Tab. **31**, 1 (1984).
2.194  P.C. Zalm: Radiat. Eff. Lett. **86**, 29 (1983); J. Vac. Sci. Technol. B **2**, 151 (1984).
2.195  U. Littmark, S. Fedder: Nucl. Instrum. Methods **194**, 607 (1982).
2.196  H. Gnaser, H. Oechsner: Surf. Sci. **251/252**, 696 (1991); Nucl. Instrum. Methods B **58**, 438 (1991).
2.197  N. Laegreid, G.K. Wehner: J. Appl. Phys. **32**, 365 (1961).
2.198  C.H. Weijsenfeld, A. Hoogendoorn, M. Koedam: Physica **27**, 763 (1961).
2.199  D. Rosenberg, G.K. Wehner: J. Appl. Phys. **33**, 1842 (1962).
2.200  R.V. Stuart, G.K. Wehner: J. Appl. Phys. **33**, 2345 (1962).
2.201  H. Oechsner: Doctoral Thesis, Universität Würzburg (1963).
2.202  R.P. Webb, D.E. Harrison: J. Appl. Phys. **53**, 5243 (1982).
2.203  M.M. Jakas, D.E. Harrison: Nucl. Instrum. Methods B **14**, 535 (1986).
2.204  P.K. Rol, J.M. Fluit, F.P. Viehböck, M. de Jong: in Proc. 4th Int. Conf. Ionization Phenomena in Gases, ed. by N.R. Nilsson (North-Holland, Amsterdam, 1960), p. 275.
2.205  V.A. Molchanov, V.G. Tel'kovskii, V.M. Chickerov: Sov. Phys.-Doklady **6**, 222 (1961)
2.206  O. Almén, G. Bruce: Nucl. Instrum. Methods **11**, 257 (1961).
2.207  M.T. Robinson: in *Sputtering by Particle Bombardment I*, ed. by R. Behrisch (Springer, Berlin, Heidelberg, 1981), p. 73.
2.208  H.E. Roosendaal: in *Sputtering by Particle Bombardment I*, ed. by R. Behrisch (Springer, Berlin, Heidelberg, 1981), p. 219.
2.209  A.L. Southern, W.R. Willis, M.T. Robinson: J. Appl. Phys. **34**, 153 (1963).
2.210  J.M. Fluit, P.K. Rol, J. Kistemaker: J. Appl. Phys. **34**, 690 (1963).
2.211  G.D. Magnuson, C.E. Carlston: J. Appl. Phys. **34**, 3267 (1963).
2.212  G.D. Odintsov: Sov. Phys. Solid State **5**, 813 (1963).
2.213  D. Onderdelinden: Appl. Phys. Lett. **8**, 189 (1966); Can. J. Phys. **46**, 739 (1968).
2.214  G.K. Wehner: J. Appl. Phys. **26**, 1056 (1955); Phys. Rev. **102**, 690 (1956).
2.215  G.S. Anderson, G.K. Wehner: J. Appl. Phys. **31**, 2305 (1960).
2.216  V.E. Yurasova: Sov. Phys. Tech. Phys. **3**, 1806 (1958).

2.217  M.W. Thompson: Philos. Mag. **4**, 139 (1959).
2.218  V.E. Yurasova, N.V. Pleshivtsev, I.V. Orfanov: Sov. Phys. JETP **37**, 689 (1960).
2.219  V.A. Molchanov, V.G. Tel'kovskii, V.M. Chickerov: Sov. Phys.–Doklady **6**, 486 (1961).
2.220  M.T. Robinson, A.L. Southern: J. Appl. Phys. **38**, 2969 (1967).
2.221  R.H. Silsbee: J. Appl. Phys. **28**, 1246 (1957).
2.222  G. Leibfried: J. Appl. Phys. **30**, 1388 (1959).
2.223  C. Lehmann, P. Sigmund: Phys. Stat. Sol. **16**, 507 (1966).
2.224  T. Lenskjaer, F. Nyholm, S.D. Pedersen, N.B. Petersen: Phys. Lett. **47A**, 63 (1974).
2.225  W. Szymczak, K. Wittmaack: Nucl. Instrum. Methods **194**, 561 (1982).
2.226  W. Szymczak, K. Wittmaack: Nucl. Instrum. Methods B **82**, 220 (1993).
2.227  M. Hou, W. Eckstein: Nucl. Instrum. Methods B **13**, 324 (1986).
2.228  W. Eckstein, M. Hou: Nucl. Instrum. Methods B **31**, 386 (1988).
2.229  M. Hou, W. Eckstein: Phys. Rev. B **42**, 5959 (1990).
2.230  W. Eckstein, M. Hou: Nucl. Instrum. Methods B **53**, 270 (1991).
2.231  Y. Yamamura, W. Takeuchi: Nucl. Instrum. Methods B **29**, 461 (1987).
2.232  M Hautala, J. Likonen: Nucl. Instrum. Methods B **33**, 526 (1988).
2.233  J. Likonen, M. Hautala: Appl. Phys. A **45**, 137 (1988).
2.234  W.O. Hofer, H. Gnaser: Nucl. Instrum. Methods B **18**, 605 (1987).
2.235  H. Gnaser, W.O. Hofer: Appl. Phys. A **48**, 261 (1989).
2.236  J.P. Baxter, G.A. Schick, J. Singh, P.H. Kobrin, N. Winograd: J. Vac. Sci. Technol. A **4**, 1218 (1986).
2.237  R. Maboudian, Z. Postawa, M. El-Maazawi, B.J. Garrison, N. Winograd: Phys. Rev. B **42**, 7311 (1990).
2.238  M. El-Maazawi, R. Maboudian, Z. Postawa, N. Winograd: Phys. Rev. B **43**, 12 078 (1991).
2.239  A. Wucher, M. Watgen, C. Mößner, H. Oechsner, B.J. Garrison: Nucl. Instrum. Methods B **67**, 531 (1992).
2.240  S.W. Rosencrance, J.S. Burnham, D.E. Sanders, C. He, B.J. Garrison, N. Winograd, Z. Postawa, A.E. DePristo: Phys. Rev. B **52**, 6006 (1995).
2.241  S.W. Rosencrance, N. Winograd, B.J. Garrison, Z. Postawa: Phys. Rev. B **53**, 2378 (1996).
2.242  C. He, Z. Postawa, S.W. Rosencrance, R. Chatterjee, B.J. Garrison, N. Winograd: Phys. Rev. Lett. **75**, 3950 (1995).
2.243  B.J. Garrison, C.T. Reimann, N. Winograd, D.E. Harrison: Phys. Rev. B **36**, 3561 (1987).
2.244  B.J. Garrison, N. Winograd, D.M. Deaven, C.T. Reimann, D.Y. Lo, T.A. Tombrello, D.E. Harrison, M.H. Shapiro: Phys. Rev. B **37**, 7197 (1988).
2.245  D.E. Sanders, K.B.S. Prasad, J.B. Burnham, B.J. Garrison: Phys. Rev. B **50**, 5358 (1994).
2.246  N. Winograd, M. El-Maazawi, R. Maboudian, Z. Postawa, D.N. Bernardo, B.J. Garrison: J. Chem. Phys. **96**, 6314 (1992).
2.247  D.E. Harrison, J.P. Johnson, N.S. Levy: Appl. Phys. Lett. **8**, 33 (1966).
2.248  B.U.R. Sundqvist: in *Sputtering by Particle Bombardment III*, ed. by R. Behrisch, K. Wittmaack (Springer, Berlin, Heidelberg, 1991), p. 257.
2.249  H.M. Urbassek, W.O. Hofer: K. Dan. Vidensk. Selsk. Mat. Fys. Medd. **43**, 97 (1993).
2.250  H.F. Winters, J.W. Coburn: Surf. Sci. Rep. **14**, 161 (1992).
2.251  H. Oechsner: in *Secondary Ion Mass Spectrometry SIMS III*, ed. by A. Benninghoven, I. Giber, I. Laszlo, M. Riedel, H.W. Werner (Springer, Berlin, Heidelberg, 1982), p. 106.

2.252  K.J. Snowdon, W. Heiland: Z. Phys. A **318**, 275 (1984).
2.253  C.M. Loxton, I.S.T. Tsong, D.A. Reed: Nucl. Instrum. Methods B **2**, 465 (1984).
2.254  K. Snowdon: Nucl. Instrum. Methods B **9**, 132 (1985).
2.255  H. Oechsner: in *The Physics of Ionized Gases*, ed. by M.M. Popovic, P. Krstic (World Scientific, Singapore, 1985), p. 571.
2.256  P. Sigmund, H.M. Urbassek, D. Matragrano: Nucl. Instrum. Methods B **14**, 495 (1986).
2.257  H.M. Urbassek: Nucl. Instrum. Methods B **18**, 587 (1987).
2.258  A.E. de Vries: Nucl. Instrum. Methods B **27**, 173 (1987).
2.259  H.H. Andersen: Vacuum **39**, 1095 (1989).
2.260  I.S. Bitensky, E.S. Parilis, I.A. Wojciechowski: Nucl. Instrum. Methods B **67**, 595 (1992).
2.261  D.E. Harrison, C.B. Delaplain: J. Appl. Phys. **47**, 2252 (1976).
2.262  B.J. Garrison, N. Winograd, D.E. Harrison: J. Chem. Phys. **69**, 1440 (1978); Phys. Rev. B **18**, 6000 (1978).
2.263  N. Winograd, K.E. Foley, B.J. Garrison, D.E. Harrison: Phys. Lett. **73A**, 253 (1979).
2.264  F. Karetta, H.M. Urbassek: Appl. Phys. A **55**, 364 (1992).
2.265  A. Wucher, B.J. Garrison: Phys. Rev. B **46**, 4855 (1992).
2.266  A. Wucher, B.J. Garrison: Surf. Sci. **260**, 257 (1992); Nucl. Instrum. Methods B **67**, 518 (1992).
2.267  A. Wucher, B.J. Garrison: J. Chem. Phys. **105**, 5999 (1996).
2.268  G. Betz, W. Husinsky: Nucl. Instrum. Methods B **102**, 281 (1995).
2.269  M.H. Shapiro, T.A. Tombrello: Nucl. Instrum. Methods B **84**, 453 (1994).
2.270  J.W. Hartman, M.H. Shapiro, T.A. Tombrello: Nucl. Instrum. Methods B **124**, 31 (1997); Nucl. Instrum. Methods B **132**, 406 (1997).
2.271  W. Ens, R. Beavis, K.G. Standing: Phys. Rev. Lett. **50**, 27 (1983).
2.272  W. Begemann, S. Dreihöfer, K.H. Meiwes-Broer, H.O. Lutz: Z. Phys. D **3**, 183 (1986).
2.273  W. Begemann, K.H. Meiwes-Broer, H.O. Lutz: Phys. Rev. Lett. **56**, 2248 (1986).
2.274  N.K. Dzhemilev, A.M. Goldenberg, I.V. Veriovkin, S.V. Verkhoturov: Int. J. Mass Spectrom. Ion Proc. **107**, R19 (1991); Int. J. Mass Spectrom. Ion Proc. **141**, 209 (1995).
2.275  N.K. Dzhemilev, A.M. Goldenberg, I.V. Veriovkin, S.V. Verkhoturov: Nucl. Instrum. Methods B **114**, 245 (1996).
2.276  R.E. Honig: J. Appl. Phys. **29**, 549 (1958).
2.277  G. Blaise, G. Slodzian: C. R. Acad. Sci. B **266**, 1525 (1968).
2.278  I. Katakuse, T. Ichihara, Y. Fujita, T. Matsuo, T. Sakurai, H. Matsuda: Int. J. Mass Spectrom. Ion Proc. **67**, 229 (1985).
2.279  W. Gerhard: Z. Phys. B **22**, 31 (1975).
2.280  W. Gerhard, H. Oechsner: Z. Phys. B **22**, 41 (1975).
2.281  V. Bonacic-Koutecky, L. Cespiva, P. Fantucci, J. Koutecky: J. Chem. Phys. **98**, 7981 (1993); Z. Phys. D **26**, 287 (1993).
2.282  C. Jackschath, I. Rabin, W. Schulze: Z. Phys. D **22**, 517 (1992).
2.283  G.P. Können, A. Tip, and A.E. de Vries: Radiat. Eff. **21**, 269 (1974); Radiat. Eff. **26**, 23 (1975).
2.284  C.B. Cooper, J.R. Woodyard: Phys. Lett. **79A**, 124 (1980).
2.285  C.B. Cooper, H.A. Hamed: Surf. Sci. **143**, 215 (1984).
2.286  K. Franzreb, A. Wucher, H. Oechsner: Phys. Rev. B **43**, 14 396 (1991); Surf. Sci. **279**, L225 (1992).
2.287  P. Wurz, W. Husinsky, G. Betz: Appl. Phys. A **52**, 213 (1991).

2.288  H. Gades, H.M. Urbassek: Nucl. Instrum. Methods B **103**, 131 (1995).

2.289  W. Husinsky, G. Nicolussi, G. Betz: Nucl. Instrum. Methods B **82**, 323 (1993).

2.290  S.R. Coon, W.F. Calaway, J.W. Burnett M.J. Pellin, D.M. Gruen, D.R. Spiegel, J.M. White: Surf. Sci. **259**, 275 (1991).

2.291  S.R. Coon, W.F. Calaway, M.J. Pellin, J.M. White: Surf. Sci. **298**, 161 (1993).

2.292  S.R. Coon, W.F. Calaway, M.J. Pellin: Nucl. Instrum. Methods B **90**, 518 (1994).

2.293  Z. Ma, S.R. Coon, W.F. Calaway, M.J. Pellin, D.M. Gruen, E.I. von Nagy-Felsobuki: J. Vac. Sci. Technol. A **12**, 2425 (1994).

2.294  Z. Ma, W.F. Calaway, M.J. Pellin, E.I. von Nagy-Felsobuki: Nucl. Instrum. Methods B **94**, 197 (1994).

2.295  T. Lill, W.F. Calaway, Z. Ma, M.J. Pellin: Surf. Sci. **322**, 361 (1995).

2.296  A. Wucher, M. Wahl, H. Oechsner: Nucl. Instrum. Methods B **82**, 337 (1993); Nucl. Instrum. Methods B **83**, 73 (1993).

2.297  M. Wahl, A. Wucher: Nucl. Instrum. Methods B **94**, 36 (1994).

2.298  A. Wucher, M. Wahl: Nucl. Instrum. Methods B **115**, 581 (1996).

2.299  A. Wucher, M. Wahl: in *Secondary Ion Mass Spectrometry SIMS X*, ed. by A. Benninghoven, B. Hagenhoff, H.W. Werner (Wiley, Chichester, 1997), p. 65.

2.300  T.J. Raeker, A.E. DePristo: Int. Rev. Phys. Chem. **10**, 1 (1991).

2.301  H.M. Urbassek: Nucl. Instrum. Methods B **31**, 541 (1988); Radiat. Eff. Def. Solids **109**, 293 (1989).

2.302  P. Fayet, J.P. Wolf, L. Wöste: Phys. Rev. B **33**, 6792 (1986).

2.303  R. de Jonge, T. Baller, M.G. Tenner, A.E. de Vries, K.J. Snowdon: Nucl. Instrum. Methods B **17**, 213 (1986); Europhys. Lett. **2**, 449 (1986).

2.304  A. Wucher: Phys. Rev. B **49**, 2012 (1994).

2.305  C.E. Klots: J. Chem. Phys. **83**, 5854 (1985); Z. Phys. D **21**, 335 (1991).

2.306  F. Bernhard, H. Oechsner, E. Stumpe: Nucl. Instrum. Methods **132**, 329 (1976).

2.307  T. Mousel, W. Eckstein, H. Gnaser: Nucl. Instrum. Methods B, submitted.

2.308  T. Takagi: *Ionized-Cluster Beam Deposition and Epitaxy* (Noyes, Park Ridge, 1988).

2.309  H. Haberland, M. Karrais, M. Mall, Y. Thurner: J. Vac. Sci. Technol. A **10**, 3266 (1992).

2.310  I. Yamada: Nucl. Instrum. Methods B **99**, 240 (1995).

2.311  A.D. Appelhans, J.E. Delmore: Anal. Chem. **61**, 1087 (1989).

2.312  M.J. Van Stipdonk, R.D. Harris, E.A. Schweikert, Rapid Comm. Mass Spectrom. **10**, 1987 (1996).

2.313  K. Iltgen, C. Bendel, A. Benninghoven, E. Niehuis: J. Vac. Sci. Technol. A **15**, 460 (1997).

2.314  M.G. Blain, S. Della-Negra, H. Joret, Y. Le Beyec, E.A. Schweikert: Phys. Rev. Lett. **63**, 1625 (1989).

2.315  M. Benguerba, A. Brunelle, S. Della-Negra, J. Depauw, H. Joret, Y. Le Beyec, M.G. Blain, E.A. Schweikert, G. Ben Assayag, P. Sudraud: Nucl. Instrum. Methods B **62**, 8 (1991).

2.316  K. Boussofiane-Baudin, G. Bolbach, A. Brunelle, S. Della-Negra, P. Håkansson, Y. Le Beyec: Nucl. Instrum. Methods B **88**, 160 (1994).

2.317  T. Yamaguchi, J. Matsuo, M. Akizuki, C.E. Ascheron, G.H. Takaoka, I. Yamada: Nucl. Instrum. Methods B **99**, 237 (1995).

2.318  I. Yamada, J. Matsuo, Z. Insepov, D. Takeuchi, M. Akizuki, N. Toyoda: J. Vac. Sci. Technol. A **14**, 781 (1996).

2.319  H.H. Andersen, H.L. Bay: Radiat. Eff. **19**, 139 (1973); J. Appl. Phys. **45**, 953 (1974); J. Appl. Phys. **46**, 2416 (1975).
2.320  D.A. Thompson, S.S. Johar: Appl. Phys. Lett. **34**, 342 (1979); Nucl. Instrum. Methods **170**, 281 (1980); Radiat. Eff. **55**, 91 (1981).
2.321  S.S. Johar, D.A. Thompson: Surf. Sci. **90**, 319 (1979).
2.322  D.A. Thompson: Radiat. Eff. **56**, 105 (1981); J. Appl. Phys. **52**, 982 (1981).
2.323  A.R. Oliva-Florio, E.V. Alonso, R.A. Baragiola: J. Ferron, M.M. Jakas: Radiat. Eff. Lett. **50**, 3 (1979).
2.324  P.C. Zalm, L.J. Beckers: J. Appl. Phys. **56**, 220 (1984).
2.325  D.J. Oostra, R.P. van Ingen, A. Haring, A.E. de Vries, F.W. Saris: Phys. Rev. Lett. **61**, 1392 (1988).
2.326  M.W. Matthew, R.J. Beuhler, M. Ledbetter, L. Friedman: Nucl. Instrum. Methods B **14**, 448 (1986); J. Phys. Chem. **90**, 3152 (1986).
2.327  R. Beuhler, L. Friedman: Chem. Rev. **86**, 521 (1986).
2.328  P. Sigmund: J. Phys. (Coll. C2) **50**, C2-175 (1989).
2.329  V.I. Shulga, P. Sigmund: Nucl. Instrum. Methods B **47**, 236 (1990); Nucl. Instrum. Methods B **62**, 23 (1991).
2.330  Z. Pan, P. Sigmund: Nucl. Instrum. Methods B **51**, 344 (1990).
2.331  P. Sigmund, I.S. Bitensky, J. Jensen: Nucl. Instrum. Methods B **112**, 1 (1996).
2.332  H.H. Andersen: K. Dan. Vidensk. Selsk. Mat. Fys. Medd. **43**, 127 (1993).
2.333  Y. Le Beyec: Int. J. Mass Spectrom. Ion Proc. **174**, 101 (1998).
2.334  M. Szymonski, A.E. de Vries: Phys. Lett. **63A**, 359 (1977).
2.335  S. Ahmad, B.W. Farmery, M.W. Thompson: Nucl. Instrum. Methods **170**, 327 (1980).
2.336  D. Pramanik, D.N. Seidman: Nucl. Instrum. Methods **209/210**, 453 (1983).
2.337  G. Carter: Radiat. Eff. Lett. **43**, 193 (1979); Radiat. Eff. Lett. **50**, 105 (1980).
2.338  Y. Yamamura, Y. Kitazoe: Radiat. Eff. **39**, 251 (1978).
2.339  M.W. Thompson, R.S. Nelson: Philos. Mag. **7**, 2015 (1962).
2.340  R. Kelly: Radiat. Eff. **32**, 91 (1977); Surf. Sci. **90**, 280 (1979).
2.341  P. Sigmund: Appl. Phys. Lett. **25**, 169 (1974); Appl. Phys. Lett. **27**, 52 (1975).
2.342  P. Sigmund, C. Claussen: J. Appl. Phys. **52**, 990 (1981).
2.343  C. Claussen: Nucl. Instrum. Methods **194**, 567 (1982).
2.344  M. Urbassek, P. Sigmund: Appl. Phys. A **35**, 19 (1984).
2.345  P. Sigmund, M. Szymonski: Appl. Phys. A **33**, 141 (1984).
2.346  K. Besocke, S. Berger, W.O. Hofer, U. Littmark: Radiat. Eff. **66**, 35 (1982).
2.347  W.O. Hofer, K. Besocke, B. Stritzker: Appl. Phys. A **30**, 83 (1983).
2.348  Y. Yamamura: Nucl. Instrum. Methods B **45**, 707 (1990).
2.349  M.H. Shapiro, T.A. Tombrello: Phys. Rev. Lett. **65**, 92 (1990); Nucl. Instrum. Methods B **58**, 161 (1991).
2.350  Z. Pan: Nucl. Instrum. Methods B **66**, 325 (1992).
2.351  H. Hsieh, R.S. Averback, H. Sellers, C.P. Flynn: Phys. Rev. B **45**, 4417 (1992).
2.352  H. Haberland, Z. Insepov, M. Moseler: Phys. Rev. B **51**, 11 061 (1995).
2.353  T.A. Tombrello: Nucl. Instrum. Methods B **99**, 225 (1995).
2.354  G. Betz, W. Husinsky: Nucl. Instrum. Methods B **122**, 311 (1997).
2.355  K. Bromann, H. Brune, C. Félix, W. Harbich, R. Monot, J. Buttet, K. Kern: Surf. Sci. **377–379**, 1051 (1997).
2.356  H.H. Andersen, A. Brunelle, S. Della-Negra, J. Depauw, D. Jacquet, Y. Le Beyec, J. Chaumont, H. Bernas: Phys. Rev. Lett. **80**, 5433 (1998).

## Chapter 3

3.1   G. Betz, G.K. Wehner: in *Sputtering by Particle Bombardment II*, ed. by R. Behrisch (Springer, Berlin, Heidelberg, 1983), p. 11.

3.2   H.H. Andersen: in *Ion Implantation and Beam Processing*, ed. by J.S. Williams, J.M. Poate (Academic, Sydney, 1984), p. 127.

3.3   P. Sigmund, N.Q. Lam: K. Dan. Vidensk. Selsk. Mat. Fys. Medd. **43**, 255 (1993).

3.4   N. Andersen, P. Sigmund: K. Dan. Vidensk. Selsk. Mat. Fys. Medd. **39** (3) (1974).

3.5   P. Sigmund: J. Vac. Sci. Technol. **17**, 396 (1980).

3.6   P. Sigmund, A. Oliva: Nucl. Instrum. Methods B **82**, 269 (1993).

3.7   N.Q. Lam, H. Wiedersich: J. Nucl. Mater. **103/104**, 433 (1981); Radiat. Eff. Lett. **67**, 107 (1982).

3.8   N.Q. Lam, H. Wiedersich: Nucl. Instrum. Methods B **18**, 471 (1987).

3.9   N.Q. Lam: Surf. Interf. Anal. **12**, 65 (1988); Mater. Res. Soc. Symp. Proc. Vol. 100, ed. by M.J. Aziz, L.E. Rehn, B. Stritzker (MRS, Pittsburgh 1988), p. 29.

3.10  N.Q. Lam: Scanning Microscopy, Suppl. **4**, 311 (1990).

3.11  M. Szymonski: Nucl. Instrum. Methods **194**, 523 (1982).

3.12  M.W. Sckerl, P. Sigmund, N.Q. Lam: Nucl. Instrum. Methods B **120**, 221 (1996).

3.13  M.W. Sckerl, N.Q. Lam, P. Sigmund: Nucl. Instrum. Methods B **140**, 75 (1998).

3.14  P. Sigmund: in *Sputtering by Particle Bombardment I*, ed. by R. Behrisch (Springer, Berlin, Heidelberg, 1981), p. 9.

3.15  P. Sigmund, A. Oliva, G. Falcone: Nucl. Instrum. Methods **194**, 541 (1982); Nucl. Instrum. Methods B **9**, 354 (1985).

3.16  P. Sigmund: Nucl. Instrum. Methods B **18**, 375 (1987).

3.17  R. Kelly: Nucl. Instrum. Methods **149**, 553 (1978); Surf. Sci. **100**, 85 (1980).

3.18  G. Betz: Surf. Sci. **92**, 283 (1980).

3.19  R. Shimizu: Nucl. Instrum. Methods B **18**, 486 (1987).

3.20  U. Conrad, H.M. Urbassek: Nucl. Instrum. Methods B **61**, 295 (1991).

3.21  P. Sigmund, M.W. Sckerl: Nucl. Instrum. Methods B **82**, 242 (1993).

3.22  H.M. Urbassek, U. Conrad: Nucl. Instrum. Methods B **73**, 151 (1993).

3.23  M. Vicanek, J.J. Jimenéz-Rodríguez, P. Sigmund: Nucl. Instrum. Methods B **36**, 124 (1989).

3.24  J.P. Biersack, W. Eckstein: Appl. Phys. A **34**, 73 (1984).

3.25  W. Eckstein, J.P. Biersack: Appl. Phys. A **37**, 95 (1985).

3.26  W. Eckstein, W. Möller: Nucl. Instrum. Methods B **7/8**, 727 (1987).

3.27  R. Kelly: Nucl. Instrum. Methods B **18**, 388 (1987).

3.28  H. Gades, H.M. Urbassek: Nucl. Instrum. Methods B **69**, 232 (1992); Nucl. Instrum. Methods B **88**, 218 (1994).

3.29  D.P. Jackson: Radiat. Eff. **18**, 185 (1973).

3.30  M.H. Shapiro, P.K. Haff, T.A. Tombrello, D.E. Harrison, R.P. Webb: Radiat. Eff. **89**, 234 (1985).

3.31  B.J. Garrison, N. Winograd, D. Lo, T.A. Tombrello, M.H. Shapiro, D.E. Harrison: Surf. Sci. **180**, L129 (1987).

3.32  A. Oliva, R. Kelly, G. Falcone: Nucl. Instrum. Methods B **19/20**, 101 (1987).

3.33  N.Q. Lam, K. Johannessen: Nucl. Instrum. Methods B **71**, 371 (1992).

3.34  M. Szymonski, R.S. Bhattacharya, H. Overeinijnder, A.E. de Vries: J. Phys. D: Appl. Phys. **11**, 751 (1978).

3.35  M. Szymonski, R.S. Bhattacharya: Appl. Phys. **20**, 207 (1979).

3.36  M. Szymonski: Appl. Phys. **23**, 89 (1980).

3.37  H. Oechsner, J. Bartella: Proc. 7th Int. Conf. Atomic Collisions in Solids, Vol. II, ed. by I.U. Bulgakov, A.F. Tulinov (Moscow State University Press, 1981), p. 55.

3.38  R.P. Schorn, M.A. Zaki Ewiss, E. Hintz: Appl. Phys. A **46**, 291 (1988).

3.39  D. Grischkowsky, M.L. Yu, A.C. Balant: Surf. Sci. **127**, 315 (1983).

3.40  E. Dullni: Nucl. Instrum. Methods B **2**, 610 (1984).

3.41  W. Husinsky, G. Betz, I. Girgis: J. Vac. Sci. Technol. A **2**, 698 (1984).

3.42  W. Husinsky, G. Betz: Nucl. Instrum. Methods B **15**, 165 (1986).

3.43  W. Husinsky, P. Wurz, B. Strehl, G. Betz: Nucl. Instrum. Methods B **18**, 452 (1987).

3.44  M. Saidoh, H. Gnaser, W.O. Hofer: Appl. Phys. A **40**, 197 (1986).

3.45  H. Gnaser, M. Saidoh, J. von Seggern, W.O. Hofer: Nucl. Instrum. Methods B **15**, 169 (1986).

3.46  G.P. Chen, J. von Seggern, H. Gnaser, W.O. Hofer: Appl. Phys. A **49**, 711 (1989).

3.47  R. Kelly: in *Materials Modification by High-Fluence Beams*, ed. by R. Kelly, M.F. da Silva (Kluwer, Dordrecht, 1988), p. 305; Mater. Sci. Engin. A **115**, 11 (1989).

3.48  R. Kelly: Nucl. Instrum. Methods B **39**, 43 (1989).

3.49  I. Bertóti, R. Kelly, M. Mohai, A. Tóth: Nucl. Instrum. Methods B **80/81**, 1219 (1993).

3.50  Z. Qu: Surf. Sci. **176**, L873 (1986).

3.51  R. Kelly, J.B. Sanders: Surf. Sci. **57**, 143 (1976).

3.52  S. Dzioba, R. Kelly: J. Nucl. Mater. **76/77**, 175 (1978).

3.53  P. Sigmund: J. Appl. Phys. **50**, 7261 (1979).

3.54  A. Gras-Marti: Phys. Stat. Sol. (a) **76**, 621 (1983).

3.55  I. Abril, R. Garcia-Molina, A. Gras-Marti: Phys. Stat. Sol. (a) **96**, 161 (1986).

3.56  P. Sigmund: Nucl. Instrum. Methods B **34**, 15 (1988); J. Vac. Sci. Technol. A **7**, 585 (1989).

3.57  Z.L. Liau, J.W. Mayer, W.L. Brown, J.M. Poate: J. Appl. Phys. **49**, 5295 (1978).

3.58  Z.L. Liau, B.Y. Tsaur, J.W. Mayer: J. Vac. Sci. Technol. **16**, 121 (1979).

3.59  M.L. Roush, F. Davarya, O.F. Goktepe, T.D. Andreadis: Nucl. Instrum. Methods **209/210**, 67 (1983).

3.60  O.F. Goktepe, T.D. Andreadis, M. Rosen, G.P. Mueller, M.L. Roush: Nucl. Instrum. Methods B **13**, 434 (1986).

3.61  W. Möller, W. Eckstein: Nucl. Instrum. Methods B **2**, 814 (1984).

3.62  U. Conrad, H.M. Urbassek: Nucl. Instrum. Methods B **83**, 125 (1993).

3.63  J.W. Mayer, B.Y. Tsaur, S.S. Lau, L.S. Hung: Nucl. Instrum. Methods **182/183**, 1 (1981).

3.64  B.M. Paine, R.S. Averback: Nucl. Instrum. Methods B **7/8**, 666 (1985).

3.65  R.S. Averback: Nucl. Instrum. Methods B **15**, 675 (1986).

3.66  Y.T. Cheng: Mater. Sci. Rep. **5**, 45 (1990).

3.67  U. Littmark, W.O. Hofer: in *Thin Film and Depth Profile Analysis*, ed. by H. Oechsner (Springer, Berlin, Heidelberg, 1984), p. 159.

3.68  T. Ishitani, R. Shimizu: Appl. Phys. **6**, 241 (1975).

3.69  P.K. Haff, Z.E. Switkowski: J. Appl. Phys. **48**, 3383 (1977).

3.70  H.H. Andersen: Appl. Phys. **18**, 131 (1979).

3.71  U. Littmark, W.O. Hofer: Phys. Lett **71A**, 457 (1979).

3.72  W.O. Hofer, U. Littmark: Nucl. Instrum. Methods **168**, 329 (1980).

3.73  P. Sigmund, A. Gras-Marti: Nucl. Instrum. Methods **168**, 389 (1980); Nucl. Instrum. Methods **182/183**, 25 (1981).

3.74  S. Matteson: Appl. Phys. Lett. **39**, 288 (1981).

3.75  G. Carter, R. Collins, D.A. Thompson: Radiat. Eff. **55**, 99 (1981).
3.76  P. Sigmund: Appl. Phys. A **30**, 43 (1983).
3.77  J.A. Peinador, I. Abril, J.J. Jimenez-Rodriguez, A. Gras-Marti: Phys. Rev. B **44**, 2061 (1991).
3.78  H. Gades, H.M. Urbassek: Phys. Rev. B **51**, 14559 (1995); Nucl. Instrum. Methods B **115**, 485 (1996).
3.79  G.V. Kornich, G. Betz, B.V. King: Nucl. Instrum. Methods B **115**, 461 (1996).
3.80  Y.T. Cheng, M. van Rossum, M.A. Nicolet, W.L. Johnson: Appl. Phys. Lett. **45**, 185 (1984).
3.81  W.L. Johnson, Y.T. Cheng, M. van Rossum, M.A. Nicolet: Nucl. Instrum. Methods B **7/8**, 657 (1985).
3.82  P. Shewmon: *Diffusion in Solids* (Minerals, Metals & Materials Soc., Warrendale, 1989).
3.83  R. Kelly, A. Miotello: Nucl. Instrum. Methods B **59/60**, 517 (1991).
3.84  A. Miotello, R. Kelly: Surf. Sci. **268**, 340 (1992); Surf. Sci. **314**, 275 (1994); Nucl. Instrum. Methods B **122**, 458 (1997).
3.85  R. Kelly, A. Miotello: Surf. Coat. Technol. **51**, 343 (1992); Nucl. Instrum. Methods B **122**, 374 (1997).
3.86  K. Wittmaack: Vacuum **34**, 119 (1984).
3.87  K. Wittmaack: in *Sputtering by Particle Bombardment III*, ed. by R. Behrisch and K. Wittmaack, (Springer, Berlin, Heidelberg, 1991).
3.88  S. Hofmann: Prog. Surf. Sci. **36**, 35 (1991).
3.89  P.C. Zalm: Rep. Prog. Phys. **58**, 1321 (1995).
3.90  B.V. King, I.S.T. Tsong: Ultramicroscopy **14**, 75 (1984); J. Vac. Sci. Technol. A **2**, 1443 (1984).
3.91  B.V. King, S.G. Puranik, M.A. Sobhan, R.J. MacDonald: Nucl. Instrum. Methods B **39**, 153 (1989).
3.92  M.P. Macht, R. Willecke, V. Naundorf: Nucl. Instrum. Methods B **43**, 507 (1989).
3.93  Y.T. Cheng, S.J. Simko, M.C. Militello, A.A. Dow, G.W. Auner, M.H. Alkaisi, K.R. Padmanabhan: Nucl. Instrum. Methods B **64**, 38 (1992).
3.94  K. Wittmaack: Surf. Interf. Anal. **21**, 323 (1994); Phil. Trans. R. Soc. Lond. A **354**, 2731 (1996).
3.95  K. Wittmaack: J. Appl. Phys. **53**, 4817 (1982).
3.96  P. Williams: Appl. Phys. Lett. **36**, 758 (1980).
3.97  P. Williams, J.E. Baker: Nucl. Instrum. Methods **182/183**, 15 (1981).
3.98  K. Wittmaack, D.B. Poker: Nucl. Instrum. Methods B **47**, 224 (1990).
3.99  J.B. Clegg, I.G. Gale: Surf. Interf. Anal. **17**, 190 (1991).
3.100 M.G. Dowsett, R.D. Barlow, H.S. Fox, R.A.A. Kubiak, R. Collins: J. Vac. Sci. Technol. B **10**, 336 (1992).
3.101 S.M. Hues, P. Williams: Nucl. Instrum. Methods B **15**, 206 (1986).
3.102 P.C. Zalm, C.J. Vriezema: Nucl. Instrum. Methods B **67**, 495 (1992).
3.103 P. Wynblatt, R.C. Ku: in *Interfacial Segregation*, ed. by W.C. Johnson, J.M. Blakely (Amer. Soc. Metals, Metals Park, Ohio, 1979), p.115.
3.104 R. Kelly: Surf. Interf. Anal. **7**, 1 (1985).
3.105 R. Kelly, A. Oliva: in *Erosion and Growth of Solids Stimulated by Atom and Ion Beams*, ed. by G. Kiriakidis, G. Carter, J.L. Whitton (Nijhoff, Dordrecht, 1986), p. 41.
3.106 R. Kelly, A. Oliva: Nucl. Instrum. Methods B **13**, 283 (1986).
3.107 H. Shimizu, M. Ono, K. Nakayama: J. Appl. Phys. **46**, 460 (1975).
3.108 G.Betz, P. Braun, W. Färber: J. Appl. Phys. **48**, 1404 (1977).

3.109  D.G. Swartzfager, S.B. Ziemecki, M.J. Kelley: J. Vac. Sci. Technol. **19**, 185 (1981).

3.110  R.S. Li, T. Koshikawa, K. Goto: Surf. Sci. **121**, L561 (1982).

3.111  R.S. Li, T. Koshikawa: Surf. Sci. **151**, 459 (1985).

3.112  R.S. Li, L.X. Tu, Y.Z. Sun: Surf. Sci. **163**, 67 (1985); Appl. Surf. Sci. **26**, 77 (1986).

3.113  L.E. Rehn, N.Q. Lam: J. Mater. Engin. **9**, 205 (1987).

3.114  H.H. Brongersma, M.J. Sparnaay, T.M. Buck: Surf. Sci. **71**, 657 (1978).

3.115  Y.S. Ng, T.T. Tsong, S.B. McLane: Phys. Rev. Lett. **42**, 588 (1979).

3.116  N.Q. Lam, H.A. Hoff, H. Wiedersich, L.E. Rehn: Surf. Sci. **149**, 517 (1985).

3.117  N.Q. Lam, H.A. Hoff: Surf. Sci. **193**, 353 (1988).

3.118  H.A. Hoff, N.Q. Lam: Surf. Sci. **204**, 233 (1988).

3.119  S. Tang, N.Q. Lam: Surf. Sci. **223**, 179 (1989).

3.120  A. Oliva, R. Kelly, G. Falcone: Surf. Sci. **166**, 403 (1986).

3.121  L.E. Rehn, S. Danyluk, H. Wiedersich: Phys. Rev. Lett. **43**, 1764 (1979).

3.122  S.M. Myers: Nucl. Instrum. Methods **168**, 265 (1980).

3.123  G.J. Dienes, A.C. Damask: J. Appl. Phys. **29**, 1713 (1958).

3.124  W. Wagner, L.E. Rehn, H. Wiedersich, V. Naundorf: Phys. Rev. B **28**, 6780 (1983).

3.125  H. Wiedersich: Nucl. Instrum. Methods B **7/8**, 1 (1985); in *Physics of Radiation Effects in Crystals*, ed. by R.A. Johnson, A.N. Orlov (North-Holland, Amsterdam, 1986), p. 225.

3.126  N.Q. Lam, S.J. Rothman: in *Radiation Damage in Metals*, ed. by N.L. Peterson, S.D. Harkness (Amer. Soc. for Metals, Metals Park, OH, 1976), p. 125.

3.127  R. Sizmann: J. Nucl. Mater. **69/70**, 386 (1978).

3.128  L.E. Rehn, V.T. Boccio, H. Wiedersich: Surf. Sci. **128**, 37 (1983).

3.129  D. Marton, J. Fine, G.P. Chambers: Phys. Rev. Lett. **61**, 2697 (1988).

3.130  J. Fine, T.D. Andreadis, F. Davarya: Nucl. Instrum. Methods **209/210**, 521 (1983).

3.131  H. Wiedersich, P.R. Okamoto, N.Q. Lam: J. Nucl. Mater. **83**, 98 (1979).

3.132  R.A. Johnson, N.Q. Lam: Phys. Rev. B **13**, 4364 (1976); Phys. Rev. B **15**, 1794 (1977).

3.133  W.L. Patterson, G.A. Shirn: J. Vac. Sci. Technol. **4**, 343 (1967).

3.134  M.L. Tarng, G.K. Wehner: J. Appl. Phys. **42**, 2449 (1971).

3.135  H. Shimizu, M. Ono, K. Nakayama: Surf. Sci. **36**, 817 (1973).

3.136  H.W. Pickering: J. Vac. Sci. Technol. **13**, 618 (1976).

3.137  Z. Qu, T.S. Xie: Surf. Sci. **194**, L127 (1988).

3.138  R.S. Li: Surf. Sci. **193**, 373 (1988); Appl. Surf. Sci. **35**, 409 (1988/89).

3.139  J. du Plessis, G.N. van Wyk, E. Taglauer: Surf. Sci. **220**, 381 (1989).

3.140  H.F. Winters, J.W. Coburn: Appl. Phys. Lett. **28**, 176 (1976).

3.141  P.K. Haff: Appl. Phys. Lett. **31**, 259 (1977).

3.142  P.S. Ho: Surf. Sci. **72**, 253 (1978).

3.143  R. Webb, G. Carter, R. Collins: Radiat. Eff. **39**, 129 (1978).

3.144  N.J. Chou, M.W. Shafer: Surf. Sci. **92**, 601 (1980).

3.145  W. Möller, W. Eckstein, J.P. Biersack: Comp. Phys. Comm. **51**, 355 (1988).

3.146  W. Eckstein: *Computer Simulation of Ion Solid Interactions* (Springer, Berlin, Heidelberg, 1991).

3.147  H.F. Winters, P. Sigmund: J. Appl. Phys. **45**, 4760 (1974).

3.148  E. Taglauer, W. Heiland: Appl. Phys. **33**, 950 (1978).

3.149  H.F. Winters: J. Vac. Sci. Technol. **20**, 493 (1982).

3.150  B. Baretzky, E. Taglauer: Surf. Sci. **162**, 996 (1985).

3.151   B. Baretzky, W. Möller, E. Taglauer: Nucl. Instrum. Methods B **18**, 496 (1987); Vacuum **43**, 1207 (1992).

3.152   R.S. Li, C.F. Li, W.L. Zhang: Appl. Phys. A **50**, 169 (1990).

3.153   R.S. Li, C.F. Li, G. Liu, X.S. Zhang, D.S. Bao: J. Appl. Phys. **70**, 5351 (1991).

3.154   J. Bartella, H. Oechsner: Surf. Sci. **126**, 581 (1983).

3.155   J. Bartella: Doctoral Thesis, Universität Kaiserslautern (1985).

3.156   S.T. Nakagawa, Y. Yamamura: Nucl. Instrum. Methods B **33**, 780 (1988).

3.157   Z. Cao: Doctoral Thesis, Universität Kaiserslautern (1997).

3.158   G.R. Castro, A. Ballesteros: Surf. Sci. **204**, 415 (1988).

3.159   K. Lawiczak-Jablonska, M. Heinonen, A. Sulyok: Surf. Sci. **222**, 129 (1989).

3.160   M. Schmid, A. Biedermann, C. Slama, H. Stadler, P. Weigand, P. Varga: Nucl. Instrum. Methods B **82**, 259 (1993).

3.161   P. Weigand, W. Hofer, P. Varga: Surf. Sci. **287/288**, 350 (1993).

3.162   P. Weigand, B. Jelinek, W. Hofer, P. Varga: Surf. Sci. **295**, 57 (1993); Surf. Sci. **301**, 306 (1994); Surf. Sci. **307–309**, 416 (1994).

3.163   M. Schmid, A. Biedermann, H. Stadler, C. Slama, P. Varga: Appl. Phys. A **55**, 468 (1992).

3.164   M. Schmid, A. Biedermann, H. Stadler, P. Varga: Phys. Rev. Lett. **69**, 925 (1992).

3.165   P. Varga, M. Schmid, W. Hofer: Surf. Rev. Lett. **3**, 1831 (1996).

3.166   R.R. Olson, G.K. Wehner: J. Vac. Sci. Technol. **14**, 319 (1977).

3.167   R.R. Olson, M.E. King, G.K. Wehner: J. Appl. Phys. **50**, 3677 (1979).

3.168   H.H. Andersen, F. Besenbacher, P. Goddisken: in *Symposium on Sputtering*, ed. by P. Varga, G. Betz, F.P. Viehböck (Technical University Vienna, 1980), p. 446.

3.169   H.H. Andersen, V. Chernysh, B. Stenum, T. Sørensen, H.J. Whitlow: Surf. Sci. **123**, 39 (1982).

3.170   H.H. Andersen, B. Stenum, T. Sørensen, H.J. Whitlow: Nucl. Instrum. Methods **209/210**, 487 (1983).

3.171   H.H. Andersen, B. Stenum, T. Sørensen, H.J. Whitlow: Nucl. Instrum. Methods B **2**, 601, 623 (1984).

3.172   S. Ichimura, H. Shimizu, H. Murakami, Y. Ishida: J. Nucl. Mat. **128/129**, 601 (1984).

3.173   H.J. Kang, Y. Matsuda, R. Shimizu: Surf. Sci. **127**, L179 (1983).

3.174   M.F. Dumke, T.A. Tombrello, R.A. Weller, R.M. Housley, E.H. Cirlin: Surf. Sci. **124**, 407 (1983).

3.175   K.M. Hubbard, R.A. Weller, D.L. Weathers, T.A. Tombrello: Nucl. Instrum. Methods B **36**, 395 (1989); Nucl. Instrum. Methods B **40/41**, 278 (1989).

3.176   M.H. Shapiro, K.R. Bengtson, T.A. Tombrello: Nucl. Instrum. Methods B **103**, 123 (1995).

3.177   T. Aoyama, M. Tanemura, F. Okuyama: Appl. Surf. Sci. **100/101**, 351 (1996).

3.178   M. Tanemura, T. Aoyama, A. Otani, M. Ukita, F. Okuyama, T.K. Chini: Surf. Sci. **376**, 163 (1997).

3.179   A. Wucher, W. Reuter: J. Vac. Sci. Technol. A **6**, 2316 (1988).

3.180   W. Bock: Doctoral Thesis, Universität Kaiserslautern (1996).

3.181   M. Szymonski: Phys. Lett **82A**, 203 (1981).

3.182   M. Vicanek, U. Conrad, H.M. Urbassek: Phys. Rev. B **47**, 617 (1993).

3.183   T. Mousel, W. Eckstein, H. Gnaser: Nucl. Instrum. Methods B, submitted.

3.184   H. Gnaser, H. Oechsner: Phys. Rev. B **47**, 14093 (1993).

3.185   H. Gnaser, H. Oechsner: Nucl. Instrum. Methods B **82**, 347 (1993).

3.186   G.K. Wehner, D. Rosenberg: J. Appl. Phys. **31**, 177 (1960).

3.187  G.K. Wehner, Y.H. Kim, D.H. Kim, A.M. Goldmann: Appl. Phys. Lett. **52**, 1187 (1988).

3.188  R.P. Webb, D.E. Harrison: J. Appl. Phys. **53**, 5243 (1982).

3.189  M.M. Jakas, D.E. Harrison: Nucl. Instrum. Methods B **14**, 535 (1986).

3.190  D.E. Harrison: Crit. Rev. Solid State Mater. Sci. **14**, S1 (1988).

3.191  C.B. Cooper, J.R. Woodyard: Phys. Lett. **79A**, 124 (1980).

3.192  H. Gnaser, H. Oechsner: Surf. Sci. **251/252**, 696 (1991); Nucl. Instrum. Methods B **58**, 438 (1991).

3.193  T. Lill, W.F. Calaway, M.J. Pellin, D.M. Gruen: Phys. Rev. Lett. **73**, 1719 (1994).

3.194  T. Lill, W.F. Calaway, M.J. Pellin: J. Appl. Phys. **78**, 505 (1995); Nucl. Instrum. Methods B **100**, 361 (1995).

3.195  H.M. Urbassek: Nucl. Instrum. Methods B **122**, 427 (1997).

3.196  J. M. Fluit, L. Friedman, A.J.H. Boerboom, J. Kistemaker: J. Chem. Phys. **35**, 1143 (1961).

3.197  P. de Bievre, I.L. Barnes: Int. J. Mass Spectrom. Ion Proc. **65**, 211 (1985).

3.198  V.I. Shulga, P. Sigmund: Nucl. Instrum. Methods B **103**, 383 (1995).

3.199  G.K. Wehner: Appl. Phys. Lett. **30**, 185 (1977).

3.200  O. Arai, Y. Tazawa, T. Shimamura, T. Kobayashi: Japan J. Appl. Phys. **18**, 1231 (1979).

3.201  W.A. Russell, D.A. Papanastassiou, T.A. Tombrello: Radiat. Eff. **52**, 41 (1980).

3.202  J. Okano, T. Ochiai, H. Nishimura: Appl. Surf. Sci. **22/23**, 72 (1985).

3.203  H. Gnaser, I.D. Hutcheon: Surf. Sci. **195**, 499 (1988).

3.204  L.M. Baumel, M.R. Weller, R.A. Weller, T.A. Tombrello: Nucl. Instrum. Methods B **34**, 427 (1988).

3.205  H. Gnaser, H. Oechsner: Phys. Rev. Lett. **63**, 2673 (1989).

3.206  H. Gnaser, H. Oechsner: Nucl. Instrum. Methods B **48**, 544 (1990).

3.207  D.L. Weathers, S.J. Spicklemire, T.A. Tombrello, I.D. Hutcheon, H. Gnaser: Nucl. Instrum. Methods B **73**, 135 (1993).

3.208  P.A.J. Ackermans, M.A.P. Creuwels, H.H. Brongersma, P.J. Scanlon: Surf. Sci. **227**, 361 (1990).

3.209  W. Bieck, H. Gnaser, H. Oechsner: Nucl. Instrum. Methods B **101**, 335 (1995).

3.210  M.H. Shapiro, P.K. Haff, T.A. Tombrello, D.E. Harrison: Nucl. Instrum. Methods B **12**, 137 (1985).

3.211  M.H. Shapiro, T.A. Tombrello, D.E. Harrison: Nucl. Instrum. Methods B **30**, 152 (1988).

3.212  D.Y. Lo, T.A. Tombrello, M.H. Shapiro: Nucl. Instrum. Methods B **40/41**, 270 (1989).

3.213  U. Conrad, H.M. Urbassek: Surf. Sci. **278**, 414 (1992).

3.214  J.J. Jiménez-Rodríguez, U. Conrad, H.M. Urbassek, I. Abril, A. Gras-Martí: Nucl. Instrum. Methods B **67**, 527 (1992).

3.215  W. Eckstein: Nucl. Instrum. Methods B **83**, 329 (1993).

3.216  L.P. Zheng, R.S. Li, M.Y. Li, J.Y. Lin, Z.Y. Zou: Nucl. Instrum. Methods B **62**, 61 (1991).

3.217  L.P. Zheng, M.Y. Li: Nucl. Instrum. Methods B **114**, 28 (1996).

3.218  H. Gades, H.M. Urbassek: Nucl. Instrum. Methods B**102**, 261 (1995).

3.219  V.I. Shulga, P. Sigmund: Nucl. Instrum. Methods B **119**, 359 (1996).

3.220  V.I. Shulga: Radiat. Eff. Def. Solids **142**, 351 (1997).

3.221  W. Eckstein, R. Dohmen: Nucl. Instrum. Methods B **129**, 327 (1997).

3.222  P.K. Haff, Z.E. Switkowski: Appl. Phys. Lett. **29**, 549 (1976).

3.223  C.C. Watson, P.K. Haff: J. Appl. Phys. **51**, 691 (1980).

3.224  P.K. Haff, C.C. Watson, T.A. Tombrello: J. Geophys. Res. **86**, 9553 (1981).
3.225  P. Sigmund: Nucl. Instrum. Methods B **82**, 192 (1993).

**Chapter 4**

4.1    R.W. Cahn, E.J. Kramer (Eds.): *Processing of Semiconductors*, Materials Science and Technology, Vol. 16 (VCH, Weinheim, 1996).
4.2    S.M. Sze (Ed.): *VLSI Technology* (McGraw-Hill, New York, 1988).
4.3    J.F. Ziegler (Ed.): *Handbook of Ion Implantation* (North-Holland, Amsterdam, 1992).
4.4    *The National Technology Roadmap for Semiconductors* (Semiconductor Industry Association, San Jose, 1997, http://notes.sematech.org/97pelec.htm).
4.5    J.W. Corbett: Surf. Sci. **90**, 205 (1979).
4.6    J.W. Corbett, J.P. Karins, T.Y. Tan: Nucl. Instrum. Methods **182/183**, 457 (1981).
4.7    R.J. MacDonald, D. Haneman: J. Appl. Phys. **37**, 3048 (1966).
4.8    J.W. Mayer, L. Erikson, S.T. Picraux, J.A. Davies: Can. J. Phys. **46**, 663 (1968).
4.9    R.S. Nelson, D.J. Mazey: Can. J. Phys. **46**, 689 (1968).
4.10   F.F. Morehead, B.L. Crowder, R.S. Title: J. Appl. Phys. **43**, 1112 (1972).
4.11   J.C. Bean, G.E. Becker, P.M. Petroff, T.E. Seidel: J. Appl. Phys. **48**, 907 (1977).
4.12   J.S. Williams, K.T. Short, R.G. Elliman, M.C. Ridgway, R. Goldberg: Nucl. Instrum. Methods B **48**, 431 (1990).
4.13   G. Bai, M.A. Nicolet: J. Appl. Phys. **70**, 649 (1991).
4.14   J.B. Malherbe: Crit. Rev. Solid State Mater. Sci. **19**, 55 (1994); Crit. Rev. Solid State Mater. Sci. **19**, 129 (1994).
4.15   F.L. Vook, H.J. Stein: Radiat. Eff. **2**, 23 (1969).
4.16   F.F. Morehead, B.L. Crowder: Radiat. Eff. **6**, 27 (1970).
4.17   J.R. Dennis, E.B. Hale: J. Appl. Phys. **49**, 1119 (1978).
4.18   W. Wesch: Nucl. Instrum. Methods B **68**, 342 (1992).
4.19   S.U. Campisano, S. Coffa, V. Raineri, F. Priolo, E. Rimini: Nucl. Instrum. Methods B **80/81**, 514 (1993).
4.20   M. Kitabatake, P. Fons, J.E. Greene: J. Vac. Sci. Technol. A **8**, 3726 (1990); J. Vac. Sci. Technol. A **9**, 91 (1991).
4.21   M.V. Ramana Murty, H.A. Atwater: Phys. Rev. B **45**, 1507 (1992).
4.22   B. Strickland, C. Roland: Phys. Rev. B **51**, 5061 (1995).
4.23   M.J. Caturla, T. Diaz de la Rubia, G.H. Gilmer: Nucl. Instrum. Methods B **106**, 1 (1995).
4.24   T. Diaz de la Rubia, G.H. Gilmer: Phys. Rev. Lett. **74**, 2507 (1995).
4.25   M.J. Caturla, T. Diaz de la Rubia, L.A. Marques, G.H. Gilmer: Phys. Rev. B **54**, 16683 (1996).
4.26   G.H. Gilmer, T. Diaz de la Rubia, D.M. Stock, M. Jaraiz: Nucl. Instrum. Methods B **102**, 247 (1995).
4.27   H. Hensel, H.M. Urbassek: Phys. Rev. B **57**, 4756 (1998).
4.28   R.S. Averback, M. Ghaly: Nucl. Instrum. Methods B **127/128**, 1 (1997).
4.29   H.E. Farnsworth, R.E. Schlier, T.H. George, R.M. Burger: J. Appl. Phys. **29**, 1150 (1958).
4.30   U.F. Geanola: J. Appl. Phys. **28**, 868 (1957).
4.31   J.L. Zilko, J.E. Greene: J. Appl. Phys. **51**, 1549 (1980).
4.32   B.R. Appleton, S.J. Pennycook, R.A. Zuhr, N. Herbots, T.S. Noggle: Nucl. Instrum. Methods B **19/20**, 975 (1987).
4.33   T. Itoh (Ed.): *Ion Beam Assisted Film Growth* (Elsevier, Amsterdam, 1989).

4.34  J.Y. Tsao, E. Chason, K.M. Horn, D.K. Brice, S.T. Picraux: Nucl. Instrum. Methods B **39**, 72 (1989).

4.35  S.T. Picraux, D.K. Brice, K.M. Horn, J.Y. Tsao, E. Chason: Nucl. Instrum. Methods B **48**, 414 (1990).

4.36  D.K. Brice, J.Y. Tsao, S.T. Picraux: Nucl. Instrum. Methods B **44**, 68 (1989).

4.37  J.W. Corbett, J.C. Bourgoin: in *Point Defects in Solids*, Vol. 2, ed. by J.H. Crawford, L.M. Slifkin (Plenum, New York, 1975), p. 1.

4.38  G.K. Wehner, R.M. Warner, P.D. Wang, Y.H. Kim: J. Appl. Phys. **64**, 6754 (1988).

4.39  E. Chason, P. Bedrossian, K.M. Horn, J.Y. Tsao, S.T. Picraux: Appl. Phys. Lett. **57**, 1793 (1990).

4.40  C.-H. Choi, R. Ai, S.A. Barnett: Phys. Rev. Lett. **67**, 2826 (1991).

4.41  M.V. Ramana Murty, H.A. Atwater, A.J. Kellock, J.E.E. Baglin: Appl. Phys. Lett. **62**, 2566 (1993).

4.42  N.-E. Lee, G.A. Tomasch, J.E. Greene: Appl. Phys. Lett. **65**, 3236 (1994).

4.43  H.J.W. Zandvliet, H.B. Elswijk, E.J. van Loenen, I.S.T. Tsong: Phys. Rev. B **46**, 7581 (1992).

4.44  H. Feil, H.J.W. Zandvliet, M.-H. Tsai, J.D. Dow, I.S.T. Tsong: Phys. Rev. Lett. **69**, 3076 (1992).

4.45  H.J.W. Zandvliet: Surf. Sci. **377–379**, 1 (1997).

4.46  H.J.W. Zandvliet, I.S.T. Tsong: in *Morphological Organizations in Epitaxial Growth and Removal*, ed. by Z. Zhang, M.G. Lagally (World Scientific, Singapore, 1998), in press.

4.47  N. Kitamura, M.G. Lagally, M.B. Webb: Phys. Rev. Lett. **71**, 2082 (1993).

4.48  H.J.W. Zandvliet, H.K. Louwsma, P.E. Hegeman, B. Poelsema: Phys. Rev. Lett. **75**, 3890 (1995).

4.49  S.J. Chey, J.E. van Nostrand, D.G. Cahill: Phys. Rev. B **52**, 16696 (1995).

4.50  S.J. Chey, D.G. Cahill: Surf. Sci. **380**, 377 (1997).

4.51  P. Bellon, S.J. Chey, J.E. van Nostrand, M. Ghaly, D.G. Cahill, R.S. Averback: Surf. Sci. **339**, 135 (1995).

4.52  H.J.W. Zandvliet, E. de Groot: Surf. Sci. **371**, 79 (1997).

4.53  X.-S. Wang, R.J. Pechman, J.H. Weaver: Appl. Phys. Lett. **65**, 2818 (1994).

4.54  X.-S. Wang, R.J. Pechman, J.H. Weaver: J. Vac. Sci. Technol. B **13**, 2031 (1995).

4.55  R.J. Pechman, X.-S. Wang, J.H. Weaver: Phys. Rev. B **51**, 10929 (1995).

4.56  X.-S. Wang, R.J. Pechman, J.H. Weaver: Surf. Sci. **364**, L511 (1996).

4.57  P. Bedrossian, T. Klitsner: Phys. Rev. B **44**, 13783 (1991).

4.58  P. Bedrossian, T. Klitsner: Phys. Rev. Lett. **68**, 646 (1992).

4.59  P. Bedrossian: Surf. Sci. **301**, 223 (1994).

4.60  P. Bedrossian, J.E. Houston, J.Y. Tsao, E. Chason, S.T. Picraux: Phys. Rev. Lett. **67**, 124 (1991).

4.61  Y.W. Mo, B.S. Swartzentruber, R. Kariotis, M.B. Webb, M.G. Lagally: Phys. Rev. Lett. **63**, 2393 (1989).

4.62  Y.W. Mo, J. Kleiner, M.B. Webb, M.G. Lagally: Phys. Rev. Lett. **66**, 1998 (1991).

4.63  D.J. Chadi: Phys. Rev. Lett. **59**, 1691 (1987).

4.64  B.S. Swartzentruber, C.M. Matzke, D.L. Kendall, J.E. Houston: Surf. Sci. **329**, 83 (1995).

4.65  R.J. Hamers, U.K. Köhler, J.E. Demuth: Ultramicroscopy **31**, 10 (1989).

4.66  A.J. Hoeven, J.M. Lenssinck, D. Dijkkamp, E.J. van Loenen, J. Dieleman: Phys. Rev. Lett. **63**, 1830 (1989).

4.67 Y.W. Mo, R. Kariotis, B.S. Swartzentruber, M.B. Webb, M.G. Lagally: J. Vac. Sci. Technol. B **8**, 232 (1990); J. Vac. Sci. Technol. A **8**, 201 (1990).

4.68 E. Chason, J.Y. Tsao, K.M. Horn, S.T. Picraux, H.A. Atwater: J. Vac. Sci. Technol. A **8**, 2507 (1990).

4.69 E. Chason, T.M. Mayer, B.K. Kellerman, D.T. McIlroy, A.J. Howard: Phys. Rev. Lett. **72**, 3040 (1994).

4.70 Z. Zhang, J. Detch, H. Metiu: Phys. Rev. B **48**, 4972 (1993).

4.71 P. Meakin: Phys. Rep. **235**, 189 (1993).

4.72 J. Lapujoulade: Surf. Sci. Rep. **20**, 191 (1994).

4.73 R. Kunkel, B. Poelsema, L.K. Verheij, G. Comsa: Phys. Rev. Lett. **65**, 733 (1990).

4.74 R.G. Elliman, J. Linnros, W.L. Brown: Mater. Res. Soc. Symp. Proc., Vol. 100, ed. by M.J. Aziz, L.E. Rehn, B. Stritzker (MRS, Pittsburgh, 1988), p. 363.

4.75 Z.H. Lu, D.F. Mitchell, M.J. Graham: Appl. Phys. Lett. **65**, 552 (1994).

4.76 M. Ishii, Y. Hirose, T. Sato, T. Ohwaki, Y. Taga: J. Vac. Sci. Technol. A **15**, 820 (1997).

4.77 I. Konomi, A. Kawano, Y. Kido: Surf. Sci. **207**, 427 (1989).

4.78 Y. Kido, H. Nakano: Surf. Sci. **239**, 254 (1990).

4.79 S. Kalbitzer, H. Oetzmann: Radiat. Eff. **47**, 57 (1980).

4.80 A.H. Al-Bayati, K.G. Orrman-Rossiter, R. Badheka, D.G. Armour: Surf. Sci. **237**, 213 (1990).

4.81 D.G. Armour, A.H. Al-Bayati: Nucl. Instrum. Methods B **67**, 279 (1992).

4.82 R.S. Williams: Solid State Comm. **41**, 153 (1982).

4.83 L.J. Huang, W.M. Lau, H.T. Tang, W.N. Lennard, I.V. Mitchell, P.J. Schultz, M. Kasrai: Phys. Rev. B **50**, 18 453 (1994).

4.84 W.M. Lau, I. Bello, L.J. Huang, X. Feng, M. Vos, I.V. Mitchell: J. Appl. Phys. **74**, 7101 (1993).

4.85 W. Bock, H. Gnaser, H. Oechsner: Surf. Sci. **282**, 333 (1993).

4.86 H. Gnaser, B. Heinz, W. Bock, H. Oechsner: Phys. Rev. B **52**, 14 086 (1995).

4.87 H. Gnaser: Appl. Surf. Sci. **100/101**, 316 (1996).

4.88 N.P. Lieske: J. Phys. Chem. Solids **45**, 821 (1984).

4.89 J.C. Fernandez, W.S. Yang, H.D. Shih, F. Jona, D.W. Jepsen, P.M. Marcus: J. Phys. C: Solid State Phys. **14**, L55 (1981).

4.90 R.J. Culbertson, Y. Kuk, L.C. Feldman: Surf. Sci. **167**, 127 (1986).

4.91 J.E. Rowe, H. Ibach: Phys. Rev. Lett. **31**, 102 (1973).

4.92 A. Koma, R. Ludeke: Surf. Sci. **55**, 735 (1976).

4.93 R. Ludeke, A. Koma: Phys. Rev. B **13**, 739 (1976).

4.94 N. Ishimaru, H. Ueba, C. Tatsuyama: Surf. Sci. **193**, 193 (1988).

4.95 A. Taoufik, A. Chouiyakh, B. Lang: Radiat. Eff. **104**, 117 (1987).

4.96 R.L. Jacobson, G.K. Wehner: J. Appl. Phys. **36**, 2674 (1965).

4.97 W. Eckstein, J.P. Biersack: Z. Phys. B **63**, 471 (1986).

4.98 A.U. MacRae, G.W. Gobeli: J. Appl. Phys. **35**, 1629 (1964).

4.99 A. Kahn: Surf. Sci. Rep. **3**, 193 (1983).

4.100 M.W. Puga, G. Xu, S.Y. Tong: Surf. Sci **164**, L789 (1985).

4.101 W.E. Spicer, P. Pianetta, I. Lindau, P.W. Chye: J. Vac. Sci. Technol. **14**, 885 (1977).

4.102 J.L. Singer, J.S. Murday, L.R. Cooper: Surf. Sci. **108**, 7 (1981).

4.103 B.D. Weaver, D.R. Frankl, R. Blumenthal, N. Winograd: Surf. Sci. **222**, 464 (1989).

4.104 R. Ludeke, L. Esaki: Surf. Sci. **47**, 132 (1975).

4.105 H. Lüth, G.J. Russell: Surf. Sci. **45**, 329 (1974).

4.106 J. van Laar, A. Huijser, T.L. Rooy: J. Vac. Sci. Technol. **14**, 894 (1977).

4.107   C. von Festenberg: Z. Phys. **227**, 453 (1969).
4.108   W. Gudat, D.E. Eastman: J. Vac. Sci. Technol. **13**, 831 (1976).
4.109   D.E. Eastman, T.C. Chiang, P. Heimann, F.J. Himpsel: Phys. Rev. Lett. **45**, 656 (1980).
4.110   R. Ludeke, A. Koma: J. Vac. Sci. Technol. **13**, 241 (1976).
4.111   J.R. Chelikowsky, M.L. Cohen: Phys. Rev. B **20**, 4150 (1979).
4.112   A. Vaseashta, A. Elshabini-Riad, L.C. Burton: Mater. Sci. Eng. B **9**, 489 (1991).
4.113   Y.G. Wang, S. Ashok: Nucl. Instrum. Methods B **39**, 461 (1989).
4.114   C. Shwe, M. Gal: Appl. Phys. Lett. **62**, 516 (1993).
4.115   D.K. Sadana: Nucl. Instrum. Methods B **7/8**, 375 (1985).
4.116   M. Kawabe, N. Kanzaki, K. Masuda, S. Namba: Appl. Opt. **17**, 2556 (1978).
4.117   N.A. Bert, K.Y. Pogrebitskii, I.P. Soshnikov, Y.N. Yur'ev: Sov. Phys. Techn. Phys. **37**, 449 (1992).
4.118   Y. Sekino, M. Owari, M. Kudo, Y. Nihei: Japan J. Appl. Phys. **25**, 538 (1986).
4.119   S.K. Jindal, S.S. Lin, L.D. Brown, H.L. Marcus, M.A. Schmerling: Scan. Microsc. **2**, 1879 (1988).
4.120   H.J. Kang, Y.M. Moon, T.W. Kang, J.Y. Leem, J.J. Lee, D.S. Ma: J. Vac. Sci. Technol. A **7**, 3251 (1989).
4.121   Y. Kido, J. Kawmoto: J. Appl. Phys. **58**, 3377 (1985).
4.122   N.P. Tognetti, R.G. Elliman, G. Carter: Vacuum **33**, 165 (1983).
4.123   K.S. Jones, C.J. Santana: J. Mater. Res. **6**, 1048 (1991).
4.124   X. Wang: Appl. Surf. Sci. **33/34**, 88 (1988).
4.125   R.H. Williams, I.T. McGovern: Surf. Sci. **51**, 14 (1975).
4.126   C. Jardin, D. Robert, B. Achard, B. Gruzza, C. Pariset: Surf. Interf. Anal. **10**, 301 (1987).
4.127   S.J. Pearton, U.K. Chakrabarti, A.P. Perley, K.S. Jones: J. Appl. Phys. **68**, 2760 (1990).
4.128   J.B. Malherbe, W.O. Barnard: Surf. Sci. **255**, 309 (1991).
4.129   C.M. Demanet, J.B. Malherbe, N.G. van der Berg, V. Sankar: Surf. Interf. Anal. **23**, 433 (1995).
4.130   M.M. Sung, S.H. Lee, S.M. Lee, D. Marton, S.S. Perry, J.W. Rabalais: Surf. Sci. **382**, 147 (1997).
4.131   S. Valeri, G.C. Gazzadi, A. Rota, A. di Bona: Appl. Surf. Sci. **120**, 323 (1997).
4.132   T.W. Sigmon: Nucl. Instrum. Methods B **7/8**, 402 (1985).
4.133   L.O. Bubulac: J. Cryst. Growth **86**, 723 (1988).
4.134   G.L. Destéfanis: J. Cryst. Growth **86**, 700 (1988); Nucl. Instrum. Methods **209/210**, 567 (1983).
4.135   G. Bahir, E. Finkman: J. Vac. Sci. Technol. A **7**, 348 (1989).
4.136   E. Belas, P. Höschl, R. Grill, J. Franc, P. Moravec, K. Lischka, H. Sitter, A. Toth: Semicond. Sci. Technol. **8**, 1695 (1993).
4.137   Y.C. Lu, R.S. Feigelson, R.K. Route: J. Appl. Phys. **67**, 2583 (1990).
4.138   A.M. Mazzone: Philos. Mag. Lett. **60**, 131 (1989).
4.139   K. Nordlund, J. Keinonen, A. Kuronen: Phys. Scr. **T54**, 34 (1994).
4.140   V. Konoplev, A. Gras-Marti: Philos. Mag. A **71**, 1265 (1995).
4.141   P. Sigmund: Appl. Phys. Lett. **14**, 114 (1969).
4.142   J.P. Biersack: Nucl. Instrum. Methods **182/183**, 199 (1981).
4.143   J.F. Ziegler, J.P. Biersack, U. Littmark: *The Stopping and Range of Ions in Matter*, Vol. 1 (Pergamon, New York, 1985).
4.144   H. Gant, W. Mönch: Surf. Sci. **105**, 217 (1981).
4.145   I.K. Robinson: J. Vac. Sci. Technol. A **6**, 1966 (1988).

4.146  R. Kelly, N.Q. Lam: Radiat. Eff. **10**, 247 (1971).
4.147  D.A. Thompson, A. Golanski, K.H. Haugen, D.V. Stevanovic, G. Carter, C.E. Christodoulides: Radiat. Eff. **52**, 69 (1980).
4.148  L.A. Christel, J.F. Gibbons, T.W. Sigmon: J. Appl. Phys. **52**, 7148 (1981).
4.149  J. Narayan, D. Fathy, O.S. Oen, O.W. Holland: J. Vac. Sci. Technol. A **2**, 1303 (1984).
4.150  L.H. Dubois, G.P. Schwartz: J. Vac. Sci. Technol. B **2**, 101 (1984).
4.151  Y.D. Zheng, Y.H. Chang, B.D. McCombe, R.F.C. Farrow, T. Temofonte, F.A. Shirland: Appl. Phys. Lett. **49**, 1187 (1986).
4.152  H. Lüth: Vacuum **38**, 223 (1988); Surf. Sci. **126**, 126 (1983); Surf. Sci. **168**, 773 (1986).
4.153  X. Yin, H.-M. Chen, F.H. Pollak, Y. Cao, P.A. Montano, P.D. Kirchner, G.D. Pettit, J.M. Woodall: J. Vac. Sci. Technol. B **9**, 2114 (1991).
4.154  T.S. Jones, M.Q. Ding, N.V. Richardson, C.F. McConville: Appl. Surf. Sci. **45**, 85 (1990); Surf. Sci. **247**, 1 (1991).
4.155  G.R. Bell, C.F. McConville, T.S. Jones: Appl. Surf. Sci. **104/105**, 17 (1996).
4.156  V. Martinelli, L. Siller, M. G. Betti, C. Mariani, U. del Pennino: Surf. Sci. **391**, 73 (1997).
4.157  W.M. Lau, R.N.S. Sodhi, B.J. Flinn, K.H. Tan, G.M. Bancroft: Appl. Phys. Lett. **51**, 177 (1987).
4.158  R.A. Kubiak, E.H.C. Parker, R.M. King, K. Wittmaack: J. Vac. Sci. Technol. A **1**, 34 (1983).
4.159  P. Pianetta, I. Lindau, C.M. Garner, W.E. Spicer: Phys. Rev. B **18**, 2792 (1978).
4.160  W. Ranke, K. Jacobi: Surf. Sci. **47**, 525 (1975).
4.161  A. Barcz, M. Croset, M.L. Mercandalli: Surf. Sci. **95**, 511 (1980).
4.162  S. Valeri, M. Lolli, P. Sberveglieri: Surf. Sci. **238**, 63 (1990).
4.163  A. Sakalas, S. Zhukauskas: Solid State Commun. **70**, 363 (1989).
4.164  S. Valeri, A. di Bona: Surf. Sci. **251/252**, 995 (1991).
4.165  G.E. McGuire: Surf. Sci. **76**, 130 (1978).
4.166  T.D. Bussing, P.H. Holloway, Y.X. Wang, J.F. Moulder, J.S. Hammond: J. Vac. Sci. Technol. B **6**, 1514 (1988).
4.167  P.H. Holloway: Appl. Surf. Sci. **26**, 550 (1986).
4.168  K. Orrman-Rossiter, A.H. Al-Bayati, D.G. Armour: Surf. Sci. **225**, 341 (1990).
4.169  S. Valeri, M.G. Lancellotti, A. di Bona, G. Granozzi, G.A. Rizzi: Appl. Surf. Sci. **56**, 205 (1992).
4.170  A. van Oostrom: J. Vac. Sci. Technol. **13**, 224 (1976).
4.171  I.L. Singer, J.S. Murday, J. Comas: J. Vac. Sci. Technol. **18**, 161 (1982).
4.172  Y.X. Wang, P.H. Holloway: J. Vac. Sci. Technol. A **2**, 567 (1984); J. Vac. Sci. Technol. B **2**, 613 (1984).
4.173  S. Valeri, M. Lolli: Surf. Interf. Anal. **16**, 59 (1990).
4.174  J.B. Malherbe, W.O. Barnard, I. Le R. Strydom, C.W. Louw: Surf. Interf. Anal. **18**, 491 (1992).
4.175  J.L. Sullivan, W. Yu, S.O. Saied: Surf. Interf. Anal. **22**, 515 (1994).
4.176  W. Yu, J.L. Sullivan, S.O. Saied: Surf. Sci. **307–309**, 691 (1994).
4.177  G.S. Anderson, G.K. Wehner: Surf. Sci. **2**, 367 (1964).
4.178  G.S. Anderson: J. Appl. Phys. **37**, 3455 (1966).
4.179  H.C. Casey, G.L. Pearson: in *Point Defects in Solids*, Vol. 2, ed. by J.H. Crawford, L.M. Slifkin (Plenum, New York, 1975), p. 163.
4.180  J.C. Tsang, A. Kahn, P. Mark: Surf. Sci. **97**, 119 (1980).
4.181  C.W. Tu, A.R. Schlier: Appl. Surf. Sci. **11/12**, 355 (1982).
4.182  B. Gruzza, B. Achard, C. Pariset: Surf. Sci. **162**, 202 (1985).

4.183  R.N.S. Sodhi, W.M. Lau, S.I.J. Ingrey: Surf. Interf. Anal. **12**, 321 (1988).
4.184  S. Abdellaoui, B. Gruzza, C. Pariset, M. Bouslama, C. Jardin, D. Robert: Surf. Sci. **208**, L21 (1989).
4.185  W.O. Barnard, J.B. Malherbe, G. Myburg: S. Afr. J. Phys. **14**, 22 (1991).
4.186  J.B. Malherbe: Appl. Surf. Sci. **70/71**, 322 (1993).
4.187  J. Morais, T.A. Fazan, R. Landers: Phys. Stat. Sol. **141**, K19 (1994).
4.188  F.A. Stevie, P.M. Kahora, D.S. Simons, P. Chi: J. Vac. Sci. Technol A **6**, 76 (1988).
4.189  A. Karen, K. Okuno, F. Soeda, A. Ishitani: J. Vac. Sci. Technol. A **9**, 2247 (1991).
4.190  M. Gauneau, R. Chaplain, A. Rupert, Y. Toudic, D. Riviere, R. Callec: Nucl. Instrum. Methods B **80/81**, 543 (1993).
4.191  H. Gnaser, C. Kallmayer, H. Oechsner: J. Vac. Sci. Technol. B **13**, 19 (1995).
4.192  R. Carin, J.P. Deville, J. Werckmann: Surf. Interf. Anal. **16**, 65 (1990).
4.193  D. Cvetko, L. Floreano, A. Morgante, M. Peloi, F. Tommasini: Surf. Sci. **307–309**, 519 (1994).
4.194  W. Yu, J.L. Sullivan, S.O. Saied: Surf. Sci. **352–354**, 781 (1996).
4.195  W. Yu, J.L. Sullivan, S.O. Saied, G.A.C. Jones: Nucl. Instrum. Methods B **135**, 250 (1998).
4.196  A.I. Zagorenko, V.I. Zaporozchenko: Surf. Interf. Anal. **14**, 438 (1989).
4.197  R. Kaplan: J. Appl. Phys. **56**, 1636 (1984).
4.198  J.J. Bellina, J. Ferrante, M.V. Zeller: J. Vac. Sci. Technol. A **4**, 1692 (1986).
4.199  J.J. Bellina, M.V. Zeller: Appl. Surf. Sci. **25**, 380 (1986).
4.200  N. Laidani, M. Bonelli, A. Miotello, L. Guzman, L. Calliari, M. Elena, R. Bertoncello, A. Glisenti, R. Capelleti, P. Ossi: J. Appl. Phys. **74**, 2013 (1993).
4.201  A. Miotello, L. Calliari, R. Kelly, N. Laidani, N. Bonelli, L. Guzman: Nucl. Instrum. Methods B **80/81**, 931 (1993).
4.202  M. Bettini, H.J. Richter: Surf. Sci. **80**, 334 (1979).
4.203  U. Solzbach, H.J. Richter: Surf. Sci. **97**, 191 (1980).
4.204  A. Ebina, K. Asano, T. Takahashi: Phys. Rev. B **22**, 1980 (1980).
4.205  P. Morgen, J.A. Wilson: Nucl. Instrum. Methods B **26**, 585 (1987).
4.206  H. Neff, K.Y. Lay, M.S. Su, P. Lange, K.J. Bachmann: Surf. Sci. **189/190**, 661 (1987).
4.207  G.C. Morris, B.J. Wood: Mater. Forum **15**, 44 (1991).
4.208  E.K. Chieh, Z.A. Munir: J. Mater. Sci. **26**, 4268 (1991).
4.209  C.M. Stahle, C.R. Helms: J. Vac. Sci. Technol. A **10**, 3239 (1992).

## Chapter 5

5.1   R.H. Sloane, R. Press: Proc. R. Soc. Lond. A **168**, 284 (1938).
5.2   R.F.K. Herzog, F.P. Viehböck: Phys. Rev. **76**, 855 (1949).
5.3   R.E. Honig: J. Appl. Phys. **29**, 549 (1958).
5.4   R. Castaing, G. Slodzian: J. Microscopie **1**, 395 (1962).
5.5   H. Liebl: J. Appl. Phys. **38**, 5277 (1967).
5.6   A. Benninghoven: Z. Naturf. **22a**, 841 (1967); Z. Phys. **220**, 159 (1969); Z. Phys. **230**, 403 (1970).
5.7   K. Wittmaack: in *Inelastic Ion–Surface Collisions*, ed. by H.H. Tolk, J.C. Tully, W. Heiland, C.W. White (Academic, New York, 1977), p. 153.
5.8   N. Winograd: Prog. Solid State Chem. **13**, 285 (1982).
5.9   P. Williams: in *Applied Atomic Collision Physics*, Vol. 4, ed. by S. Datz (Academic, Orlando, 1983), p. 327.
5.10  A. Benninghoven, F.G. Rüdenauer, H.W. Werner: Secondary Ion Mass Spectrometry (Wiley, New York, 1987).

5.11    M.L. Yu: in *Sputtering by Particle Bombardment III*, ed. by R. Behrisch, K. Wittmaack (Springer, Berlin, Heidelberg, 1991), p. 91.
5.12    R.G. Wilson, F.A. Stevie, C.W. Magee: *Secondary Ion Mass Spectrometry* (Wiley, New York, 1989).
5.13    N.H. Tolk, I.S.T. Tsong, C.W. White: Anal. Chem. **49**, 16A (1977).
5.14    P. Williams, I.S.T. Tsong, S. Tsuji: Nucl. Instrum. Methods **170**, 591 (1980).
5.15    G. Slodzian, J.F. Hennequin: Comp. Rend. Acad. Sc. (Paris) B **263**, 1246 (1966).
5.16    C.A. Andersen: Int. J. Mass Spectrom. Ion Phys. **2**, 61 (1969); Int. J. Mass Spectrom. Ion Phys. **3**, 413 (1970).
5.17    J. Maul, K. Wittmaack: Surf. Sci. **47**, 358 (1975).
5.18    K. Wittmaack: Surf. Sci. **112**, 168 (1981).
5.19    K. Wittmaack: in *Secondary Ion Mass Spectrometry SIMS XI*, ed. by G. Gillen, R. Lareau, I. Bennett, F. Stevie (Wiley, Chichester, 1998), p. 11.
5.20    W. Reuter, J.G. Clabes: Anal. Chem. **60**, 1404 (1988).
5.21    V.E. Krohn: J. Appl. Phys. **33**, 3523 (1962).
5.22    P. Williams, R.K. Lewis, C.A. Evans, P.R. Hanley: Anal. Chem. **49**, 1399 (1977).
5.23    P. Vallerand, M. Baril: Int. J. Mass Spectrom. Ion Phys. **24**, 241 (1977).
5.24    M. Bernheim, J. Rebiere, G. Slodzian: J. Microsc. Spectrosc. Electron. **5**, 261 (1980).
5.25    K. Wittmaack: Surf. Sci. **126**, 573 (1983).
5.26    J. Los, J.J.C Geerlings: Phys. Rep. **190**, 133 (1990).
5.27    B.H. Cooper, E.R. Behringer: in *Low Energy Ion–Surface Interactions*, ed. by J.W. Rabalais (Wiley, Chichester, 1994), p. 263.
5.28    W. Heiland: in *Low Energy Ion–Surface Interactions*, ed. by J.W. Rabalais (Wiley, Chichester, 1994), p. 313.
5.29    W.O. Hofer: Scanning Microscopy, Suppl. **4**, 256 (1990).
5.30    R. Baragiola: in *Low Energy Ion–Surface Interactions*, ed. by J.W. Rabalais (Wiley, Chichester, 1994), p. 187.
5.31    G. Betz, P. Varga (Eds.): Proc. 8th Int. Workshop on Inelastic Ion–Surface Collisions [Nucl. Instrum. Methods B **58**, 301–530 (1991)].
5.32    M. Bernheim, J.P. Gauyacq (Eds.): Proc. 9th Int. Workshop on Inelastic Ion–Surface Collisions [Nucl. Instrum. Methods B **78**, 3–345 (1993)].
5.33    A.V. Barnes, N.H. Tolk, P. Nordlander (Eds.): Proc. 10th Int. Workshop on Inelastic Ion–Surface Collisions [Nucl. Instrum. Methods B **100**, 209–449 (1995)].
5.34    W. Heiland, E. Taglauer (Eds.): Proc. 11th Int. Workshop on Inelastic Ion–Surface Collisions [Nucl. Instrum. Methods B **125**, 1–336 (1997)].
5.35    M.L. Yu, N.D. Lang: Nucl. Instrum. Methods B **14**, 403 (1986).
5.36    W.F. van der Weg, P.K. Rol: Nucl. Instrum. Methods **38**, 274 (1965)
5.37    M.L. Yu: Phys. Rev. Lett. **40**, 574 (1978).
5.38    M.L. Yu, N.D. Lang: Phys. Rev. Lett. **50**, 127 (1983).
5.39    S. Prigge, E. Bauer: Adv. Mass Spectrom. **8A**, 543 (1981).
5.40    M.L. Yu: Phys. Rev. Lett. **47**, 1325 (1981).
5.41    M.L. Yu: Phys. Rev. B **24**, 1147 (1981); Phys. Rev. B **24**, 5625 (1981).
5.42    M.L. Yu: Phys. Rev. B **26**, 4731 (1982).
5.43    N.D. Lang: Phys. Rev. B **27**, 2019 (1983).
5.44    A. Blandin, A. Nourtier, D.W. Hone: J. Phys. **37**, 396 (1976).
5.45    J.K. Nørskov, B.I. Lundqvist: Phys. Rev. B **19**, 5661 (1979).
5.46    R. Brako, D.M. Newns: Surf. Sci. **108**, 253 (1981).
5.47    M.L. Yu: Phys. Scripta **T6**, 67 (1983); J. Vac. Sci. Technol. A **1**, 500 (1983).
5.48    M. Bernheim, F. Le Bourse: Nucl. Instrum. Methods B **27**, 94 (1987).

5.49   G.A. v.d. Schootbrugge, A.G.J. de Witt, J.M. Fluit: Nucl. Instrum. Methods
       **132**, 321 (1976).
5.50   A.R. Bayly, R.J. MacDonald: Radiat. Eff. **34**, 169 (1977).
5.51   R.F. Garrett, R.J. MacDonald, D.J. O'Connor: Nucl. Instrum. Methods **218**,
       333 (1983).
5.52   M.J. Vasile: Phys. Rev. B **29**, 3785 (1984).
5.53   H. Gnaser: Phys. Rev. B **54**, 16 456 (1996).
5.54   H. Gnaser: Phys. Rev. B **54**, 17 141 (1996).
5.55   Z. Sroubek: Nucl. Instrum. Methods **218**, 336 (1983).
5.56   Z. Sroubek: Phys. Rev. B **25**, 6046 (1982); Appl. Phys. Lett. **45**, 850 (1984).
5.57   Z. Sroubek: Appl. Phys. Lett. **42**, 514 (1983).
5.58   Z. Sroubek: Nucl. Instrum. Methods **194**, 533 (1982).
5.59   K. Wittmaack: Surf. Sci. **53**, 626 (1975).
5.60   I.F. Urazgil'din: Phys. Rev. B **47**, 4139 (1993); Nucl. Instrum. Methods B
       **78**, 271 (1993).
5.61   D.V. Klushin, M.Y. Gusev, I.F. Urazgil'din: Nucl. Instrum. Methods B **90**,
       542 (1994); Nucl. Instrum. Methods B **100**, 316 (1995).
5.62   M.L. Yu: Phys. Rev. B **29**, 2311 (1984).
5.63   T.R. Lundquist: J. Vac. Sci. Technol. **15**, 684 (1978); Surf. Sci. **90**, 548
       (1979).
5.64   A.R. Krauss, D.M. Gruen: Surf. Sci. **92**, 14 (1980).
5.65   R.G. Hart, C.B. Cooper: Surf. Sci. **94**, 105 (1980)
5.66   H.J. Barth, E. Mühling, W. Eckstein: Surf. Sci. **166**, 458 (1986).
5.67   W. Eckstein: Nucl. Instrum. Methods B **27**, 78 (1987).
5.68   Y.L. Wang: Phys. Rev. B **38**, 8633 (1988).
5.69   B.J. Garrison: Surf. Sci. **167**, L225 (1986).
5.70   A. Wucher, H. Oechsner: Surf. Sci. **199**, 567 (1988).
5.71   G. Slodzian: Surf. Sci. **48**, 161 (1975); Phys. Scripta **T6**, 54 (1983).
5.72   P. Williams: Surf. Sci. **90**, 588 (1979); Appl. Surf. Sci. **13**, 241 (1982).
5.73   L. Landau: Z. Phys. Sov. **2**, 46 (1932).
5.74   C. Zener: Proc. R. Soc. Lond. A **137**, 696 (1932).
5.75   E.C.G. Stueckelberg: Helv. Phys. Acta **5**, 369 (1932).
5.76   M.L. Yu, K. Mann: Phys. Rev. Lett. **57**, 1476 (1986).
5.77   K. Mann, M.L. Yu: Phys. Rev. B **35**, 6043 (1987).
5.78   M.L. Yu: Nucl. Instrum. Methods B **18**, 542 (1987).
5.79   E.E. Nikitin: *Theory of Elementary Atomic and Molecular Processes in
       Gases* (Clarendon, Oxford, 1974).
5.80   C. Coudray: Radiat. Eff. Def. Solids **110**, 343 (1989); Nucl. Instrum. Methods
       B **78**, 278 (1993).
5.81   C.A. Andersen, J.R. Hinthorne: Anal. Chem. **45**, 1421 (1973).
5.82   A.E. Morgan, H.W. Werner: Anal. Chem. **48**, 699 (1976); Anal. Chem. **49**,
       927 (1977).
5.83   D.S. Simons, J.E. Baker, C.A. Evans: Anal. Chem. **48**, 1341 (1976).
5.84   M.L. Yu: Nucl. Instrum. Methods B **15**, 151 (1986).
5.85   A.E. Morgan, H.A.M. de Grefte, N. Warmoltz, H.W. Werner, H.J. Tolle:
       Appl. Surf. Sci. **7**, 372 (1981).
5.86   A.E. Morgan, H.A.M. de Grefte, H.J. Tolle: J. Vac. Sci. Technol. **18**, 164
       (1981).
5.87   P. Sander, U. Kaiser, R. Jede, D. Lipinsky, O. Ganschow, A. Benninghoven:
       J. Vac. Sci. Technol. A **3**, 1946 (1985).
5.88   V.R. Deline, W. Katz, C.A. Evans, P. Williams: Appl. Phys. Lett. **33**, 832
       (1978).
5.89   H. Gnaser: Nucl. Instrum. Methods **218**, 312 (1983).

5.90   H. Oechsner, Z. Sroubek: Surf. Sci. **127**, 10 (1983).
5.91   J.L. Alay, W. Vandervorst: Phys. Rev. B **50**, 15 015 (1994).
5.92   H. Gnaser: Surf. Interf. Anal. **25**, 737 (1997).
5.93   H.A. Storms, K.F. Brown, J.D. Stein: Anal. Chem. **49**, 2023 (1977).
5.94   Y. Gao: J. Appl. Phys. **64**, 3760 (1988).
5.95   C.W. Magee, W.L. Harrington, E.M. Botnick: Int. J. Mass Spectrom. Ion Proc. **103**, 45 (1990).
5.96   K. Wittmaack: Phys. Scripta **T6**, 71, (1983).
5.97   G. Carter, J.S. Colligon, J.H. Leck: Proc. Phys. Soc. **79**, 299 (1962).
5.98   F. Schulz, K. Wittmaack: Radiat. Eff. **29**, 31 (1976).
5.99   G. Carter, M.J. Nobes, I.V. Katardjiev, I. Abril, A. Gras-Marti, J.J. Jimenez-Rodriguez, J.A. Peinador: Vacuum **44**, 783 (1993).
5.100  J.E. Chelgren, W. Katz, V.R. Deline, C.A. Evans, R.J. Blattner, P. Williams: J. Vac. Sci. Technol. **16**, 324 (1979).
5.101  C.E. Christodoulides, W.A. Grant, J.S. Williams: Nucl. Instrum. Methods **149**, 219 (1978).
5.102  N. Menzel, K. Wittmaack: Nucl. Instrum. Methods **191**, 235 (1981).
5.103  R. Valizadeh, J.A. van den Berg, R. Badheka, A. Al Bayati, D.G. Armour, D. Sykes: Nucl. Instrum. Methods B **64**, 609 (1992).
5.104  J. Scholtes: Fresenius J. Anal. Chem. **353**, 499 (1995).
5.105  J. Biersack, S. Berg, C. Nender: Nucl. Instrum. Methods B **59/60**, 21 (1991).
5.106  W. Eckstein, M. Hou, V.I. Shulga: Nucl. Instrum. Methods B **119**, 477 (1996).
5.107  G.S. Tompa, W.E. Carr, M. Seidl: Appl. Phys. Lett. **48**, 1048 (1986); Appl. Phys. Lett. **49**, 1511 (1986).
5.108  G.S. Tompa, W.E. Carr, M. Seidl: Surf. Sci. **198**, 431 (1988).
5.109  J. Hölzl, F.K. Schulte: in *Solid Surface Physics*, Springer Tracts in Modern Physics, Vol. 85 (Springer, Berlin, Heidelberg, 1979), p. 1.
5.110  A.P. Jansen, P. Akhter, C.J. Harland, J.A. Venables: Surf. Sci. **93**, 453 (1980).
5.111  G. Bachmannn, H. Oechsner, J. Scholtes: Fresenius Z. Anal. Chem. **329**, 195 (1987).
5.112  G. Blaise, G. Slodzian: Surf. Sci. **40**, 708 (1973).
5.113  R.E. Weber, W.T. Peria: Surf. Sci. **14**, 13 (1969).
5.114  J.E. Ortega, E.M. Oellig, J. Ferron, R. Miranda: Phys. Rev. B **36**, 6213 (1987).
5.115  C.A. Papageorgopoulos, M. Kamaratos: Surf. Sci. **221**, 263 (1989).
5.116  R.G. Wilson, S.W. Novak, S.P. Smith, S.D. Wilson, J.C. Norberg: in *Secondary Ion Mass Spectrometry SIMS VI*, ed. by A. Benninghoven, A.M. Huber, H.W. Werner (Wiley, Chichester, 1988), p. 133.
5.117  H. Hotop, W. C. Lineberger: J. Phys. Chem. Ref. Data **4**, 539 (1975); J. Phys. Chem. Ref. Data **14**, 731 (1985).
5.118  H.B. Michaelson: J. Appl. Phys. **48**, 4729 (1977).
5.119  H. Gnaser, I.D. Hutcheon: Phys. Rev. B **35**, 877 (1987).
5.120  H. Gnaser, I.D. Hutcheon: Phys. Rev. B **38**, 11 112 (1988).
5.121  H. Gnaser, I.D. Hutcheon: in *Secondary Ion Mass Spectrometry SIMS VI*, ed. by A. Benninghoven, A.M. Huber, H.W. Werner (Wiley, Chichester, 1988), p. 29.
5.122  H. Gnaser: Radiat. Eff. Def. Solids **109**, 265 (1989).
5.123  J.C. Huneke, J.T. Armstrong, G.J. Wasserburg: Geochim. Cosmochim. Acta **47**, 1635 (1983).
5.124  P. Williams: Scanning Electron Microscopy **II**, 553 (1985).
5.125  G. Slodzian, J.C. Lorin, A. Havette: J. Phys. Lett. **41**, L558 (1980).

5.126  J.C. Lorin, A. Havette, G. Slodzian: in *Secondary Ion Mass Spectrometry SIMS III*, ed. by A. Benninghoven, I. Giber, I. Laszlo, M. Riedel, H.W. Werner (Springer, Berlin, Heidelberg, 1982), p. 141.

5.127  E.U. Engström, A. Lodding, H. Odelius, U. Södervall: Mikrochim. Acta I, 387 (1987).

5.128  U. Södervall, H. Odelius, A. Lodding, E.U. Engström: Scanning Microscopy 1, 471 (1987).

5.129  U. Södervall, E.U. Engström, H. Odelius, A. Lodding: in *Secondary Ion Mass Spectrometry SIMS VI*, ed. by A. Benninghoven, A.M. Huber, H.W. Werner (Wiley, Chichester, 1988), p. 83.

5.130  A.J.T. Jull: Inter. J. Mass Spectrom. Ion Phys. 41, 135 (1982).

5.131  N. Shimizu, S.R. Hart: J. Appl. Phys. 53, 1303 (1982).

5.132  S.A. Schwarz: J. Vac. Sci. Technol. A 5, 308 (1987).

5.133  G. Hortig, M. Müller: Z. Phys. 221, 119 (1969).

5.134  I. Katakuse, T. Ichihara, Y. Fujita, T. Matsuo, T. Sakurai, H. Matsuda: Int. J. Mass Spectrom. Ion Proc. 67, 229 (1985); Int. J. Mass Spectrom. Ion Proc. 74, 33 (1986).

5.135  I. Katakuse, T. Ichihara, T. Matsuo, T. Sakurai, H. Matsuda: Int. J. Mass Spectrom. Ion Proc. 91, 99 (1989).

5.136  K. Wittmaack: Phys. Lett. 69A, 322 (1979).

5.137  H.M. Urbassek, W.O. Hofer: K. Dan. Vidensk. Selsk. Mat. Fys. Medd. 43, 97 (1993).

5.138  D.M. Wood: Phys. Rev. Lett. 46, 749 (1981).

5.139  M. Wahl, A. Wucher: Nucl. Instrum. Methods B 94, 36 (1994).

5.140  A. Wucher, M. Wahl: Nucl. Instrum. Methods B 115, 581 (1996).

5.141  M.M. Kappes: Chem. Rev. 88, 369 (1988).

5.142  V. Bonacic-Koutecky, L. Cespiva, P. Fantucci, J. Koutecky: Z. Phys. D 26, 287 (1993).

5.143  C. Jackschath, I. Rabin, W. Schulze: Z. Phys. D 22, 517 (1992).

5.144  H. Gnaser and H. Oechsner: Fres. J. Anal. Chem. 341, 54 (1991); Surf. Interf. Anal. 17, 646 (1991); Surf. Interf. Anal. 21, 257 (1994).

5.145  H. Gnaser: J. Vac. Sci. Technol. A 12, 452 (1994).

5.146  H. Gnaser, H. Oechsner: Surf. Sci. Lett. 302, L289 (1994).

5.147  H.M. Urbassek: Nucl. Instrum. Methods B 18, 587 (1987).

5.148  H.S.W. Massey, E.H.S. Burhop, H.B. Gilbody: *Electronic and Ionic Impact Phenomena*, Vol. 3 (Oxford University Press, London, 1971).

5.149  S. Bloom, H. Margenau: Phys. Rev. 85, 670 (1952).

5.150  J.-F. Hennequin: J. Phys. 29, 483 (1968); J. Phys. 29, 1053 (1968).

5.151  K. Wittmaack: Nucl. Instrum. Methods 170, 565 (1980).

5.152  S.N. Schauer, P. Williams, R.N. Compton: Phys. Rev. Lett. 65, 625 (1990).

5.153  H. Gnaser, H. Oechsner: Nucl. Instrum. Methods B 82, 518 (1993).

5.154  S. Yang, K.J. Taylor, M.J. Craycraft, J. Conceicao, C.L. Pettiette, O. Cheshnovsky, R.E. Smalley: Chem. Phys. Lett. 144, 431 (1988).

5.155  R.L. Hettich, R.N. Compton, R.H. Ritchie: Phys. Rev. Lett. 67, 1242 (1991).

5.156  H.-G. Weikert, L.S. Cederbaum, F. Tarantelli, A.I. Boldyrev: Z. Phys. D 18, 299 (1991).

5.157  M.K. Scheller, L.S. Cederbaum: J. Phys. B: At. Mol. Opt. Phys. 25, 2257 (1992).

5.158  M.K. Scheller, L.S. Cederbaum: Chem. Phys. Lett. 216, 141 (1993); J. Chem. Phys. 99, 441 (1993).

5.159  T. Sommerfeld, M.K. Scheller, L.S. Cederbaum: Chem. Phys. Lett. 209, 216 (1993); J. Phys. Chem. 98, 8914 (1994).

5.160  J.D. Watts, R.J. Bartlett: J. Chem. Phys. 97, 3445 (1992).

5.161  M.K. Scheller, R.N. Compton, L.S. Cederbaum: Science **270**, 1160 (1995).
5.162  A.D. Bekkerman, N.K. Dzhemilev, S.E. Maksimov, V.V. Solomko, S.V. Verkhoturov, I.V. Veryovkin: Vacuum **47**, 1073 (1996).
5.163  H.S.W. Massey: *Negative Ions* (Cambridge University Press, Cambridge, 1976).
5.164  B.K. Janousek, J.I. Brauman: in *Gas Phase Ion Chemistry*, Vol. 2, ed. by M. T. Bowers (Academic, New York, 1979), p. 53; P.S. Drzaic, J. Marks, J.I. Brauman: in *Gas Phase Ion Chemistry Vol. 3*, ed. by M.T. Bowers (Academic, New York, 1984), p. 167.
5.165  D.R. Bates: Adv. At. Mol. Opt. Phys. **27**, 1 (1991).
5.166  H. Gnaser: Phys. Rev. A **56**, R2518 (1997).
5.167  Y.K. Bae, M.J. Coggiola, J.R. Peterson: Phys. Rev. A **29**, 2888 (1984).
5.168  E.W. Thomas: Prog. Surf. Sci. **10**, 383 (1980).
5.169  K.J. Snowdon, W. Heiland: Z. Phys. A **318**, 275 (1984).
5.170  E.W. Thomas, L. Efstathiou: Nucl. Instrum. Methods B **2**, 479 (1984).
5.171  H. Müller, R. Hausmann, H. Brenten, V. Kempter: Surf. Sci. **303**, 56 (1994).
5.172  T. Sommerfeld, L.S. Cederbaum: Phys. Rev. Lett. **80**, 3723 (1998).
5.173  A.J. Eccles, J.A. van den Berg, A. Brown, J.C. Vickerman: Appl. Phys. Lett. **49**, 188 (1986).
5.174  S.T. de Zwart, T. Fried, D.O. Boerma, R. Hoekstra, A.G. Drentje, A.L. Boers: Surf. Sci. **177**, L939 (1986).
5.175  D.L. Weathers, T.A. Tombrello, M.H. Prior, R.G. Stokstad, R.E. Tribble: Nucl. Instrum. Methods B **42**, 307 (1989).
5.176  I.S. Bitensky, M.N. Murakhmetov, E.S. Parilis: Sov. Phys. Tech. Phys. **24**, 618 (1979).
5.177  T. Neidhart, F. Pichler, F. Aumayr, H.P. Winter, M. Schmid, P. Varga: Phys. Rev. Lett. **74**, 5280 (1995).
5.178  T. Neidhart, F. Pichler, F. Aumayr, H.P. Winter, M. Schmid, P. Varga: Nucl. Instrum. Methods B **98**, 465 (1995).
5.179  P. Varga, T. Neidhart, M. Sporn, G. Libiseller, M. Schmid, F. Aumayr, H.P. Winter: Phys. Scripta **T73**, 307 (1996).
5.180  M. Sporn, G. Libiseller, T. Neidhart, M. Schmid, F. Aumayr, H.P. Winter, P. Varga, M. Grether, D. Niemann, N. Stolterfoht: Phys. Rev. Lett. **79**, 945 (1997).
5.181  D.H.G. Schneider, M.A. Briere: Phys. Scripta **53**, 228 (1996).
5.182  T. Schenkel, M.A. Briere, H. Schmidt-Böcking, K. Bethge, D.H. Schneider: Phys. Rev. Lett. **78**, 2481 (1997).
5.183  T. Schenkel, A.V. Barnes, A. Hamza, A. Schach von Wittenau, D.H. Schneider: Nucl. Instrum. Methods B **125**, 153 (1997).
5.184  F. Aumayr, H.P. Winter: Nucl. Instrum. Methods B **90**, 523 (1994); Comm. At. Mol. Phys. **29**, 275 (1994).
5.185  A. Arnau, F. Aumayr, P.M. Echenique, M. Grether, W. Heiland, J. Limburg, R. Morgenstern, P. Roncin, S. Schippers, R. Schuch, N. Stolterfoht, P. Varga, T.J.M. Zouros, H.P. Winter: Surf. Sci. Rep. **27**, 113 (1997).
5.186  J. Burgdörfer, P. Lerner, F.W. Meyer: Phys. Rev. A **44**, 5647 (1991).
5.187  S. Della-Negra, J. Depauw, H. Joret, V. Le Beyec, E.A. Schweikert: Phys. Rev. Lett. **60**, 948 (1988).
5.188  T. Schenkel, M.A. Briere, H. Schmidt-Böcking, K. Bethge, D.H. Schneider: Mater. Sci. Forum **248/249**, 413 (1997).
5.189  T. Schenkel, A.V. Barnes, A.V. Hamza, D.H. Schneider, J.C. Banks, B.L. Doyle: Phys. Rev. Lett. **80**, 4325 (1998).

5.190  T. Schenkel, A.V. Barnes, A.V. Hamza, D.H. Schneider, Eur. Phys. J. D **1**, 297 (1998).

## Appendix

A.1   H. Ibach (Ed.): *Electron Spectroscopy for Surface Analysis* (Springer, Berlin, Heidelberg, 1977).

A.2   H. Oechsner (Ed.): *Thin Film and Depth Profile Analysis* (Springer, Berlin, Heidelberg, 1984).

A.3   G. Ertl, J. Küppers: *Low-Energy Electrons and Surface Chemistry* (VCH, Weinheim, 1985).

A.4   D.P. Woodruff, T.A. Delchar: *Modern Techniques of Surface Science* (Cambridge University Press, Cambridge, 1986).

A.5   L.C. Feldman, J.W. Mayer: *Fundamentals of Surface and Thin Film Analysis* (North-Holland, New York, 1986).

A.6   M. Henzler, W. Göpel: *Oberflächenphysik des Festkörpers* (Teubner, Stuttgart, 1991).

A.7   D.J. O'Connor, B.A. Sexton, R.S.C. Smart (Eds.): *Surface Analysis Methods in Materials Science* (Springer, Berlin, Heidelberg, 1992).

A.8   H. Lüth: *Surfaces and Interfaces of Solid Materials* (Springer, Berlin, Heidelberg, 1995).

A.9   D. Briggs, M.P. Seah: *Practical Surface Analysis*, Vol 1: Auger and X-ray Photoelectron Spectroscopy (Wiley, Chichester, 1994).

A.10  D. Briggs, M.P. Seah: *Practical Surface Analysis*, Vol 2: Ion and Neutral Spectroscopy (Wiley, Chichester, 1996).

A.11  A. Benninghoven, F.G. Rüdenauer, H.W. Werner: *Secondary Ion Mass Spectrometry* (Wiley, New York, 1987).

A.12  R.G. Wilson, F.A. Stevie, C.W. Magee: *Secondary Ion Mass Spectrometry* (Wiley, New York, 1989).

A.13  M.L. Yu: in *Sputtering by Particle Bombardment III*, ed. by R. Behrisch, K. Wittmaack (Springer, Berlin, Heidelberg, 1991), p. 91.

A.14  H. Gnaser, J. Fleischhauer, W.O. Hofer: Appl. Phys. A **37**, 211 (1985).

A.15  H. Oechsner: in *Thin Film and Depth Profile Analysis*, ed. by H. Oechsner (Springer, Berlin, Heidelberg, 1984), p. 63.

A.16  H. Oechsner: Int. J. Mass Spectrom. Ion Proc. **143**, 271 (1995).

A.17  W. Bieck, H. Gnaser, H. Oechsner: Appl. Phys. Lett. **63**, 845 (1993).

A.18  N. Winograd, B.J. Garrison: in *Ion Spectroscopies for Surface Analysis*, ed. by A.W. Czanderna, D.M. Hercules (Plenum, New York, 1991), p. 45.

A.19  M.G. Payne, L. Deng, N. Thonnard: Rev. Sci. Instrum. **65**, 2433 (1994).

A.20  C. He, C.H. Becker: Surf. Interf. Anal. **24**, 79 (1996).

A.21  G.K. Nicolussi, M.J. Pellin, K.R. Lykke, J.L. Trevor, D.E. Mencer, A.M. Davis: Surf. Interf. Anal. **24**, 363 (1996).

A.22  K. Wittmaack: in *Sputtering by Particle Bombardment III*, ed. by R. Behrisch, K. Wittmaack (Springer, Berlin, Heidelberg, 1991), p. 161.

A.23  S. Hüfner: *Photoelectron Spectroscopy* (Springer, Berlin, Heidelberg, 1996).

A.24  B. Feuerbach, B. Fitton, R.F. Willis (Eds.): *Photoemission and the Electronic Properties of Surfaces* (Wiley, New York, 1978).

A.25  W.K. Chu, J.W. Mayer, M.A. Nicolet: *Backscattering Spectrometry* (Academic, New York, 1978).

A.26  J.F. van der Veen: Surf. Sci. Rep. **5**, 199 (1985).

A.27  E.S. Mashkova, V.A. Molchanov: *Medium-Energy Ion Reflection from Solids* (North-Holland, Amsterdam, 1985).

A.28  H. Niehus, W. Heiland, E. Taglauer: Surf. Sci. Rep. **17**, 213 (1993).

A.29    M.A. Van Hove, W.H. Weinberg, C.M. Chan: *Low-Energy Electron Diffraction* (Springer, Berlin, Heidelberg, 1986).

A.30    H. Raether: *Excitation of Plasmon and Interband Transitions by Electrons* (Springer, Berlin, Heidelberg, 1980).

A.31    H. Ibach, D.L. Mills: *Electron Energy Loss Spectroscopy and Surface Vibrations* (Academic, New York, 1982).

A.32    J.A. Stroscio, W.J. Kaiser: *Scanning Tunneling Microscopy* (Academic, San Diego, 1993).

A.33    H.J. Güntherodt, R. Wiesendanger: *Scanning Tunneling Microscopy I* (Springer, Berlin, Heidelberg, 1994).

A.34    L. Reimer: *Scanning Electron Microscopy* (Springer, Berlin, Heidelberg, 1985).

A.35    L. Reimer: *Transmission Electron Microscopy* (Springer, Berlin, Heidelberg, 1993).

A.36    S. Amelinckx, D. van Dyck, J. van Landuyt, G. van Tendeloo: *Handbook of Microscopy* (Wiley, Chichester, 1996).

A.37    R. Feidenhans'l: Surf. Sci. Rep. **10**, 105 (1989).

A.38    M.T. Robinson: K. Dan. Vidensk. Selsk. Mat. Fys. Medd. **43**, 27 (1993).

A.39    J.R. Beeler: *Radiation Effects Computer Simulations* (North-Holland, Amsterdam, 1983).

A.40    H.H. Andersen: Nucl. Instrum. Methods B **18**, 321 (1987).

A.41    W. Eckstein: *Computer Simulation of Ion Solid Interactions* (Springer, Berlin, Heidelberg, 1991).

A.42    R.M. Nieminen: K. Dan. Vidensk. Selsk. Mat. Fys. Medd. **43**, 81 (1993).

A.43    W.G. Hoover: *Molecular Dynamics* (Springer, Berlin, Heidelberg, 1986).

A.44    F.F. Abraham: Adv. Phys. **35**, 1 (1986).

A.45    M.P. Allen, D.J. Tildesley: *Computer Simulation of Liquids* (Clarendon, Oxford, 1987).

A.46    M.T. Robinson, I.M. Torrens: Phys. Rev. B **9**, 5008 (1974).

A.47    M.T. Robinson: Phys. Rev. B **40**, 10717 (1989).

A.48    M. Hou, W. Eckstein, M.T. Robinson: Nucl. Instrum. Methods B **82**, 234 (1993).

A.49    J.P. Biersack, L.G. Haggmark: Nucl. Instrum. Methods **174**, 257 (1980).

A.50    J.P. Biersack, W. Eckstein: Appl. Phys. A **34**, 73 (1984).

A.51    J.P. Biersack: Nucl. Instrum. Methods B **27**, 21 (1987).

A.52    W. Möller, W. Eckstein, J.P. Biersack: Comp. Phys. Comm. **51**, 355 (1988).

A.53    Y. Yamamura: Nucl. Instrum. Methods **194**, 515 (1982); Nucl. Instrum. Methods B **33**, 493 (1988); Nucl. Instrum. Methods B **45**, 707 (1990).

A.54    D.E. Harrison: Radiat. Eff. **70**, 1 (1983); Crit. Rev. Solid State Mater. Sci. **14**, S1 (1988).

A.55    V.I. Shulga: Radiat. Eff. **70**, 65 (1983); Radiat. Eff. **82**, 169 (1984); Radiat. Eff. **85**, 1 (1985).

A.56    R. Smith, R.P. Webb: Nucl. Instrum. Methods B **67**, 373 (1992).

A.57    M.H. Shapiro: Radiat. Eff. Defects Solids **142**, 259 (1997).

A.58    H.M. Urbassek: Nucl. Instrum. Methods B **122**, 427 (1997).

# Index

# Springer Tracts in Modern Physics